NO LONGER THE PROPERTY OF BUTLER UNIVERSITY LIBRARIES

Designing Experiments & Games of Chance

Blaise Pascal, death mask, *(Société de Port-Royal, Paris)*

Designing Experiments & Games of Chance

The Unconventional Science of Blaise Pascal

WILLIAM R. SHEA

Science History Publications/USA
2003

First published in the United States of America
by Science History Publications/USA
a division of Watson Publishing International
P.O. Box 493, Canton, MA 02021-0493
www.shpusa.com

© 2003 Watson Publishing International

Library of Congress Cataloging-in-Publication Data
Shea, William R.
　　Designing experiments & games of chance : the unconventional science of Blaise Pascal / William R. Shea.
　　　　p.　　cm.
　　Includes bibliographical references and index.
　　ISBN 0-88135-376-0 (alk. paper)
　　1. Games of chance (Mathematics)　2. Experimental design.　3. Pascal, Blaise, 1623–1662—Contributions in mathematics.　4. Pascal, Blaise, 1623–1662—Contributions in physics.　5. Science, Renaissance.　I. Title: Designing experiments and games of chance.　II. Title.

QA271.S58 2003
519.2'7—dc22   2003059201

If the publishers have unwittingly infringed the copyright in an illustration reproduced, they will gladly pay an appropriate fee on being satisfied as to the owner's title.

All rights reserved. No part of this book may be used or reproduced in any manner whatsoever without written permission of the copyright holder except in the case of brief quotations embodied in critical articles and reviews.

Designed and typeset by Publishers' Design and Production Services, Inc.

Designed, typeset, and printed in the USA.

To
Marcelo Sánchez Sorondo
priest, scholar and friend

Contents

Introduction   ix

CHAPTER 1
A Precocious Genius   1

CHAPTER 2
Filling the Void: From Galileo to Torricelli   17

CHAPTER 3
Creating Emptiness: Pascal's Ingenious Methods   41

CHAPTER 4
The Battle of the Void: The Jesuits Against Pascal   65

CHAPTER 5
The Great Experiment on a Mountain in France   99

CHAPTER 6
Why Pumps Work: The Treatise on the Equilibrium of Liquids   129

CHAPTER 7
Submerged in a Sea of Air: The Treatise on the Weight of the Mass of Air   155

Contents

CHAPTER 8
Pascal on the Advancement of Learning in Science and Religion 187

CHAPTER 9
The Use of Logic and the Role of Experiments 209

CHAPTER 10
The Marvelous Arithmetical Triangle 241

CHAPTER 11
Mastering Games of Chance 257

CHAPTER 12
Putting the Arithmetical Triangle to Work 291

CHAPTER 13
The Brilliance and Pride of a Gambler and a Mathematician 313

Conclusion 333

Selected Bibliography 341

Index 347

Introduction

During his comparatively brief life (he died at thirty-nine, the age Mozart was to die), Blaise Pascal devoted his unusual talents to mathematics, physics, and religion. His religious views are well known, and the general interest in this aspect of his intellectual and moral pursuit may be responsible for an undue neglect of his mathematical and scientific achievements. Those who admired Pascal principally for his religious genius tended to belittle—as Pascal's family did after his death—the importance of his work in other fields. Many who are familiar with his *Pensées*, which are fragments of an intended *Apology for Christianity*, lack a just appreciation of the originality of his thought in mathematics and physics. The present book is an attempt to fill this gap by providing a general description and a critical analysis of Pascal's work in these fields. After an introductory chapter on his early life, we immediately turn, in Chapter 2, to the origins of the debate over the possibility of a vacuum, a problem that Pascal was to tackle with remarkable insight and ingenuity, as we describe in Chapter 3 and Chapter 4. In Chapter 5, we explain why Pascal ascended a mountain with a barometer to prove that we live submerged under a sea of air. In Chapter 6 and Chapter 7, we examine his physics of liquids and his explanation of pumps. In Chapter 8, we consider his views on the advancement of learning in science and religion, and, in Chapter 9, we outline his novel philosophy of science. In the next four chapters, we analyze his pioneering work on probability and the odds of winning at games of chance before turning to an incident about a curve called the cycloid that brought out the best in Pascal's mathematical ability but the worst in his human nature.

The problems that Pascal tackled and the way he went about finding solutions are familiar to most of us. This is why some readers may be surprised by the words "Unconventional Science" in the subtitle of this book.

Introduction

They are meant as a reminder of the radically different way of looking at the world that was developed by Pascal and his contemporaries. In recent years, historians of science have drawn attention to the plurality of strands, including scholastic and hermetic elements, that went into the texture of modern science, but we should not loose sight of the profound change that occurred in the seventeenth century. Scientific investigation, in the sense of the formulation of hypotheses on the basis of empirical data stated with sufficient rigor to allow the hypotheses to be stringently tested, was not widespread before that time. We shall see how this was the case when we consider Pascal's running argument with the Jesuit professor, Fr. Etienne Noël. A competent, if somewhat bumptious Aristotelian, Noël believed that quantity was irrelevant because it could only tell us how much of a substance we have, not what kind. In this perspective, science rested on ordinary experience and aimed at studying the habitual behavior of bodies. For instance, heavy objects naturally fall downwards and this was taken as an indication that they have an innate form that makes them move in that way. There seemed no need for technical concepts such as those of mathematics, and there was no question of a quantitative structuring of experience. This is where Pascal, walking in the footsteps of Galileo and Torricelli, broke new ground. He developed a new way of thinking, which is usually described by the short-hand term *the experimental method*, meaning the active questioning of nature under conditions defined by the experimenter rather than the mere observation of the phenomena that spontaneously present themselves.

Pascal's experiments on the vacuum were not only novel, they went against the conventions that Fr. Noël and the academic establishment took for granted. They were surprised by his approach, and told him in no uncertain terms that the systematic simplification of the natural context was wrong-headed. We are now so well acquainted with Pascal's viewpoint, that what surprises us is that he should not have met with instant and universal approval. In the field of mathematics, Pascal's approach was equally unconventional. His analysis of the arithmetical triangle had been hailed from afar, but he used it in a way that went much beyond his predecessors. Odds and games of chance were never to be the same after Pascal, and there would be no modern science if he had not tamed Lady Luck.

Pascal's starting point was a concrete fact or a practical question. It was the way mercury fell in an evacuated tube that led him to investigate the properties of the vacuum, and it was the problem of a friend who had to place a bet that made him work out the odds at tossing dice. The same

Introduction

reliance of experiment and experience can be seen in Pascal's own life. His deeply religious faith was anchored in a mystical experience that led him to carry out what he considered an experiment in Christian living, and he would have considered it paradoxical that the method he developed was to transform a civilization organized around Christianity into one centred on science. He applied his probability calculus to the problem of personal destiny and reached the unconventional conclusion that the fulfilment of human science lies beyond terrestrial life. Pascal never attempted to cover all fields of knowledge, but he takes us on a stimulating intellectual journey where we discover the new world of experimental science and learn to play for high stakes.

Acknowledgments

This book was completed shortly after I was appointed Galileo Professor of the History of Science at the University of Padua where Galileo spent what he fondly recalled as "the best eighteen years of my life." I am grateful to my new colleagues for their warm welcome, their scholarly support, and their administrative help. I owe special thanks to Vincenzo Milanesi, the Rector of the University, Irene Favaretto, the Pro-Rector, Eugenio Calimani, the Dean of the Faculty of Science, and Giovanni Costa, the Director of the Center for the History and Philosophy of Science. They kindly made me feel that studying experiments on the void and wrestling with difficult odds at games of chance was neither useless nor uninteresting.

This book owes more than I can express to the contributions of several historians of science whose works are quoted in the following pages. I will limit myself to mentioning here W.E. Knowles Middleton, who introduced me to the richness of the vacuum and whose *History of the Barometer* was a constant source of inspiration. For friendly advice and constructive criticism, I was fortunate to be able to rely on Catherine Chevalley, Francis Debeauvais, Georges Frick, Paolo Galluzzi, Raffaella Simili, Walter Tega, and René Voltz. I am particularly grateful to my dedicated and patient publisher, Neale Watson, who saved me from many an egregious error and taught me much that I needed to known about the art of scientific communication. I am also indebted to the Social Sciences and Humanities Research Council of Canada for funding the project that eventually led to this book. My thanks also to Simonetta Bacchin and the re-

Introduction

markable secretariat of the Faculty of Science of the University of Padua for their competence and their invariable friendliness. I also want to record my profound appreciation for the way Graziella Borgato made the move to Padua so easy and so pleasant. Finally, I wish to say how much I owe to my wife, Evelyn. Without her encouragement and her probing questions the pages of this book would have remained void, and I would have been overcome by odds that I could never master alone.

William R. Shea
University of Padua
1 September 2003

CHAPTER 1

A Precocious Genius

Shadowed by Illness

Blaise Pascal was born on 19 June 1623 in the city of Clermont, the only son of Etienne Pascal and Antoinette Begon. He had an elder sister, Gilberte, born in 1620, and a younger one, Jacqueline, who was two years his junior.[1] His grandfather was a tax official who became private secretary to Queen Louise, the wife of King Henri III, before becoming the Treasurer of France for the Province of Auvergne. Blaise's father, Etienne, was a Presiding Judge at the taxation court in Montferrand, the twin city of Clermont, and, after a period of retirement in Paris, was appointed head of the taxation office in Rouen by Cardinal Richelieu.

Pascal was later to have serious medical problems, which were foreshadowed at an early age. When he was two years old, he became extremely thin and his parents feared for his life. The symptoms were most unusual: he could not look at water without having a fit, and although he was happy when his mother or his father came to him separately, he could not bear seeing them together. In an age prone to see the hidden

---

[1] Our main source is the *Life of Pascal* written after his death by his sister, Gilberte Périer. It was published for the first time, in a pirated edition, in Amsterdam in 1684, and two years later in Paris, the year before Gilberte died. A second version was discovered and published by Léon Brunschvicg in the twentieth century. Both are included in the standard edition of Pascal's works, the *Œuvres Complètes de Pascal*, edited by Pierre Mesnard. Paris: Desclée de Brouwer, 1964, vol. I, pp. 571–602 and pp. 603–642. Four volumes of this new edition have appeared to date and cover the period 1623–1662, but do not include the *Pensées* or the *Provinciales*. This edition will be quoted as *Œuvres de Pascal* followed by the volume in Roman and the page in Arabic numbers.

hands of witches, it was rumored that a curse had been placed on him by an old woman who belonged to the poor that his family helped.[2] Examined and threatened by the boy's father, the woman broke down and confessed that she had indeed placed a spell on the boy because the judge had refused to arrange some legal matter for her. She declared that the spell was deadly and could only be canceled by some other death, but she promptly added that any animal would do. Pascal's father offered a horse, but she replied that a cat would be more than enough. She was immediately given one, but as she walked out of the room and down the staircase, some servants insulted her and in her excitement she flung the cat out of the window. Although it did not fall from any great height, it landed on its back and was pronounced dead upon striking the ground. This was considered most amazing, because it was well known that a cat that is dropped upside down invariably lands on its four feet.[3] After the old woman recovered her composure, she went out into the garden to gather some herbs, which she mixed with flour to make a poultice.[4] No sooner was it applied to Blaise's stomach than he fell into a coma.

The doctors, summoned to his bedside, declared that the boy was dead, but his father refused to have him made ready for burial. He summoned the old woman, and when she entered the room, he went wild, rushed at her, and slapped her face. She meekly replied that she was only getting what she deserved because she had forgotten to mention that the boy would become unconscious and would only revive around midnight. The family anxiously awaited the outcome, but when the clock struck twelve, Blaise did not stir. Then around two in the morning, he opened his eyes and saw his mother and father together without dread, "which proved that he was cured," comments Gilberte.[5] A few days later his fear of water

---

[2] The story was told by Gilberte Périer while Pascal was still alive but it was only published in the nineteenth century. Mesnard gives it in *Œuvres de Pascal*, I, pp. 507–508. An enlarged and moralized version is to be found in the *Memoirs* that Marguerite Périer dictated around 1723 when she was already 76 years old and somewhat given to hyperbole (*Œuvres de Pascal*, I, pp. 1091–1093).

[3] I do not recommend carrying out the experiment, but the physical explanation is quite straightforward: by pulling in its paws, the cat is able to rotate very fast for the same reason that a skater can speed up his rate of spinning by bringing his arms in.

[4] Marguerite Périer adds that the witch told Pascal's father that three different kinds of herbs had to be gathered by a child under seven years of age. The local apothecary offered the services of his daughter and brought her early on the next morning (*Œuvres de Pascal*, I, p. 1092).

[5] *Œuvres de Pascal*, I, 508.

subsided and he was soon splashing about in his bath. As we can see, this bizarre incident tells us more about belief in witchcraft in the seventeenth century than about the nature of the aliment that may have been at the origin of Pascal's subsequent illness.[6]

Pascal's mother died in 1626 and shortly thereafter his father resigned as President of the taxation court to concentrate on the education of his three children. The family moved to Paris in 1631, and from that time until 1640, when Pascal's father was sent to Rouen as tax officer by Cardinal Richelieu, Blaise studied at home under his supervision. The father did not want to teach his son subject matters that were beyond his ability: he introduced him to Latin when he was twelve years old, once Blaise had understood that the rules of grammar can be used as keys to unlock the meaning of language. The boy was thus spared seeing grammar as a tedious collection of rules to be memorized and, as his sister Gilberte put it, "when he had to learn it, he knew why." In the colleges run by the Jesuits, boys were taught Latin and Greek at a much earlier age.

Pascal's father was not only a distinguished lawyer but an amateur scientist, and he welcomed every opportunity to interest his son in curious facts of nature. One day, when the family was at table, a piece of china was accidentally struck with a knife and began to hum. When Pascal placed his hand on the plate, he noticed that the noise stopped immediately. This puzzled him and he began experimenting with other forms of sound, and even wrote a paper (which Gilberte graces with the word *treatise*) on his results when he was only eleven.

Euclid Mastered

Pascal's father was acquainted with the leading mathematicians in France and he made important contributions of his own. He studied the curve that

---

[6] We find the following clinical remark in Gilberte's account: "He was as thin as children who have *chartre*," a condition that is described by Furetière in his *Dictionnaire universel* (1690) as "an illness that leads to passivity and gradual loss of weight, and is also known as *phtisis*," namely tuberculosis (cited in *Œuvres de Pascal*, I, p. 1091, n. 1). Marin Mersenne, who suffered from skin rashes all his life, mentions women in his native village who treated such complaints by blowing on the infected area two or three times in the form of a cross and reciting the Our Father (*Quaestiones in Genesim* (1623), col. 599, cited in French translation in Armand Beaulieu, *Mersenne, le Grand Minime*. Bruxelles: Fondation Nicolas-Claude Fabri de Peiresc, 1995, p. 7).

the Ancients called the *conchoid of the circle* with such success that it was renamed *the snail of Pascal*, as it is still known today. In modern notation it can be written as $r = a + b\cos\theta$. But Etienne Pascal did not want his son to become engrossed in mathematics before he had mastered Latin and Greek, and he put his books on the subject out of the boy's reach. "What is mathematics and what is it all about"? Blaise persisted in asking his father, who finally replied that it was "a general method of tracing precise figures and finding their ratios," but that he was not to think about it. The result of such provocation for a gifted child may well be imagined. Blaise thought about it furiously. With a piece of charcoal, he drew figures on the floor-tiles, and he tried to draw perfect circles or trace triangles with equal sides. Since he did not even know the names of geometrical properties, he invented his own mathematical terms, calling, for instance, a circle a *round* and a line a *bar*.

One day, when Blaise was absorbed in studying a geometrical figure, his father walked into the room and asked him what he was doing. In his homespun terminology, Blaise explained that he had just found how to show that the sum of the three interior angles of a triangle is equal to two right angles. His father immediately recognized this as a Proposition of Book I of Euclid's *Elements*, and asked what had made him look for such a proof. Blaise answered that it was because he had found such and such a geometrical property, and when his father asked him again how he had discovered this, he mentioned yet another geometrical property until, step by step, he was led back to his initial definitions and axioms.[7] Etienne Pascal was so moved by his son's astonishing performance that he rushed to the house of his close friend, Jacques Le Pailleur, and told him, "with tears in his eyes," what he had just witnessed.

Gilberte does not describe her brother's do-it-yourself method but the Proposition that he reached is number 32 of Book I of the *Elements* and, if he followed the same route as Euclid, he was led to prove that the sum of the three interior angles of a triangle is equal to two right angles after realizing that if any side of a triangle is produced, the exterior angle is equal to the sum of the two interior and opposite angles. This property follows from Proposition 29 of Book I, which uses the postulate of parallel lines to show that a straight line that falls on two parallel straight lines makes the exterior angle equal to the interior and opposite angle on the

---

[7]*Œuvres de Pascal*, I, p. 606, and pp. 574–575 for the first version.

same side, and hence that the two interior angles on the same side are together equal to two right angles.

It was clear that there was no point in keeping Blaise from mathematics, and he was given a copy of Euclid's *Elements*. Tallemant des Réaux, the author of a collection of anecdotes about famous people of his acquaintance, goes further and claims that Pascal read the first six books of Euclid entirely on his own. This is the kind of embellishment that someone who has no talent for mathematics might wish to ascribe to those who seem to him to penetrate the mysteries of geometry without effort, and it should be taken with a grain a salt. But Blaise was undoubtedly brilliant and he was admitted, probably when he was no more than fourteen or fifteen, into a select discussion group that met in the convent of the friar Marin Mersenne, near the Place Royale (now Place des Vosges) in Paris.

Mersenne had become the clearing-house of the world of science and mathematics, and in addition to bringing scholars and scientists together each week, he wrote to all and sundry, in his cramped handwriting, on what was new and significant. In the absence of learned societies, professional journals, and other means of exchanging ideas, Mersenne made a valuable contribution to the development and dissemination of scientific ideas. He soon became an outspoken admirer of Pascal's mathematical gifts, and he was later to advertise his work on the vacuum. It is at Mersenne's weekly meetings that Pascal met such remarkable personalities as Gilles Personne de Roberval, the professor of mathematics at the Collège Royal; Ismael Boulliau, an eminent astronomer; Pierre Gassendi, the leading French exponent of atomism; Claude Mydorge, a distinguished mathematician with whom Descartes had worked on optics; Pierre Petit, a royal engineer; and Pierre Carcavy, the future librarian of Colbert and Louis XIV.[8] The most important contact for Blaise was Girard Desargues, who had just published a slim volume with the forbidding title, *Brouillon project d'une atteinte aux évènemens des rencontres du Cone avec un Plan*, which can be rendered as *Rough Draft of an Attempt to Deal with the Outcome of the Meeting of a Cone with a Plane*. Why he studiously

---

[8]Michel de Marolles, who met Pascal at the Academy, states in his *Mémoires* published in 1656 that the meetings were held on Saturday in the apartment of Jacques Le Pailleur. Further evidence is provided by the list of written or intended mathematical works that Pascal submitted to the *Parisian Academy of Mathematical Sciences* in 1654, and in which he says that he was educated in their midst (*Œuvres de Pascal*, II, p. 1031). Le Pailleur died on 4 November 1654. See Jean Mesnard, "Pascal à l'Académie Le Pailleur" in René Taton (ed.), *L'Oeuvre Scientifique de Pascal*, Paris: Presses Universitaires de France, 1964, pp. 7–16.

avoided the obvious title, *Conics*, which Apollonius of Perga had given to his classical work on the subject, is unknown. Desargues laid the foundations of a new kind of geometry, but lost most of his readers in a maze of botanical terms, by calling *trunk* a straight line with points on it, *knots* points on a line through which pass other lines, *branch* each of these lines, and *twig* a segment on one of the lines.[9]

The idea on which Desargues' work is based is a simple observation: a circle, when viewed obliquely, has the shape of an oval. For instance, the shadow of a lampshade looks like a circle on the ceiling, and like an oval where it falls on the wall. This illustrates the fact that shapes and sizes change according to the way light rays fall on them. Throughout these changes, certain properties remain the same, and it is these that Desargues studied. He realized that a conic section stays a conic section no matter how many times it undergoes a projection. In order to use this feature, Desargues took the original step of assuming that a parabola has a locus "at infinity," and that parallel lines meet at "a point at infinity." The theory of perspective, developed in the Renaissance, had made such ideas plausible by considering rays of light from the sun as parallel, and treating them as a cylinder or a parallel pencil of lines, unlike rays from a terrestrial source that they treated as a cone or a point pencil. In Desargues' view, the cylinder is merely a cone whose vertex is infinitely distant, and a parallel pencil of lines is simply a family of lines, all of which go through the same point at infinity.

It is in this way that the young Pascal became interested in geometry, the most concrete of the mathematical disciplines, and specifically in one of its most concrete branches, what we now know as projective geometry. Whereas his older contemporary, Descartes, favored the abstract approach of algebraical analysis, Pascal preferred drawing figures that could be visualized in space, and he enjoyed studying the transformations that result from moving them about. The conic sections were particularly interesting in this respect since the circle, the ellipse, the parabola, and the hyperbola result from different ways of slicing a cone. Following Desargues, Pascal identified parallel and concurrent straight lines because the first can

---

[9]Desargues did not expect a large readership and had only 50 copies printed in 1639. All were lost, but a transcript made by Philippe de la Hire in 1679 was discovered in Paris in the nineteenth century. In 1951, an original copy was found in the Bibliothèque Nationale in Paris and was published by René Taton in *L'oeuvre mathématique de G. Desargues*, Paris: Presses Universitaires de France, 1951.

be derived from the second when they are projected to infinity. Likewise, he considered a straight line as the limiting case of a circle whose center is located at infinity, and a cylinder the limiting case of a cone whose summit is at an infinite distance. The notion of infinity, which was necessary for this operation, remained incomprehensible, but it ushered in the idea of paradox that we will later find associated with Pascal's notion of scientific proof.

Pascal's First Poster

Pascal was only fifteen when Desargues' book appeared, but in less than a year he published his own original developments in a short but remarkable paper, the *Essay on Conics*, which consisted of a large printed page 35 × 43 centimeters. It is what we would now call a poster,[10] and it contains a proposition, described as the *Mystic Hexagon* that has since been known as Pascal's theorem. In essence, it states that the opposite sides of a hexagon inscribed in a conic section intersect in three collinear points. Pascal expresses this in a somewhat different way by saying that if A, B, C, D, E, and F are successive vertices of a hexagon in a conic, and if P is the intersection point of AB and DE, and Q the point of intersection of BC and EF (see Figure 1 where the conic is the egg-shaped figure, and ABCDEF is a plane figure with six sides),[11] then the lines PQ, CD, and FA "are of the same order" or, as we would put it, are members of a pencil, whether a point pencil or a parallel pencil. Pascal deduced several applications from this proposition, the first being the construction of the tangent to a conic section, but these developments do not appear in the poster.

Marin Mersenne was so impressed that he wrote to a correspondent in London that an eighteen-year-old mathematician (Pascal was actually seventeen at the time) had been able to capture the essential property of conic

---

[10]The *Essay on Conics* is published in the *Œuvres de Pascal*, II, pp. 228–235, and is translated in D.J. Struik (ed.), *A Source Book in Mathematics 1200–1800*. Princeton University Press, 1986, pp. 163–167. Pascal's poster suffered the same fate as Desargues' book and soon got lost. It was rediscovered by Colin Maclaurin in 1727. See René Taton, "L'oeuvre de Pascal en géométrie projective," *Revue d'histoire des sciences* 15(1962), pp. 197–252, reprinted in *L'Oeuvre Scientifique de Pascal*. Paris: P.U.F., 1964, pp. 17–72; René Taton, "*L'Essay pour les Coniques de Pascal*," *Revue d'Histoire des Sciences* 8(1955), pp. 1–18.

[11]I use the figure as redrawn by Carl B. Boyer in his *History of Mathematics*, Princeton: Princeton University Press, 1985, p. 396.

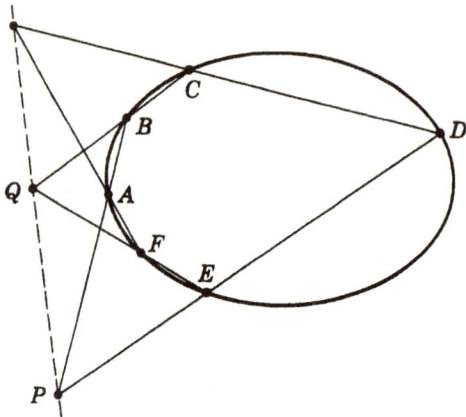

FIGURE 1   *Diagram of Pascal's theorem written when he was seventeen years old.*

sections in a single proposition from which he drew 400 corollaries.[12] He also mentioned the boy-prodigy to Descartes, who replied, rather ungraciously, "I am not surprised that someone can provide demonstrations that are easier than those of Apollonius, a prolix and confused writer who only proved what is quite easy, but there are other things concerning conics that would give a sixteen-year-old child a harder time."[13] As he had promised in his poster, Pascal wrote a treatise on conics, the *Conicorum Opus Completum*, which was not published in his lifetime. The manuscript was loaned to Leibniz by Pascal's heirs when he visited Paris in 1676 and, after taking notes, he made a detailed description of the content, and returned it with the recommendation that it be printed. Unfortunately, Leibniz's advice was never followed, and only the first part, *On Conic Sections*, has survived. It was found among Leibniz's papers after his death and published in 1892 along with Leibniz's notes.

Pascal used his new method to construct tangents to a circle and tangents to a sphere. The first had been studied by Apollonius in Antiquity,

---

[12] Letter of Mersenne to Theodor Haak, 28 November 1640, *Correspondance du P. Marin Mersenne*, edited by P. Tannery, C. de Waard et al., 16 vols. Paris: Editions du C.N.R.S., 1933–1986, vol. XI, p. 429. The date is erroneously given as 18 November 1640 in the *Œuvres de Pascal*, II, p. 239.

[13] Letter of Descartes to Mersenne, 25 December 1639, *Œuvres de Pascal*, II, p. 237. Apollonius of Perga, who lived in the third century B.C., wrote a treatise on conic sections that is one of the greatest mathematical works from the ancient world.

and was known through the summary that is found in Pappus, a fourth-century mathematician. In 1600, Viète had tried to reconstruct Apollonius' lost procedure in a book that he entitled *The French Apollonius*. Pascal now proposed to generalize the whole method by formulating the problem as follows: Given any three elements (points, lines, or circles), draw a circle that passes through the given points and is tangent to the given lines and circles. The second problem, that of the construction of tangents to a sphere, "is derived," writes Pascal, "from a single property of conic sections that is useful in coping with many other difficult problems. Yet it barely fills a page."[14] Unfortunately, Pascal does not tell us what this property is, and we are left to wonder.

Pascal did not realize that the great developments in mathematics in the seventeenth century were occurring in analytic geometry. François Viète (1540–1603) had seen that the wave of the future was algebraic geometry, and the two leading mathematicians of Pascal's day, Descartes and Fermat, followed his lead. Descartes' *Geometry* was published in 1637, when Pascal was fourteen years old, but the work had less of an impact than might be expected because his co-ordinate system was not clearly explained. Indeed, the phrase, "Cartesian co-ordinates," so often used today, is an anachronism. We do not find in Descartes' *Geometry* the "X-axis" and "Y-axis" with which we are familiar from high school geometry. Descartes' innovation was only fully appreciated after Frans van Schooten published a Latin translation of the *Geometry* with commentaries and additional material in 1649. This may be the reason why Pascal failed to see the extent to which the algorithms of algebra opened a royal road for geometry.

The Calculating Machine

Blaise's father, Etienne, was appointed tax officer in Rouen in 1640 after a rebellion of local peasants had led to the destruction of the registers. The accounting system had to be completely overhauled, and the task took more time and was more onerous than anticipated. Three years later, Pascal's father was still computing well into the night, as we know from a postscript to a letter to his daughter, Gilberte: "My loving daughter will excuse me if I do not write as I should wish to do if I had time. I have never

---

[14]Pascal, *Letter to the Parisian Academy of Mathematical Sciences*, written in 1654, *Œuvres de Pascal*, II, p. 1033.

faced such difficulties as at present, and I could not cope with more. For four months now, I have not gone to bed more than six times before two o'clock in the morning."[15]

Pascal, who was still in his teens, offered to help his father and was soon confronted with the tedious calculation of fiscal charges. In order to afford relief to his father as well as himself, he began looking into the possibility of mechanical calculation, a task rendered more complicated by the fact that French currency was not computed in tens but by a method like the one that prevailed in the United Kingdom until as late as 1971. The French *livre* consisted of 20 *sols*, and each *sol* was equivalent to 12 *deniers*. The meant that the livre was worth 240 deniers, not 100, and this gave Pascal a headache. After several trials, he invented a system to add and subtract with a series of gears that went forward, but not backward. This was similar to the mechanical calculating machines that served as precursors to today's electronic calculators.

The idea itself is easy to grasp. Pascal placed side by side in an oblong box eight small drums, around the upper and lower halves of which the numbers 0 to 9 were written in descending and ascending orders, respectively (see Figure 2). The upper row of figures was for subtraction, the lower for addition. One half of each drum was shut off from outside view by a sliding metal bar. Below each drum was a wheel consisting of ten (or twenty or twelve) movable spokes inside a fixed rim numbered in ten (or more) equal sections from 0 to 9, rather like a clockface. The numbers to be added or subtracted were fed into the machine by means of wheels on the box lid that were turned with a metal hook. A simple gearing mechanism of five toothed wheels linked the ten movable spokes to their appropriate drum. To increase by one the number showing in any aperture, it was necessary to turn the appropriate drum forward 1/10th of a revolution. For instance, to add 315 + 173, first 315 was recorded on the three drums closest to the right-hand side with 5 appearing in the sighting aperture to the extreme right, 1 next to it, and 3 next to that one. Then 173 was entered by giving the drum on the extreme right three turns, moving the drum to its left by 7/10 of a revolution, and the drum to this one's left by 1/10. The total of 488 could then be read off in the appropriate slots. But, easy as this operation was, a problem clearly arose when the numbers

---

[15]Postscript to the letter of Blaise to Gilberte, 31 January 1643, *Œuvres de Pascal*, II, p. 283; the letter is translated in Emile Cailliet and John C. Blankenagel (eds.), *Great Works of Pascal*. Philadelphia: Westminster Press, 1948, pp. 39–40.

FIGURE 2 *Pascal's Calculating Machine, viewed sideways and from the top.*

to be added together involved totals that had to be carried forward: say, 412 + 179 where 9 + 2 = 11 required that 1 be carried forward. Before Pascal, there had been no previous attempt at a calculating machine capable of carrying column totals forward, and this presented a serious technical challenge.

Pascal overcame the problem of transfers to the power of ten by inventing an ingenious *sautoir* (or escapement arm) whereby a pawl attached to a counterweight regulated the motion of a wheel representing a power of ten, through contact with a tooth of that wheel. Whenever the unit wheel completed a revolution, the counterweight rose automatically, the pawl worked itself loose from the tooth of the wheel against which a spring normally held it in place, and slid away to the right where it pushed on the wheel leftwards by 1/10 of a revolution as the counterweight fell back.

Subtraction, the second operation, was not as straightforward as might seem because the wheels and gearing mechanism could not be put into reverse. Consequently, Pascal devised a system of parallel notations enabling the converse arithmetical process to be carried out with the minimum of complexity. The drums that were used for addition were also numbered in reverse order from 9 to 0, and the only further and inevitable refinement was occasioned by the problem of negative transfers.

Pascal had greater trouble with multiplication and division, the last two of the four basic operations of arithmetic. He does not seem to have been aware of Wilhelm Schickard's so-called calculating-clock, manufactured in Germany as early as 1624, and apparently able to perform the operations of multiplication and division. Nine years after Pascal's death, Leibniz invented a more versatile machine capable of the four arithmetical processes which, thanks to a *stepped* wheel, performed the operations of multiplication or division by means of repeated additions or subtractions.

Pascal's sister, Gilberte, tells us that Pascal made his discovery when he was nineteen. "He embodied," she writes, "his scientific idea in a machine that can be used to perform mathematical operations without having to think about them." Nonetheless, the invention itself required considerable thought, and it was about this time that Pascal's health began to fail, although Gilberte insists "that what exhausted him was not so much the idea of the calculator, or figuring out all the requisite movements, but instructing the workmen how to proceed."[16] Pascal admitted as much to Chancellor Séguier (whose position was roughly that of a modern prime minister): "I encountered difficulties as great as those that I wished to avoid and to which I was seeking a remedy. I could handle pen and compass but not hammer and nails, and although the workmen knew how to use their skill, they had little idea of the science on which it rests."[17]

There was more to Pascal's problems than finding people who could make cogs and wheels to his specification. A Rouen watchmaker heard about his invention and rushed to put an unauthorized copy on the mar-

---

[16]*Œuvres de Pascal*, I, pp. 576–577. On the genesis of Pascal's machine, see René Taton, "Sur l'invention de la machine arithmétique," *Revue d'histoire des sciences* 16(1963), pp. 139–160, reprinted in *L'Oeuvre Scientifique de Pascal*. Paris: P.U.F., 1964, pp. 205–227; Guy Mourlevat, *Les machines arithmétiques de Blaise Pascal*. Clermont-Ferrand: La Française d'Edition et d'Imprimerie, 1988.

[17]*Œuvres de Pascal*, II, p. 332. Pascal's letter to Chancellor Séguier is translated in Pascal, *Selections*, edited by Richard H. Popkin. London: Collier Macmillan, 1989, pp. 21–23. This work will be subsequently referred to as Popkin trans.

ket. It was inferior to the original, but Pascal was so annoyed that he stopped production and dismissed the craftsmen that he had hired. His friends remonstrated with him, appealed to his entrepreneurial spirit, and advised him on marketing techniques. Better still, they sought out potential customers and showed the machine to important people such as the Prince de Condé and the Chancellor Séguier. Both dignitaries were impressed and urged Pascal to carry on. This occurred early in 1644 and by the next year, some fifty working models later, Pascal offered the Chancellor the prototype that we can still admire today in the *Musée des techniques* in Paris. A suitable *Letter of Presentation*, which accompanied the gift, was published by Pascal along with a *Notice* that reads like a publicity blurb. Would-be purchasers are warned about the inferior quality of rival products, and told to "call on Mr Roberval, the professor of mathematics at the Collège de France, who will show them, free of charge, how easily it can be operated." Roberval's address is provided along with the opening hours, namely "before 8.00 a.m. every morning, and in the afternoon on Saturday."[18]

Sales were not brisk. The major deterrent was the price that contemporary writers mention as 100 or even 400 *livres*, a substantial sum when we consider that the monthly rent for a large apartment in the best part of Paris was about 30 *livres*. The invention was a prestige object for the wealthy, not a business machine for the average shopkeeper. One of Pascal's first customers was the new Queen of Poland, Louise-Marie de Gonzague, who took a calculator along when she left for Warsaw at the end of 1645. King Ladislaw VII was delighted with the machine and kept it in his own office. He ordered two more "pascals" and asked that they be modified to take into account local Polish currency where 3 *gros* = 2 *sols*; 20 *sols* = 1 *florin*, and 3 *florins* = 1 *thaler*.

When Descartes called on Pascal during his visit to Paris in September 1647, he was shown the calculating machine and given explanations by Roberval, who had been invited for that purpose. If Descartes was impressed, he showed no sign of it. Marin Mersenne was more generous and tried to further the interests of the young scientist. He made a point of recommending to Constantin Huygens, the secretary of Prince Frederick Henry of Orange, not to miss seeing the machine on his next trip to Paris with his son Christiaan.

---

[18]*Œuvres de Pascal*, II, p. 341.

DESIGNING EXPERIMENTS & GAMES OF CHANCE

Royal Patronage

In 1649, Pascal obtained from King Louis XIII a patent that gave him the exclusive right to manufacture his calculating machine. Any violation was to be fined 3,000 *livres*, roughly $100,000 in modern currency. Whether the threat carried any clout can be doubted, but Pascal was mainly concerned with the price of the machine, which he recognized as much too high, and he tried to make it affordable by using a simpler movement. In spite of recurrent bouts of illness, he kept on improving his machine and putting new versions on display. The Duchesse d'Aiguillon, the rich niece of Cardinal Richelieu, even invited him to give a demonstration in her sumptuous residence of Le Petit Luxembourg in the spring of 1652.[19] Meanwhile, the Abbé Bourdelot, who had engineered Pascal's meeting with the Prince de Condé, had left France after the civil unrest known as "La Fronde" to become the physician of Queen Christina in Stockholm. On 14 May 1652, he wrote Pascal a flattering letter in which he praised him as "one of the geniuses that the Queen is looking for."[20] He did not go as far as to invite Pascal to Stockholm, probably because Descartes, some three years earlier, had accepted Queen Christina's summons only to discover that she expected him to instruct her before dawn. Unused to the cold and the early hours, Descartes had caught a chill and died within three months of his arrival in Stockholm. Pascal was not interested in following him to the grave, but he welcomed the opportunity to send a copy of his machine to the Queen. The letter that accompanies the gift is a curious combination of personal vanity and self-serving praise of the young sovereign: "Had I as much health as I have zeal," he writes, "I would not have allowed other hands than my own the honour and pleasure of placing my invention at the feet of the greatest princess of the world."[21]

In his most gallant and courtly style, Pascal goes on to explain that his machine performs mathematical calculations automatically, but he soon launches into a disquisition on true greatness and the difference between material and spiritual things. The Queen, he says, possesses the two traits that he admires most: sovereign authority and superior science. Begging her not to take offense, he adds that either of these traits without the other

---

[19]"La Muse Historique," 14 April 1652, cited in *Œuvres de Pascal*, II, p. 903.
[20]*Œuvres de Pascal*, II, p. 919.
[21]*Œuvres de Pascal*, II, p. 922; Popkin trans., p. 30.

is defective: "However powerful a monarch may be, something is missing to his glory if he does not possess intellectual pre-eminence; and however enlightened a subject may be, his condition is always lowered by his dependence." But the Queen, Pascal immediately adds, is known to combine temporal power with intellectual penetration and must undoubtedly be called to reign over the whole extent of the earth.[22]

Royal patronage, it must be confessed, did not do much for the sales, and although the Jesuit professor, Jean François, acknowledged in his *Arithmétique* published in 1653 that Pascal's "wheel" performed "reliably and quickly" the four basic operations of adding, subtracting, multiplying, and dividing, he thought it was just too expensive. Interest was not revived until 1659, when a machine was sent to Christiaan Huygens in the Netherlands with a covering letter in which a friend of the Pascal family praised it as having brought relief "to merchants, bankers and businessmen."[23] The model, which consisted of five drums and hence could handle sums up to 99,999, impressed Huygens but he soon realized that he would need someone to teach him how to operate it. Then as now, the written manual was a poor substitute for personal instruction.

A New Way of Teaching the Alphabet

Having made counting easier, Pascal also wanted to find a simpler way of teaching how to read. To this end he devised a new syllabic way of teaching the alphabet, which is described in the *Grammar of Port-Royal* that some of his friends published in Paris in 1660. The method consists in teaching children to pronounce the consonants phonetically. For instance, in English the letters *f*, *r*, and *y* of the word *fry* should be pronounced phonetically to make the sound *fry* and not *ef-are-why*. Pascal's syllabic method was tried out in the elementary school of the Convent of Port Royal, where his sister Jacqueline was a teacher. She commented on two drawbacks. The first is that the letter *e* at the end of French words is not always pronounced, and the second is the problem of consonants that follow vowels at the end of a word. Although the letter *n*, for instance, should

---

[22]*Œuvres de Pascal*, II, p. 925; Popkin trans., p. 31.

[23]Chapelain described the machine as "a strange combination of wheels" in his letter to Huygens of 20 August 1659, *Œuvres de Pascal*, IV, p. 690.

be taught phonetically and pronounced *ene* in French (the same sound as *n* in English), it has a different sound when combined with *e* or *o*, as in the words *en* and *on*.[24] Whether Pascal had foreseen these difficulties is something we shall never know, since his answer is not extant and the *Grammar of Port Royal* does not mention them, let alone try to answer them.

---

[24]Letter of Jacqueline Pascal to her brother, 26 October 1655, *Œuvres de Pascal*, III, p. 439

CHAPTER 2

Filling the Void

From Galileo to Torricelli

Natural philosophers began to investigate the possibility of producing an effective vacuum, namely a space entirely void of matter, roughly at the time Shakespeare wrote *Much Ado About Nothing* at the turn of the seventeenth century. But, in order to understand Pascal's involvement in this quest, we must travel to Italy and say a few words about the contribution that Galileo made in the *Two New Sciences* that he published in 1638, when Pascal was fifteen years old.[1] A curious fact had struck Galileo. When the spring-cleaning of public cisterns and fountains was carried out, the water could not be raised above 18 *braccia*, or roughly ten meters and a half. "I have learned," he writes,

> the cause of a certain effect that I had long wondered at and despaired of understanding. I once saw a cistern that had been provided with a pump under the mistaken impression that the water might thus be drawn with less effort or in greater quantity than by means of an ordinary bucket. The stock of the pump carried its sucker and valve

---

[1] The Latin, *vacuum*, is now restricted to a physical space from which air or some other substance has been removed, but in the seventeenth century it had the broader meaning associated with the English word *void* or the French word *vide*, and can often best be rendered by the general term *empty*. For the prehistory of these experiments, see W. E. Knowles Middleton, *The History of the Barometer*. Baltimore: The Johns Hopkins Press, 1964, pp. 3–18; Charles B. Schmitt, "Experimental Evidence for and against a Void," *Isis* 58 (1967), pp. 352–366.

in the upper part so that the water was lifted by attraction and not by pushing as is the case with pumps in which the sucker is placed lower down. This pump worked perfectly so long as the water in the cistern stood above a certain level; but below this level the pump failed to work. When I first noticed this I thought the machine was out of order; but the plumber whom I called in to repair it told me the defect was not in the pump but in the water which had fallen too low to be raised through such a height. He added that it was not possible, either by a pump or by any other machine working on the principle of attraction, to lift water a hair's breadth above eighteen *braccia*. Whether the pump is large or small, this is the extreme limit of the lift.[2]

In this passage, Galileo distinguishes between a *suction-pump*, which can lift water to 18 *braccia* by what he calls attraction (we would write suction today), and a *force-pump* that can push water to a much greater height if the tube is strong enough and a sufficient force is applied. The behavior of the suction-pump initially left Galileo "full of wonder and empty of understanding" until he realized that the explanation lay in the creation of a vacuum above 18 *braccia*. What may seem surprising is that Galileo did not associate this phenomenon with the weight of the air (what we refer to as atmospheric pressure). This is one of the most fascinating near misses in the history of science, and we can learn much about science in the generation that preceded Pascal by examining how this oversight occurred.

Weighing Air

In 1614, Giovanni Battista Baliani, a prominent citizen of Genoa and an amateur scientist, heard that Galileo had determined the weight of air, and he immediately wrote to ask him how he proceeded. In his reply, Galileo described how he filled a glass flask with compressed air and weighed the flask before and after. He then had the compressed air drive out the water in another flask of the same size, weighed the water expelled, and com-

---

[2]Galileo, *Discorsi intorno a due nuove scienze* in *Opere di Galileo Galilei*, ed. A. Favaro, 20 vols. Florence: G. Barbèra, 1890–1909, VIII, pp. 63–64. In his English translation, *Two New Sciences* (Madison: University of Wisconsin Press, 1974), Stillman Drake gives the pagination of this standard edition. The Florentine *braccio* is 58.4 cm, so 18 *braccia* are slightly more than 10.51 meters.

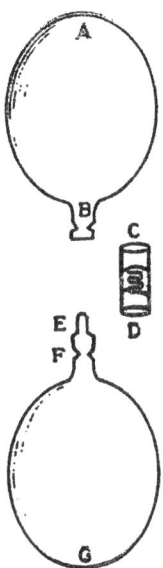

FIGURE 1 *Diagram of the two flasks that Galileo used to determine the weight of air.*

pared it to the weight of the air that had been left in the flask (see Figure 1). The water, "as far as I recall," he wrote, "weighed about 460 times as much but I am not sure of this. The operation could be repeated several times to know for sure."[3] The correct figure is closer to 800 times, and although Galileo's result was too low by almost half, it was a remarkable achievement for his day.

Pascal did not know of Galileo's correspondence with Baliani, but he could have read the description of the experiment that Galileo published in his *Two New Sciences*. Galileo took a large glass flask with a narrow neck into which he inserted a football-valve and, with a syringe, drove in two or three additional volumes of air. He then weighed the flask on a delicate balance. The valve was opened, the forced air came out, and when the flask was weighed again, it was appreciably lighter. In order to determine the weight of the air, it was necessary to measure the quantity of air

---

[3]Letter to Baliani, 12 March 1614, *Opere di Galileo*, XII, p. 36, trans. by S. Drake in *Galileo at Work*, Chicago: Chicago University Press, 19, p. 233. Aristotle, *On the Heavens*, bk. IV, ch. 4, 311 *b* 10–11.

compressed. This Galileo did in two ways. First, he allowed the excess air from the flask to escape without loss into another flask completely filled with water, and so constructed that the water displaced by the entering air could be caught in a suitable vessel, whose increase of weight he then determined. The quantity of water thus expelled was equal to the volume of air that came out of the first flask. Galileo now weighed this flask from which the forced air had gone out and noted the decrease in weight. He then compared the weights of the two equal volumes of air and water and found them to be approximately as 1:400.

Galileo also used a second method that is more elegant, and could be carried out in a single flask. This time he forced water, instead of air, into the flask until it was three-quarters full so that the air was compressed into one quarter of the space it originally occupied. He weighed the flask and then opened the valve, allowing to escape the volume of air that had previously occupied the space that was now filled by the water, namely three quarters of the capacity of the flask. Finally, he weighed the flask again. The decrease in weight indicated the weight of the air that had occupied the volume that was now filled with water.

Galileo provided key information about the experimental design. He pointed out, for instance, that he used a large glass flask with a narrow neck to which he had attached a leather collar into which he had inserted a football-valve, and that he opened the first flask by passing a rod through the second one.[4] But Baliani never got around to performing this experiment, and he does not appear to have corresponded with Galileo about the weight of air, or anything else, for over sixteen years. When he wrote again in the summer of 1630, it was to consult him about an apparently unrelated practical problem. In order to bring water to the city, Genoa had recently installed a lead siphon that led over the hills. Although the siphon had first been filled with water, when one end was opened, it behaved as if there were some leak of air at the top, and the water could not be pumped across the summit. Galileo replied to Baliani that he was sorry the work had been done without his having been consulted, as he had long before discovered that it was impossible to operate a siphon (the common name of the suction-pump) for above 20 *braccia*. He claimed that the

---

[4]Galileo, *Two New Sciences*, translated by Stillman Drake. Madison: University of Wisconsin, 1974, pp. 80–83, in the *Opere di Galileo*, vol. VIII, pp. 121–124. See Stillman Drake, *Galileo at Work*. Chicago: University of Chicago Press, 1970, pp. 231–233.

cause was "obvious, as is usually the case once the solution is found."[5] Yet he was deceived. His solution, however "obvious," is wrong, because he drew an unfortunate analogy between a rope that snaps when its load becomes too heavy and a column of water that "breaks" when it reaches 18 *braccia*.

Galileo illustrated his train of thought with the help of an example. Let us assume, he says, that an iron cable that snaps when 100 pounds are suspended from its lower end will also break when the weight of the cable itself reaches 100 pounds, and let us apply the same reasoning to a "rope of water." Since water cannot be raised in a siphon above 18 *braccia* regardless of the diameter or the length of the pipe, this must indicate that the weight of a column of water this high is the maximum possible. It appeared to Galileo that in the case of the rope the weight had to overcome both the "resistance to the void" (which would follow breakage) and the "very strong attachment of the parts," whereas in the case of a column of water only the "resistance to the void" had to be overcome.

In his letter of acknowledgment, Baliani recognized that he had failed to distinguish between a force-pump and a suction-pump, but he claimed to have grasped, in the light of their earlier discussion, that if air has weight then a vacuum is possible. What he had not suspected was how easy it is to produce. Now comes the crucial insight: if air has weight, says Baliani, its properties can be compared to those of water. If we were to stand under ten feet of water, we would not feel its weight unless it were piled up on our head while the rest of our body (in some marvelous way) remained in the open air. The same reasoning applies to the weight of the air, which we do not feel because we are completely surrounded by it.[6]

Galileo did not pursue Baliani's line of investigation in the *Two New Sciences*. Rather, he focused on facts that seemed to indicate that a vacuum was possible. For instance, he asked why two polished slabs of marble, which easily slide over each other, stick when we try to pull them apart. Since air cannot rush in instantaneously when they are separated, a vacuum must occur between the plates for a short interval of time: "Hence we must say that by force (or contrary to nature) a void is sometimes admitted, although in my opinion nothing is contrary to nature save

---

[5]Letter of Baliani to Galileo, 27 July 1630, *Opere di Galileo*, XIV, p. 125; letter of Galileo to Baliani, 6 August 1630, *Opere di Galileo*, XIV, p. 129. The estimate of 20 braccia was improved to 18 in the *Two New Sciences* (*Opere di Galileo*, VIII, p. 64).
[6]Letter to Galileo of 24 October 1630, *Opere di Galileo*, XIV, pp. 158–159.

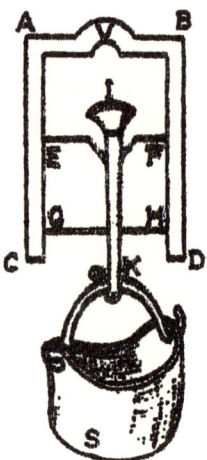

FIGURE 2 *Galileo's diagram to illustrate how to determine "the force of the vacuum."*

what is impossible, and that never happens."[7] Here Galileo's language plainly lags behind his conceptualization, for if whatever happens in nature is natural, then the expression, "contrary to nature," is a misnomer. But the philosophical issue is left in abeyance, and Galileo moves on to determine the strength of the void. His experiment, which is most likely a thought-experiment, is better understood, as he himself points out, "with the aid of a diagram rather than by mere words." A glass cylinder, ABCD, with a wooden piston, EFGH, is inverted (see Figure 2) and a wire, IK, is passed through a hole in the piston. A bucket is attached to a hook at the end of the wire. After water has been poured into the cylinder, the piston is adjusted so that EF is about seven centimeters from the top. Sand is added to the bucket until it is just heavy enough to detach the piston from the water "to which nothing held it except repugnance to the void." It follows that "by weighing the piston together with the iron wire, the container, and whatever it contains, we shall have the amount of the force of the void."[8] Once this weight is determined, it should be suspended to a glass cylinder of the same size as the cylinder of water. If the cylinder breaks,

---

[7]Galileo, *Two New Sciences*, *Opere di Galileo*, VIII, pp. 59–60.
[8]*Ibid.*, VIII, pp. 62–63.

says Galileo, we shall have compelling evidence that resistance to the void is the sole cause of the cohesion of glass, but if we have to add more weight, say four times as much, then we shall have to conclude that the void accounts for only one fifth of the resistance.

Since 18 *braccia* is the limit to which any quantity of water can be raised whether the pump is wide, narrow, or "as thin as a straw," it follows that by weighing the water contained in 18 *braccia* of a pipe we shall find "the value of the resistance of the void for cylinders of any solid material that are as large as the hollows of those pipes." Galileo does not seem to have attempted to weigh such a large amount of water. Instead he applied his theory to a hypothetical rod of copper one *braccio* long and weighing one eighth of an ounce. Assuming that the rod will break when it is made to support 50 pounds, Galileo reasons as follows: Since 1 pound = 12 ounces, 50 pounds = 600 ounces, namely 4800 times as much as the rod's weight. Hence, we can suspend a copper wire, of whatever thickness, from a height of 4801 *braccia* before it snaps under its own weight. Now 4801 *braccia* is over 2.8 kilometers! No wonder Galileo did not attempt to put his theory to an actual test. He worked out, however, the part that the *resistance to the void* would play in this case since it would have to be as much as the weight of a column of water 18 *braccia* long and the same thickness as the rod of copper. Since copper is 9 times as heavy as water, the length of this copper rod would only have to be 2 braccia, and the resistance to breakage that depends on resistance to the void would therefore be equal to the weight of this copper rod.[9] Galileo does not determine this weight in ounces but it can readily be calculated: a rod of copper one *braccio* long weighs one eight of an ounce, so a 2 *braccia* rod will weigh one quarter of an ounce. This is a mere 1/2400th of the total weight. So the vacuum plays a very modest role.

We can summarize Galileo's interesting exercise in scientific deduction by saying that although he accepted nature's abhorrence of a vacuum, he believed that it was limited and that it could be measured by the weight that is required to separate a piston from the water in a glass cylinder. In this way, he explained the strange limitation to suction-pumps, which cannot raise water above 18 *braccia*. Galileo focused exclusively on the column of water *inside* the tube, and it did not occur to him to consider the column of air *outside*. Baliani had grasped the importance of the sur-

---

[9]*Ibid.*, VIII, pp. 65–67.

rounding pressure, but all his life Galileo stuck to his explanation in terms of the breaking strength of materials, likening water in a vertical pipe closed at the top to a suspended rope or cable. Galileo's difficulties serve as a reminder that although the scientific conceptual scheme implied by the phrase "nature abhors a vacuum" may have been wrong, it was not nonsense. It accounted to some extent for a number of apparently unrelated phenomena, such as the action of pumps in common use, the adhesion of two slabs of wet marble, the action of bellows, and the inability to make a "hole" in a liquid the way we can in a solid.

Rome Faces the Void

Fifty copies of Galileo's *Two New Sciences* reached Rome at the end of 1638 and were immediately sold out at 2 *scudi* apiece, a large sum, since a well-paid official at the Vatican earned about 16 *scudi* per month. The book was "admired not to say worshipped"[10] in the Eternal City, but the mathematician Gasparo Berti doubted the claim that water could not be raised "a hair's breadth above 18 *braccia*." He thought he could show Galileo that he was wrong, and with the consent of a patron who owned a several storey building, he fixed a metal tube some 22 *braccia* long (just under thirteen meters) to the facade. The bottom end, which was fitted with a tap, was placed in a barrel of water (see Figure 3, where later features are incorporated; in the original design the end of the tube just went into a small bucket, as we see in Figure 4).[11] The tube was filled from the top, the opening closed, and the tap at the bottom opened. The water fell to a certain height, and when Berti measured it, he had the immense pleasure of finding it well above 18 *braccia*. His joy was short-lived, however, for it was pointed out to him that the height of the column of water should have been measured from the top and not from the bottom of the barrel, that is, without including (as Berti had done) the height of the water

---

[10]Letter of Jean-Jacques Bouchard, who was the secretary of Cardinal Francesco Barberini, the nephew of Urban VIII, to Vincenzo Capponi, 23 April 1639, *Opere di Galileo*, XVIII, p. 45.
[11]See Frank D. Prager, "Berti's Device and Torricelli's Barometer from 1641 to 1643," *Annali dell'Istituto e Museo di Storia della Scienza* 2 (1981), pp. 35–53. Figure 3 is taken from Gaspar Schott's beautifully illustrated *Technica Curiosa* of 1664, and Figure 4 comes from Raffaello Magiotti's letter to Mersenne of 12 March 1648 (Marin Mersenne, *La Correspondance du P. Marin Mersenne*, edited by P. Tannery, C. de Waard et al., 16 vols. Paris: Editions du C.N.R.S., 1933–1986, vol. XVI, p. 169).

FIGURE 3 *Gasparo Berti's experiment on the facade of a Roman building.*

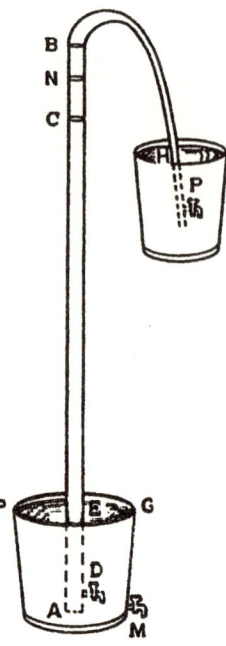

FIGURE 4   *Berti's original experiment.*

already in the barrel. When he did this, he found that Galileo had been right after all.[12]

But the matter was still worth investigating and Berti, who was a clever experimentalist, made improvements to his apparatus. He inserted a glass section BC to facilitate the observation of the water level at N (see Figure 4), but soon discovered that it would be better to fit a glass flask to

---

[12]We may note here that if he wanted "to convince Galileo," as he told his friend Raffaello Magiotti, the experiment must have been performed between 1639 and 1641, since Galileo died on 8 January 1642 (letter of Magiotti to Mersenne, 12 March 1648 in *La Correspondance du P. Marin Mersenne*, XVI, p. 170). Stillman Drake, however, argues for 1643 on the grounds that the Italian expression "convincere Galileo" means to "refute" rather than "convince" Galileo, and that the magnet used to ring the bell is not mentioned in Kircher's book, *Magnes (On Magnetism)* that he published in 1641 and re-edited 1643. Tommaso Cornelio, who describes various experiments of this kind in his *Progymnasmata Physica* of 1663, speaks of Berti as professor of mathematics at the Roman University when the experiment took place, and Berti only held this post for a few months prior to his death in 1643 (S. Drake, voice *Berti* in the *Dictionary of Scientific Bibliography*. New York: Charles Scribner's Sons, 1970, vol. II, p. 84).

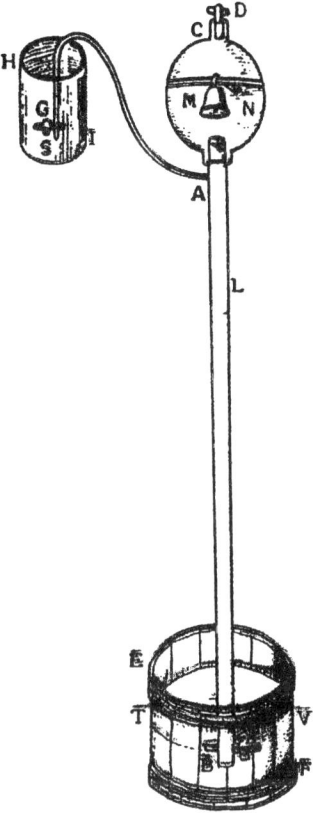

FIGURE 5 *Description of the Experiment showing the bell inside the glass container at the top.*

the top of the tube (see Figure 5). He disconnected the upper bucket and its riser from the downpipe, and then pierced the tube just below the flask to attach a smaller lead tube AS that ended with a tap G that plunged in a vessel, HI, full of water (see Figure 5).[13] The vessel had initially been placed lower than the flask in order to fill the tube AS with water poured in at C, and when the tube AS was full, the vessel was raised and placed on the windowsill as we see in the illustration.

---

[13] Figure 5 is taken from Emmanuel Maignan's *Cursus Philosophicus* published in Toulouse in 1653, and is reproduced in W. E. Knowles Middleton, *The History of the Barometer*. Baltimore: The Johns Hopkins Press, 1964, p. 11.

When the whole apparatus was full of water, tap B at the bottom of the tube was opened and the water sank to the same level as before. This tap was then closed, and tap G opened at the top. The small tube sucked up the water continually supplied by the vessel HI until the larger tube and the flask were filled "except for an air-bubble of a certain size" at the top of the flask. Since the tube was made of lead, and hence was opaque, the level of the water in the pipe was probably determined by lowering a sounding line after the tap at the bottom of the tube had been closed, and the one at the top opened. Berti went over these matters with his friends Raffaello Magiotti and the Jesuits Athanasius Kircher and Nicolò Zucchi. Kircher suggested placing a bell inside the flask to find out whether its sound could be heard. Here the problem was to get the bell to ring, and Kircher had the ingenious idea of pulling the clapper back with a magnet and then releasing it so that it fell and struck the bell (see Figure 6). The result was clearly audible and, for Zucchi and Kircher, this indicated the presence of air in the apparently empty space.[14]

This story would prove, if proof were required, that experiments, when first performed, rarely speak with the authoritative voice that they acquire in textbooks. Indeed, they may often be said to speak with a forked tongue. We now know that sound was heard, not because air was left inside the flask, but because the vibrations of the bell were communicated to the metal rod to which it was attached, and from there to the flask and the outside air. This explanation was not beyond the ken of seventeenth-century physics, and it was actually suggested by a French friar, Emmanuel Maignan, who was stationed in Rome at the time. He was not present when the experiment was carried out but, as soon as he was told that sound had been heard, he inferred that the vibratory motion of the bell had been conducted through the solid mounting to the air. Maignan offered as confirmation the fact that deaf persons can "experience" musical sounds if they feel the instrument vibrating. "Try this yourself," he writes,

> stop up your ears as well as possible and clasp firmly between your teeth the peg of a lute. Then play the instrument or, better still have

---

[14]From Athanasius Kircher, *Musurgia Universalis*, Rome, 1650, pp. 12–13, from which Figure 6 is taken. Mersenne also heard the sound of the bell when he repeated the experiment in 1648 but he did not conclude that there was air in the flask. He opted for the presence of some unspecified "subtle matter" (quoted in Cornelis de Waard, *L'expérience barométrique*. Thouars: Imprimerie Nouvelle, 1936, p. 183).

FIGURE 6 *Illustration showing how the clapper was pulled back with a magnet.*

someone else play it, and you will hear the music very well, even more loudly and agreeably than you would with open ears.[15]

How Empty Is the Void?

Whether sound can travel through a vacuum was to become an abiding subject of controversy, and twenty years later Robert Boyle was still try-

---

[15]Quoted in Middleton, *The History of the Barometer*, p. 14.

ing to find out in Oxford. He placed a bell, a watch, and a pistol in an exhausted vessel. The bell, whose clapper had been pulled aside by a magnet, produced "a very audible sound," but the ticking of the watch, which was suspended by a thread, grew fainter and fainter as the vessel was gradually evacuated by the powerful pump that Boyle had invented. The experiment with the pistol yielded similar results, for "the noise made by the striking of the flint against the steel was exceedingly languid, in comparison of what it would have been in the open air." Boyle would have liked to lower a violin in the evacuated receiver to test whether it vibrated when a string, tuned in unison, was struck outside. Unfortunately, the instrument was too large to fit into the pneumatical vessel. He also thought of placing in the receiver a pair of bellows attached to a musical pipe, but he encountered the same difficulty.[16]

If sound posed a problem, the apparent *boiling* of the water as it fell when the tap at the bottom was open raised another one. The violent bubbling caused surprise, but no one seems to have guessed that the *apparent* boiling was just that. Water boils at 100 degrees Celsius at sea level, but at the peak of Mont Blanc it boils at 85 degrees, and in a vacuum it can start bubbling at a much lower temperature. On a warm sunny day in summer in Rome, the thermometer often rises to 40 degrees Celsius, and this may explain why Raffaello Magiotti noted the bubbles in Rome while Pascal, who carried out the experiment in the winter in Rouen, does not mention them. Otto von Guericke, who repeated the experiment at Magdeburg after 1654, saw the bubbles and thought that they had penetrated through the walls of the tube. Emmanuel Maignan in Rome rejected this idea and conjectured that a small amount of air had slipped in through the mouth of the tube AS when the water was being sucked up. He later claimed to have remarked at the time that "unless the water was sustained by the balancing force of the surrounding air it could not have remained suspended."[17] If he said this, he was a step ahead of everyone else.

Sound and the presence of bubbles were bad enough, but the passage of light was an even more embarrassing objection to the possibility of the vacuum. It was generally believed at the time that light, just like sound,

---

[16]Experiment XVIII of Boyle's *New Experiments Physico-Mechanical Touching the Spring of Air* (1660) in *Works*, edited by Thomas Birch, 6 vols. (London, 1772). Facsimile Hildesheim: Georg Olms, 1965, pp. 62–64.

[17]Quoted in Middleton, *The History of the Barometer*, p. 13.

could only be transmitted through a medium such as air. A perfect vacuum, they reasoned, would be completely opaque. The philosopher Thomas Hobbes concurred and thought that it was *impossible* for the "action of a lucid body to be propagated through a vacuum."[18] Maignan, who had explained the action of a magnet by postulating that it sent out minute magnetic particles, got around the problem by applying the same reasoning to light and arguing that it was composed of corpuscles. But this seemed far-fetched to most of his contemporaries.[19]

The production of sound, the transmission of light, and the apparent boiling of water were genuine objections. In a sense, the immediate results of the experiment were more favorable to the *plenists* than to the *vacuists*, to use a terminology that became fashionable a few years later. The *plenists* could think of several ways air might get into the flask, for instance, by penetrating through the glass, or slipping in between the water and the inside wall of the tube. Berti was particularly concerned about the fact that the sound of the bell had been heard, and he believed that this tolled the death of the vacuum. He dismantled his apparatus, thanked his patron for allowing him to use the facade of his building, and turned to other things. Most of the Roman spectators soon forgot about his amusing but inconclusive experiment. When Torricelli picked up the threads in Florence a few years later, no one in Rome seemed to remember just when Berti had amused himself with a long tube of lead.[20]

---

[18]Letter to Mersenne, 17 February 1648, *Correspondance du P. Marin Mersenne*, XVI, pp. 109–110.

[19]Maignan's views are found in his *Perspectiva horaria, sive de Horographia gnomica tum theoretica, tum practica libri quatuor*. Rome, 1648, pp. 619–632. Huygens claimed in his *Traité de la lumière*, written while he was in France and read to the Académie royale des sciences in 1678 (but only published in Leiden in 1690), that the Torricellian experiment was evidence for a luminiferous aether since light went through the evacuated tube.

[20]The Rector of the Roman College, the Jesuit Nicolò Zucchi, and his colleague, Athanasius Kircher, witnessed the experiments made by Berti and offered separate accounts some years later. In 1648, Zucchi published a 20-page pamphlet, which he reprinted the next year as an appendix to his *Nova de machinis philosophia* (Rome, 1649, pp. 101–115); Kircher offered his account in his *Musurgia Universalis* of 1650. The Minim friar Emanuel Maignan later published a memorandum given to him by Berti as well as critical observations he claimed to have made at the time (*Cursus philosophicus*. Toulouse, 1653, vol. IV, pp. 1925–1936). Needless to say, this was not an age when laboratory reports were written out in fastidious detail and the date and the hour carefully noted.

The Torricellian Experiment

Evangelista Torricelli, who succeeded Galileo as Philosopher and Mathematician to the Grand Duke of Tuscany in 1642, was both a brilliant mathematician and a gifted physicist. He died prematurely in 1647, a few days after his thirty-ninth birthday, at the same age that Pascal was to pass away. Torricelli was told about Berti's experiment by his friend Raffaello Magiotti, who mentioned that if salt water had been used it would not have stood as high as ordinary water.[21] Neither Magiotti's letter to Torricelli nor any reply that might have been made has survived, but if what Magiotti claimed is true, he may have been the one who provided the stimulus that Torricelli needed. At the beginning of 1644 (but when exactly is anyone's guess) Torricelli had a brainstorm: Why not used mercury, which is about 14 times as heavy as water? Since a column of water falls to a height of ten and a half meters, Torricelli calculated that a column of mercury would drop to about 76 centimeters.[22] A tube of this length could easily be found, whereas one long enough to perform the experiment with water was hard to come by.

Torricelli borrowed a meter-long glass tube that was sealed at one end, filled it with mercury and, keeping his thumb over the end, inverted it in an open vessel of the same liquid. When he removed his thumb, the mercury fell to a height of about 76 cm, as he had worked out. Many years later, one of his students, Carlo Dati, gave an account of the experiment in which he claimed that it was not carried out by Torricelli himself, but that he imagined what would happen and instructed his friend Vincenzo Viviani, who made the apparatus, procured the mercury, and did the experiment.[23] But Torricelli had "a wonderful capacity for manual work," as Viviani himself testified, and Dati may have underplayed his involvement in the actual experiment. In any case, not only had Torricelli the apparently simple but extremely clever idea of using mercury instead of water, he also anticipated, as we shall see, virtually all subsequent developments.

---

[21]Magiotti makes this claim in his letter to Mersenne of 12 March 1648, *Correspondance du P. Marin Mersenne*, XVI, p. 170.

[22]The height at sea level on an average day is 76 centimeters. Variations about this value are quite small; for example, the highest and lowest sea level pressures recorded are 806 mm (in the middle of Siberia) and 665 mm (in a typhoon in the South Pacific). Near the earth's surface the pressure decreases with height at a rate of about 3.5 millibars for every 30 meters, a decrease of 0.345%.

[23]*Lettere a Filateli di Timauro Antiate, Della Vera Storia della Cicloide, e della Famosissima Esperienza dell'Argento Vivo*, Florence, 1663, p. 20.

In the spring of 1644, Torricelli informed a Roman friend, Michelangelo Ricci, that he was carrying out a new experiment. Ricci wrote back on 11 June to say that he was impatiently awaiting the result. He had not long to wait, for on that same day Torricelli had already written to tell him of the outcome. He based his reasoning, he said, on the realization that "we live submerged at the bottom of a sea of air, which is known by unquestioned experiments to have weight." Torricelli does not say when he had the idea that we are surrounded by a sea of air but, as he was familiar with Galileo' *Two New Sciences*, he knew that air has weight. At some time, he asked the momentous question, Might it not exert pressure on the surface of the earth and thus push up the water when the piston of a suction-pump is raised? Eighteen *braccia* (ten and a half meters) of water would thus represent the weight that is counterbalanced by the pressure of the air on the surface of the earth. This is how Baliani had reasoned in the light of what Galileo had told him about the weight of air. But let us hear Torricelli himself:

> I have already hinted to you that a philosophical [i.e., *scientific* in modern language] experiment was being performed concerning the vacuum, not simply to produce a vacuum, but to make an instrument that might show changes in the air, which is now heavier and coarser, now lighter and more subtle.[24]

By June 1644, therefore, Torricelli had arrived at the conclusion that the weight of a column of air varies with weather conditions, on the observation that air sometimes feels, as he put it, "lighter and more subtle." What is remarkable is that he wanted to make an instrument to measure this variation, an idea that does not seem to have occurred to anyone else. Torricelli was conscious of innovating, and he contrasts his theory with the positions of the Aristotelians, who maintained that a vacuum was impossible, and with that of Galileo, who thought that it was possible but difficult to produce:

> Many have said that a vacuum cannot be created, others that it can, but only with difficulty and against the repugnance of nature. I do not know of anyone who said that it could be created without difficulty and without any resistance from nature.

---

[24]Letter of 11 June to Ricci, translated in W. E. Knowles Middleton, *The History of the Barometer*, pp. 23–24. The original is published in P. Galluzzi and M. Torrini, eds., *Le Opere dei Discepoli di Galileo Galilei, Carteggio 1642–1648*. Florence: Barbèra, 1975, pp. 122–123.

Torricelli recognized that the force that kept mercury from dropping below 76 centimeters was to be sought *outside* the tube, and not *inside* where everyone else had looked for it. Here again he was aware of the boldness of his approach: "It was believed until now that this force is something internal to the tube . . . but I claim that it is external."[25] The vacuum does not suck up the mercury. The vacuum doesn't affect the mercury at all! It is the pressure of the atmosphere pushing on the vessel of mercury that drives it up the tube.

Apparent Failure

But how could Torricelli prove that the tube of mercury was really empty at the top? To show this, he used tubes of different shapes, as we can see in the sketch that he sent to Ricci (see Figure 7). In a first experiment, he filled the vessel at the bottom with mercury up to the letter C, and added some water on top to the height of D. He then filled the tube with the bulb with mercury and inverted it in the vessel. After the mercury in the tube had fallen to the height of about 76 centimeters, it was slowly raised. When the mouth of the tube reached the level of the water, the mercury fell out and water rushed in without encountering *the slightest resistance*. This made a great impression on the spectators, especially when the water surged so swiftly upward that it broke the end of the tube and splashed all over the table and the attendants. In a second experiment, Torricelli used the two tubes in Figure 7 to show that mercury always falls to the same level, regardless of the shape of the tubes. Someone might object that an unknown rarefied substance had been trapped *inside* the tube, and that this was what held up the mercury. But then there would be more of this rarefied material in the tube that ended in a bulb than in the narrower tube. The mercury should therefore have been pulled higher in the tube with a bulb, but this was not the case, for the mercury fell to the same level in each. This outcome confirmed Torricelli in his view that the earth is surrounded by a sea of air, and he even realized that since air is heavy it must weigh more at sea level than at the top of a high mountain.

The weight of the air on the surface of an open vessel of mercury is enough to keep the column in the tube from falling all the way down. Thus 76 centimeters of mercury represent the weight of mercury that the air can

---

[25]Letter of 11 June, Middleton, *History of the Barometer*, p. 23.

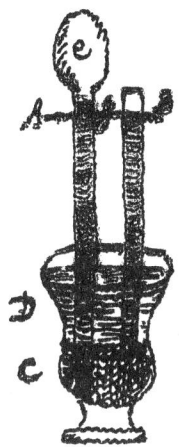

FIGURE 7 *Torricelli's experiment to show that mercury always falls to the same level, regardless of the shape of the tubes.*

maintain. But the experiment, as we have mentioned, was designed to determine the variations in the weight or pressure of the air. It seems, therefore, that Torricelli has reached his port of call or, if we prefer a military to a nautical metaphor, that he has penetrated into enemy territory and captured their citadel! Alas, the forked tongue of experiments makes itself heard once again, and Torricelli ends his letter with these wistful words:

> I have not been able to succeed in my chief intention, namely to find out with the instrument EC when the air is coarser and heavier and when more subtle and light, because the level AB changes for another cause, which I never would have suspected, namely on account of heat and cold, and that very noticeably, exactly as if the tube AE were full of air.[26]

---

[26]*Ibid.*, p. 24. But this does not mean that others did not try to make him say what they thought his *real* method was or should have been. Writing in 1663, Carlo Dati, the editor of the Torricelli-Ricci correspondence, makes the *theory precede the experiment*: Torricelli had the idea; Viviani made the experiment! Five years later, the editors of the *Saggi di Naturali Esperienze Fatte nell'Accademia del Cimento* had *the experiment precede the theory*. A little forcing with the date was also indulged to over-insure priority: in the margin next to the passage where the experiment is described, we find 1643 instead of 1644. Another instance of good rhetoric, but bad chronology, occurred as late as 1743 when Georg Matthias Bose, a German professor, gave the date of the experiment as 2 May 1643 in a public Latin oration to celebrate the centennial of the discovery.

The unsatisfactory outcome is surprising. We know that a barometer is not significantly altered by small changes in temperature. What went wrong? The most likely explanation is that the tube that Torricelli had borrowed had been rinsed with water and that a small quantity had been left behind. This would be enough to cause the mischief, but Torricelli had no inkling of this and he believed what the experiment told him. He therefore returned the tube to the storeroom and never wrote to anyone else about what he considered a failure.

Let us summarize the developments thus far. Galileo was the first to record in print that even the best-made suction-pumps cannot draw water upwards above a certain height, which he estimated at 18 *braccia*, roughly ten and a half meters. He compared the column of water in the pipe to a rope which, if long enough, breaks under its own weight. Berti, who made the experiment in Rome in a long lead pipe set up vertically, was interested in the height of the water, but neither he nor his friends seem to have realized that the column of water in the tube was pressed upwards from the bottom, and not suspended from the top. In a letter to Mersenne on 12 March 1648, Raffaello Magiotti says of the suction pumps in Florence that they work *per attractionem*, and that the water rose in Berti's pipe because "it was attracted from above."[27] Torricelli thought of using mercury, and guessed that what caused it to be upheld in the evacuated tube was not a presumed resistance to the vacuum but simply the counterbalancing weight of the atmosphere. What counted was the ratio of the weight to the surface area at the base of the column, that is, the pressure. But how could so little mercury balance so much air? Here Torricelli was hampered by Galileo's erroneous determination of the relative densities of air and water, which he had estimated as 400 to 1, when it is actually closer to 800 to 1.

Roman Disbelief

We now move to Rome, where Torricelli's friend, Michelangelo Ricci, was the last person to accept physics on faith. Torricelli rightly expected objections, and he got them, almost by return of post. After congratulating Torricelli "on examining something that no one has considered until now"[28] (which shows just how little Romans remembered of their discus-

---

[27]*Correspondance du P. Marin Mersenne*, XVI, p. 170.

[28]Middleton, *The History of the Barometer*, p. 25.

sions at the time Berti made his experiment a short time before), Ricci raised three objections that gave Torricelli the opportunity to explain his theory more fully.

First, Ricci objected that if a lid were fixed on top of the vessel of mercury in which the tube is inverted, the outside air would no longer weigh on the mercury but on the lid and, hence, would be unable to counterbalance the liquid. Torricelli had anticipated this kind of misunderstanding, and he replied that four different cases had to be distinguished: (1) if the lid touches the mercury in the vessel and is held firmly in place, the mercury in the inverted tube cannot fall because there is no room for it in the vessel; (2) if the cover does not touch the mercury in the vessel and a layer of air is left inside, the mercury in the tube drops as usual because the enclosed air has the same "condensation," that is, the same atmospheric pressure, as the external air; and (3) if the enclosed air is somewhat thinner, the mercury drops to a lower level than usual. Finally, (4) if the air is "completely rarefied," namely if a vacuum is produced, the mercury falls all the way down, provided there is enough room in the vessel to accommodate it.

The difficulty was to convince Ricci that the enclosed air has the same "condensation" as atmospheric air. Torricelli thought he could do this by comparing a column of air to a cylinder filled with wool. If a weight is placed on a cylinder of wool, the wool is compressed and the weight is felt on the bottom of the cylinder. If a knife is thrust into the cylinder so that it goes through the wool and comes out on the other side, the pressure at the bottom of the cylinder does not change, and the wool remains as compressed as before.[29]

The second objection that Ricci made is even more commonsensical. If you push the plunger of a syringe completely in and then drawn it back, you feel the same "resistance" whether you point the syringe upwards, downwards, or sideways. This shows, he claims, that the weight of the air is totally irrelevant. Torricelli turned this objection into a joke and drowned it in a squirt of wine:

> Now as to your second point. There was once a philosopher who, seeing a cask being pierced by a servant, dared to say to him that the

---

[29]The analogy had occurred to Descartes (letter to Reneri of 2 June 1631, *Œuvres*, edited by Charles Adam and Paul Tannery, 13 volumes. Paris 1897–1913. The first 11 volumes were revised and reprinted. Paris: Vrin, 1964–1974, pp. 205–208). Isaac Beeckman had also compared air to a sponge (see C. De Waard, *L'expérience barométrique*, p. 76).

wine would never come out because weight naturally presses downwards, not sideways or in other directions. But the servant made him see with his own eyes that although liquids naturally weigh downwards, they push and squirt out in every direction, even upwards, as long as they find a place to go, namely a place that offers less resistance.[30]

For his third objection, Ricci appealed to Archimedes' principle of buoyancy, which states that when a body is immersed in water, the upward force exerted on it is equal to the weight of the water that is displaced. If we apply this principle to mercury and air, we have to conclude that the buoyant force on a given amount of mercury must be equal to the weight of an equal volume of air. But how can air, which is so much lighter than mercury, raise a comparable volume of mercury? Torricelli found this objection more interesting, and he answered it without jesting. A body *immersed* in water is buoyed up by the weight of as much water as its own volume, he granted, but this does not apply to the mercury in the tube because it is *not immersed* in water. The mercury *inside the tube* is not pressed down; it is the mercury *outside* in the vessel that is pressed "by many miles of air piled up above it." This is why the mercury in the tube stays at precisely the height at which its weight is equal to that of the counterbalancing column of air. Only the vertical distance and the specific gravity need be taken into account! Torricelli knew that he was on a winning streak and he assured Ricci, "You may be as certain of this as if you had make the experiment yourself."[31]

Torricelli later tried introducing minnows, flies, and butterflies into the vacuum to see how they behaved, but they were too much affected by going up through the mercury for the experiment to be satisfactory. The Belgian François de Sluse, who studied in Rome between 1642 and 1652, made similar attempts with Ricci and Magiotti, but the outcome was equally disappointing.[32]

---

[30] Letter of 28 June to Ricci, translated in Middleton, *The History of the Barometer*, p. 27.
[31] *Ibid.*, p. 28.
[32] *Lettere a Filaleti*, Florence, 1663, p. 20, cited in Middleton, *The History of the Barometer*, p. 29. On de Sluse, see Anne-Catherine Bernès, "René-François de Sluse et l'Italie" in *Congrès de Namur* (XLIX Congrès de la Fédération des cercles d'archéologie et d'histoire de Belgique). Namur, 1990, vol. III, pp. 305–317. De Sluse was not quite thirty years old when he left Rome to return to his native Belgium.

The French Connection

Torricelli's letters to Ricci were shown to a young French mathematician, François Bonneau Du Verdus, who had studied mathematics with Roberval in Paris before leaving for Rome at the end of 1643. He was eager to impress people back home, and he promptly sent extracts of Torricelli's letters of 11 and 28 June to Mersenne in Paris. In his haste, however, he omitted the beginning and the end of the first letter, and he left out of the second letter Torricelli's reply to the second and the third objection that Ricci had raised, while enclosing a diagram that had been intended for the third one.[33]

Mersenne had enough information to repeat the experiment and realize that Torricelli invoked the weight of the air, but there were three important things about which he was in the dark: (1) Torricelli's clear statement that we live at the bottom of a sea of air, and that a vacuum can be easily produced; (2) Torricelli's "chief intention," which was to make an instrument to show the changes in atmospheric pressure; and (3) his belief that he had failed.

In the autumn of 1644, Mersenne went to Rome and stopped on the way in Florence, where Torricelli showed him his glass tube. He had reached Rome by the time the young Cardinal Giovanni-Carlo de Medici, the brother of the Grand Duke of Tuscany, made his first official visit to the Pope. The lavish entertainment that was put on for the Cardinal included a demonstration of Torricelli's experiment. The vacuum was becoming fashionable and, as we shall see in the next chapter, the French found it just as exciting.

---

[33]Letter of François du Verdus, written toward the end of July 1644, to Marin Mersenne, *Correspondance du P. Marin Mersenne*, XIII, pp. 179–181

CHAPTER 3

Creating Emptiness
Pascal's Ingenious Methods

The Early Experiments of the Pascals, Father and Son

As soon as Mersenne returned to Paris from Italy in the summer of 1645, he endeavored to replicate Torricelli's experiment with his friend Pierre Chanut, who was about to leave for Sweden to take up a diplomatic post. They were unsuccessful because of the poor quality of their equipment: Parisian glass-blowers were simply unable to make tubes that did not break when they were turned upside down after having been filled with mercury. When we read, therefore, that Torricelli had rendered the experiment easy by using mercury instead of water, we should take this—like everything we read—with a grain of salt. The experiment did become simple for those who had access to high-quality glass and first-rate glass-blowers. This was the case in Florence but not in Paris.

The engineer Pierre Petit heard about Mersenne's attempt shortly thereafter, and in the summer of 1646 he tried to create a vacuum with a two-foot-long blowpipe, which was much too short for his purpose as he soon found out (at sea level a column of mercury stands at roughly 76 centimeters and Petit's two-foot pipe was only 65.6 centimeters).[1] Petit as-

---
[1] Petit gives Mersenne as the source of his information, but he could have learned about Berti's experiments in 1643 if he traveled to Rome in that year, as is suggested by the Abbé Bourdelot in a letter to a Roman friend: "The mathematician, Mr Petit, had been called to Rome to work on fortifications" (letter to Cassiano dal Pozzo, 15 October 1643 in *Correspondance de Mersenne*, vol. XII, p. 227).

sumed that Torricelli's main intention had been the production of a vacuum, and he thought that the weight of the mercury was the decisive factor. Atmospheric pressure did not cross his mind. When he passed through Rouen on his way to Dieppe in September 1646, he called on the Pascals and told them about the experiment but without mentioning the name of its inventor.[2]

Etienne Pascal had long believed that a vacuum was possible in nature. This is important in our story because it means that his son Blaise, whom he taught, never had to work himself out of the Aristotelian conceptual scheme that ruled out the possibility of a void. The Pascals had the advantage of not being fettered by Scholasticism. They also lived in a city with one of the best glass factories in France and, furthermore, they could pay for expensive equipment. When Petit mentioned that he would like to repeat the experiment with a longer blowpipe and a greater quantity of mercury, Etienne suggested that they perform it together when Petit returned to Rouen from Dieppe. As soon as Petit arrived, the two friends went to the glassworks and obtained a tube four feet long, "the size of the little finger inside" (roughly 131 cm long and 1.5 cm wide), open at one end and "hermetically sealed, to use the terms of the art," at the other. From the druggist they procured about fifty pounds of mercury[3] to fill the tube completely and pour some in a vessel to a depth of three fingers. Above the mercury in the vessel they added another three fingers of water. Petit sealed the opening of the tube with his middle finger and carefully inverted the tube in the vessel of mercury until his finger touched the bottom. Before removing his finger, Petit peered carefully into the tube to see if any air had been trapped inside. When he found no evidence of this, he removed his finger and the mercury fell by more than eighteen inches, "something I never would have believed!" he later wrote in the letter he sent to Pierre Chanut in Stockholm.

---

[2] Writing to Antoine de Ribeyre on 12 July 1651, Blaise Pascal claimed that all he knew at the time was that the experiment had been performed in Italy (*Œuvres de Pascal*, II, p. 809). Petit's account is taken from a letter that he wrote on 19 or 26 November 1646 to Pierre Chanut, a French diplomat in Stockholm (*ibid.*, II, pp. 349–359). The text was edited by Marc-Antoine Dominicy and printed in Paris in 1647.

[3] Petit says that he asked for 40 or 50 pounds (*Œuvres de Pascal*, II, p. 351), a large amount since Magni used four or five pounds in Warsaw (letter of Desnoyers to Roberval, 17 July 1647, *Correspondance de Mersenne*, XV, p. 314). In a letter to Constantin Huygens, Mersenne mentions that he needed 30 pounds. A pound of mercury cost 45 sols in Paris, and Mersenne marveled when Descartes told him that he could get that amount in Holland for half the price (*Correspondance de Mersenne*, XVI, p. 11).

After making the experiment, Petit and Etienne Pascal began "to philosophize," and it is here that Blaise comes into the picture for the first time, as the devil's advocate. "The son of Mr Pascal," writes Petit, "objected that some people might say that air entered through the pores of the glass." Petit replied that in this case the air would have continued to penetrate through the pores and would have pushed the mercury down even lower. This did not happen, and, furthermore, much to the astonishment of the three observers, when the tube was slowly raised, the mercury remained at the same level, although the empty space above it became greater and greater. When the mouth of the tube that was being raised reached the region of water that had been poured on top of the mercury in the vessel, the mercury in the tube fell all the way down and the water rushed right up to the top of the tube.[4] This would have been impossible if there had been any air above the mercury.

Petit wanted Chanut to succeed in repeating the experiment in Sweden, and he gave him some practical advice. First, if the tube had been used for another experiment, it should be washed thoroughly and dried by heating the outside to ensure that all water vapor was expelled. Second, care should be taken to avoid mixing the mercury with however small a quantity of water. No one should blow in it before performing the experiment! Third, the vessel into which the tube was inverted should be large enough for the experimenter to bend his arm when putting the tube in. Finally, there should be at least three or four inches of mercury in the bowl. These last two recommendations could have been made after one or two experiments, but the first and the second about the dryness of the tube indicate several trials and genuine experimental acumen. It is almost certain that Torricelli's failure to achieve his goal of making a working barometer was due to the presence of water in the tube.

Experimental Caution

Pascal's father was delighted with the results, but Petit was less sanguine. As far as he was concerned, the experiment did not exclude that some air had been trapped in the tube and had expanded to fill the space left empty when the mercury dropped. These qualms were legitimate enough, since it had occurred to neither Petit nor the Pascals that the weight of the at-

---
[4]*Œuvres de Pascal*, II, pp. 352–353.

mosphere could balance the weight of the mercury. A minute amount of air could indeed have been left in the tube or have been introduced when the mercury was poured in. The air had then become "rarefied" in order to avoid the creation of a completely empty space. Such a view implied a tacit admission of nature's horror of a void, and when faced with the question, Why does mercury not fall lower? Petit and Etienne Pascal fell back on this theory and answered that "nature could not tolerate a greater void."

The discussion could not be prolonged because Petit had to leave early on the next morning. We do not know whether Petit had a sleepless night, but no sooner had he ridden out of Rouen than he realized that their answer had been wrong. Since the space occupied by the vacuum had increased when the tube was slowly raised from the bottom of the vessel, nature was clearly willing to allow a greater vacuum! In the hope of finding out why, Petit turned to the statical and hydrostatical laws with which he was familiar as an engineer. Might not the answer lie in the difference between the force required to raise the tube and the weight of the mercury in the tube? He hoped to determine this force by suspending the tube from one arm of a balance, and then adding weights to the other arm until the tube started rising. He did not carry this out immediately, and the experiment became irrelevant a short time later when the young Pascal realized that what was at issue was not the weight of mercury but the weight of air.[5] In the meantime, Petit impressed people in Paris by repeating the standard experiment with mercury in the houses of friends and acquaintances.

Science on Display

Back in Rouen, Pascal, his interest thoroughly aroused, resolved to perform a series of additional experiments to confirm the creation of a vacuum and determine the force needed to produce it. In his enthusiasm, he went in for *big science* and commissioned very long tubes, some over forty-five feet (15 meters) to repeat the experiment that Gasparo Berti had per-

---

[5]According to his own testimony, Pascal heard of Torricelli's explanation some time before the publication of his *New Experiments Concerning the Vacuum* in October 1647. Although he believed the explanation to be correct, he did not mention it because he did not have a conclusive experiment at the time (letter to Florin Périer of 15 November 1647, *Œuvres de Pascal*, II, p. 679).

formed in Rome. At first sight, this may seem fanciful, since it is hard to see how a tube as high as a five-storey building could be filled and then inverted. Some historians of science have been tempted to infer that Pascal was merely generalizing from experiments made with mercury, but their skepticism is unwarranted.[6] Pascal had found a clever way of handling these gigantic tubes by tying them upright to the mast of a ship, and then rotating the mast with a system of pulleys. These spectacular (and outrageously expensive) displays of engineering skill were performed in January and February 1647 in the yard in front of the local glass factory. It was a great hit.

One of the spectators was Jacques Pierius, a professor of philosophy at the College of Rouen, who rushed into print with a polemical pamphlet titled *Can a Void Exist in Nature?* The question mark was promptly dispatched, and the reader reassured that however ingenious the experiment left nature comfortably full.[7] Pierius claimed that the space above the mercury in the tube was filled with vapors or rarefied *spirits* that pushed the liquid down. In order to disprove the philosopher's hypothesis, Pascal sent out a general invitation to attend a new and more striking demonstration. When the spectators, who included Pierius and his students, arrived at the glassworks, they found that two 40-foot tubes had been tied to a mast. One was filled with water, the other with wine. Since water is denser than wine, Pascal knew that it would fall further down. The trick was to know by how much. That morning he had privately performed the experiment with mercury (which is 13.6 times as heavy as water) and calculated that water would stand at $31\frac{1}{9}$ feet, whereas the wine would be close to $31\frac{2}{3}$ feet. Before beginning the proceedings, like a good *metteur en scène*, Pascal asked the audience whether wine contained a greater quantity of *spirits* than water. When Pierius and the crowd answered, "Yes," Pascal egged them on to infer that if the experiment was carried out in tubes of the same length, a greater space would be left above the wine

---

[6]Alexandre Koyré voiced his doubts in "Pascal savant" (*Etudes d'histoire de la pensée scientifique*. Paris: Gallimard, 1973, pp. 362–389), but the exceptional length of the tubes is well ascertained. Pascal speaks of 46 feet (*Œuvres de Pascal*, II, p. 502), Roberval of 40 (II, p. 462), Mersenne of 45 (II, p. 488), and Pierius of 50 (II, p. 642).

[7]According to Roberval, Pierius dashed off his pamphlet within twenty-four hours of witnessing the experiment. *Can a Void Exist in Nature?* (*An Detur Vacuum in Rerum Natura?*) bears no place, date, or name of publisher, but was probably printed in Rouen (*Œuvres de Pascal*, II, 469). In 1648, Pierius published a revised version in Paris in which he mentions the experiments on the void carried out by Petit in the French capital.

because more *spirits* would be released by wine than by water. He then performed the experiment and the reverse happened: less space was left above the wine! Pascal now poured water into the tube that had contained wine, and wine into the one that had held water, and repeated the experiment. The heights that he had calculated were confirmed once more. Pascal was becoming a superb showman! Over five hundred people, including five or six Jesuits, are said to have attended one of several repeat performances.[8]

A Doctor Speaks Out

Pierius and the professors of philosophy still shook their head, but a local physician, Pierre Guiffart, joined the fray on Pascal's side. In his pamphlet, *A Discourse on the Void*, he mentions another experiment that greatly impressed him. This time Pascal placed a rope in a tube filled with water and inverted it in a vessel of mercury. When the rope was pulled, the water and the mercury rose, against their natural downward inclination, in order to fill the space vacated. Guiffart describes all this in what might be called the new *rhetoric of vacuousness*. Here is a sample: "Note this brave nothingness against which so many outstanding philosophers have been battling for so long, this dreadful void that nature fears and resists with all its might, this beautiful nothing that provides us with weapons and good arguments for its defence."[9]

Guiffart took it for granted that mercury rose because it was attracted by a vacuum. What puzzled him was why it did not rise higher in tall tubes where there is a greater vacuum than in shorter tubes where there is less. Why does mercury stand at the same height whether the experiment is performed in a 3-foot or a 50-foot tube? Because mercury does not *consent* to rise higher, replied Guiffart. He offered as evidence that when a tube full mercury is inverted in a bowl of mercury, it bobs up and down until the liquid settles down. "This shows," he writes, "that it hesitates between its desire to follow its weight and its natural horror of a void."[10]

---

[8]Letter of Pascal to Antoine de Ribeyre, 12 July 1651, *Œuvres de Pascal*, II, pp. 807; see also Roberval's letter of 20 September 1647 to Desnoyers in which he states that the experiment was repeated in public four times to shut up the "measly Aristotelians" (*Œuvres de Pascal*, II, p. 472).
[9]Pierre Guiffart, *Discours sur le vide*, Rouen, 1647 (*Œuvres de Pascal*, II, p. 428).
[10]*Ibid.*, II, pp. 428–429.

Unable to satisfy both, it oscillates for a while. Nature's horror of a void is thus real, concludes Guiffart, but it is limited, a view that had already been expressed by Galileo.

When Pascal performed experiments in public, the question period was often enlivened by similar flights of fancy. On one occasion, he was asked what would happen if the experiment with mercury were carried out in an enormously long tube that was slowly inclined in a canal that stretched some 1800 French leagues until the tube was tangent to the curvature of the earth? Since a French league is 4920 meters, such a canal would extend over 8,856 kilometers, roughly the distance from London to Chicago! Pascal did not offer to carry out the experiment, but he realized that the question was not trivial. How could a small vacuum at the top of such a long tube hold up tons and tons of mercury?[11] But whereas everyone still looked for the answer in an inside *pull*, Pascal began thinking of an outside *push*.

The Royal Professor of Mathematics Advertises the Vacuum

Meanwhile, Gilles Personne de Roberval, the Professor of Mathematics at the *Collège Royal* in Paris, had become interested in the way his bugbears, the Aristotelians, accounted for Pascal's experiments by saying that the mercury dropped because air in the tube took its place. In their view, the resistance that air normally opposes to rarefaction is overcome by a column of mercury 27 inches high. Under this force, residual air in the tube expands and fills the space that would otherwise have been left void. Roberval and Pascal (it is often hard to tell which) countered this argument by pointing out that in tubes that become wider at the top, air would have to be rarefied to a greater extent than air in narrower tubes, something that would require the application of a greater force. But experiment shows that the same force, namely the weight of a column of mercury 27 inches high, is all that is needed regardless of the shape of the tube.

An experiment described by Roberval bears on this point in an interesting way. After tilting a full tube of mercury in a vessel of mercury, he introduced a bubble of air into the tube and the mercury shot right to the

---

[11]*Ibid.*, II, p. 432.

top. When the tube was raised to the perpendicular, the column of mercury stood a little lower than usual and a larger empty space was created at the top of the tube. According to the Aristotelians, this indicated that the column of mercury, which was now lower, had less force. But who, asks Roberval, will believe that a smaller force acting on the same bubble of air can produce a greater degree of rarefaction? "Only those," he answers sarcastically, "who do not rejoice in the light of truth but in the darkness of ignorance and, like andabates, fight with their eyes closed."[12] Whether everyone in those days knew that andabates were blindfolded gladiators who fought on horseback may be doubted, but as we shall see, Roberval sealed his own eyes a short time later and became an andabates himself.

In spite of his rude words about professors of philosophy, Roberval was worried that a residue of air might have remained inside the apparently evacuated tube, and he thought that a definitive resolution of the problem would not be possible until tubes could be made large enough for birds to be placed inside. If the birds flew about, this would indicate that some air was present because their wings had to rest on something like oars that can convey no impulse to a boat if they do not use the water as a fulcrum.

The Vacuum in Clermont and Warsaw

The entertainment value of the vacuum experiment was not lost on the rest of France. Pascal's brother-in-law, Florin Périer, repeated it before a large group in Clermont, but only a handful agreed that a real vacuum had been produced. Died-in-the-wool Aristotelians, including two physicians, were adamant that the apparent void contained air. When someone in the audience declared that he would believe in the vacuum only if the mercury in the tube dropped when a hole was made in the side of the tube, Jacques Le Tenneur, who was present, promptly offered to bet 100 to 1 that it would indeed drop. We do not know if anyone accepted the wager, but the tube could not be perforated without breaking the glass and so the issue was left undecided.[13]

---

[12]Letter of Roberval of 20 September 1647 to Desnoyers (*Œuvres de Pascal*, II, p. 466). This is usually referred to as Roberval's *First Account*, to distinguish it from a *Second Account* that he wrote a year later.

[13]Letter of Le Tenneur to Mersenne, 21 October 1647, *Œuvres de Pascal*, II, p. 491.

The French were beginning to feel that they had a monopoly over vacuousness when Pierre Desnoyers, the secretary of the Queen of Poland, wrote from Warsaw to say that the Poles were astir over an experiment that an Italian priest, Valerio Magni, had just been performed at court. An account, entitled *An Ocular Demonstration of a Container without Content*, was dispatched to Roberval, asking for his comments in Latin so that King Ladislaw IV might read them.[14] Roberval complied but in the polemical tone for which he was famous. He admitted that the experiment had first been performed in Italy, but declared that all the really exciting experimental work had been done in France, notably by Pascal. He rejected the hypothesis that some residual air had been left behind in the apparent vacuum in the tube, but he was careful, nonetheless, to speak of the space above the mercury "as though empty," thereby avoiding any commitment to an effective vacuum.[15]

Pascal's New Experiments

Pascal intended to give an account of his experiments as soon as possible but he was frequently ill throughout 1647 and progress was slow. When Mersenne, whom he had informed of his plans, began saying to everyone that the work was forthcoming, Pascal felt that he had to publish at least an abridgement. His sense of urgency was increased when a description of Valerio Magni's experiments reached Paris in July. Pascal now feared that his own work might be thrust into the background, and he rushed into print to protect his intellectual property. "I thought it fit to provide this summary in advance," he writes, "because I feared that someone, who had not taken the same time, or gone to the same trouble and expense, might offer to the public, before I could, results that he would not be able to de-

---

[14]Magni spent over a year in Rome and went to Florence for a couple of months prior to returning to Poland in July 1643. In 1645 he was back in Rome, where he saw Mersenne, but he did not meet Ricci in Rome or Torricelli in Florence. Indeed their very names were unknown to him (Marin Mersenne, *La Correspondance du P. Marin Mersenne*, edited by P. Tannery, C. de Waard et al., 17 vols. Paris: Editions du C.N.R.S., 1933–1988, XIII, p. 481). Magni's starting point was what he had read in the First Day of Galileo's *Two New Sciences* concerning the "force of the vacuum."

[15]Letter of Roberval to Desnoyers or *First Account*, 20 September 1647, *Œuvres de Pascal*, II, pp. 461, 1.12; p. 464, 1.14; p. 468, 1.11; p. 472, 1.13.

scribe with the proper care and in the required order because he had not seen them himself. For no one has had tubes and siphons as long as mine."[16] The abridgement, a slim volume of 32 pages, entitled *New Experiments Concerning the Vacuum*, appeared in October 1647. It is a clean copy of the minutes of selected experiments carried out in Rouen at the beginning of the year.

Pascal stresses that his experiments were performed in a number of ways, in tubes of different shapes, and with various liquids, including mercury and water, but also wine and oil. His first goal, "is to show that even very large vessels can be emptied of all materials known in nature," and his second, "to reveal the force required to achieve this vacuum."[17] Pascal wants everyone to know how far he has outrun his rivals in the *vacuum rush*, but he protests that he does not want to take credit for what is not his own. Thus he will not describe the "Italian" experiment that he performed with Pierre Petit but, having said this much, he goes on to boast that he made it "in more ways than anyone else," with tubes twelve and even fifteen feet long, thereby conveying the impression that longer tubes provided better results. What is surprising is that Pascal does not relate his spectacular experiment with two 40-foot tubes, which he had attached to the mast of a ship and filled, one with water, the other with wine.[18] He probably omitted this particular experiment because it had been fully described in Roberval's *First Account*, namely the letter that he sent Desnoyers in Warsaw on 20 September 1647 and that had been widely disseminated in France and abroad. Pascal also leaves out *complicating factors* such as the "boiling" of water in the evacuated tube, something that had struck Magiotti, Desnoyers, and Roberval but about which he is silent.

---

[16] Pascal later wrote to Antoine de Ribeyre, "in 1647 I published the experiments that I carried out a year before in Normandy" (*Œuvres de Pascal*, II, p. 807). The *New Experiments Concerning the Vacuum* was the third work of Pascal to be published after the *Essay on Conics* (1640) and the *Notice Concerning the Adding Machine* (1645). The licence to publish was granted on 8 October 1647. The work was printed in Paris and it is dedicated to "Monsieur Pascal adviser to the King by his son, B.P." It is translated in Pascal, *Selections*, edited by Richard H. Popkin, London: Collier Macmillan, 1989, pp. 33–41.

[17] *New Experiments*, *Œuvres de Pascal*, II, p. 501; Popkin trans. p. 3.

[18] Wine is mentioned in the *New Experiments* but only for its color as in experiment 3 where a 46-foot tube is filled with "bright red wine which is easier to see" (*Œuvres de Pascal*, II, pp. 502, 504; Popkin trans., pp. 36, 37).

William R. Shea

From the Anecdotal Account to the Laboratory Report

If we compare Pascal's writings with those of his slightly older contemporaries, Mersenne and Roberval, we witness a significant shift in the way experiments are presented. With Pascal we move from an *anecdotal account* to the modern *laboratory report*, in which subjective elements of surprise, enthusiasm, or disappointment are avoided. Whereas an anecdotal account stresses what the experimenter did and saw, a laboratory report teaches what ought to have been done and seen. Each genre has it own appeal and usefulness. Mersenne and Roberval favor the anecdotal and give colorful details, and mention names and personalities, but sometimes lose sight of the main goal of the experiment. Pascal's laboratory report helps us understand what to do, not by heaping up details but by focusing on what matters and teaching how an abstract mechanism works in a given case. The experimenter in the *New Experiments* remains anonymous, and the siphon and the syringe appear to work on their own. In this perspective, an imaginary experiment (what we call a "thought-experiment") can be more useful than a real one because the results of several experiments are often better grasped when they are conflated into one ideal experiment.

Pascal acknowledges that his experiments were inspired by the one performed in Italy "four years ago," but he does not mention Torricelli, and he is silent about the role of his father. When asked later why he had not mentioned Torricelli, Pascal declared that he did not know at the time who had made the experiment, which is odd since this was known to Pierre Petit, who was his source of information. We must recall, however, that the extracts from the letters of Torricelli to Ricci of 11 and 28 June 1644, which were forwarded to Mersenne by François Du Verdus, did not identify the inventor of the mercury experiment.

Of greater interest is when Pascal learned about the idea that it is the weight of atmospheric air that keeps a column of mercury suspended. In all likelihood, this was shortly after his return to Paris from Rouen in the summer of 1647. In a letter of 25 September of that year, his sister Jacqueline mentions that he had been talking about the *column of air* with his friend Auzoult.[19] It should also be pointed out that as late as August 1647, Mersenne did not know that Torricelli had performed experiments with water and not only with mercury. He credited Pascal with having been the

---

[19]*Œuvres de Pascal*, II, p. 482.

first to find that water stands 14 times higher than mercury in an evacuated tube, and that no empty space occurs above the water until it has dropped to a height of 32 feet, "what Torricelli had foreseen but I do not believe experienced."[20]

In an apologetic account written four years later, Pascal declares: "As early as 1647, we were told about the brilliant idea that Torricelli had concerning the cause of what until then had been attributed to the abhorrence of a vacuum. But this was a mere conjecture without proof and, in order to ascertain whether it was true, I started to think of an experiment that was performed, as you know, by Mr. Périer in 1648."[21] Pascal insists that this experiment, which was carried out on the Puy-de-Dôme, was entirely his own, and that he deserves full credit for the new knowledge it provides. He would not have been able to make such a sweeping claim had he read the full version of Torricelli's letter of 11 June 1644 to Ricci where we find the two famous statements: (1) "We live submerged at the bottom of an ocean of the element of air, which has been shown, by reliable experiments, to have weight," and (2) "on the tops of high mountains the air begins to be very pure and weighs far less." These two passages had been left out by Du Verdus in the abstract he had sent Mersenne in the summer of 1644.[22]

In November 1646, when Pascal repeated the "Italian" experiment with Petit, his goal was to produce an effective vacuum, and the idea that the cause might be the weight of the atmosphere did not brush his mind. His experimental success merely confirmed him in his view "that nature does not run away from the void with as much horror as some imagine" and that "if nature abhors anything it is not a vacuum but too much matter." But Pascal was not one to throw caution to the wind: "I am satisfied," he writes in the *New Experiments*, "to exhibit a large empty space and leave to interested and learned persons to determine what happens in such a space, such as whether animals can live there or whether refraction through glass is diminished."[23] This fear of hasty generalizations may also have something to do with his friend Roberval's refusal to recognize the vacuum as real until such time as he could determine whether light could travel without a supporting medium.

---

[20]*First Preface* to the *Novarum Observationum Tomus III*, quoted in *Œuvres de Pascal*, II, p. 488.

[21]Letter of Pascal to Antoine de Ribeyre, summer of 1651, *Œuvres de Pascal*, II, pp. 812–813.

[22]As late as 1648, Pascal's father, Etienne, still wondered about the truth of Torricelli's hypothesis, which he recognized as familiar to everyone (*Œuvres de Pascal*, II, pp. 588–589).

[23]*Ibid.*, p. 501, Popkin trans., p. 35.

Pascal wrote the *New Experiments* when he was recovering from a period of illness and this may explain why it is somewhat repetitive in nature. The tone is moderate and conciliatory, and Pascal is careful to speak of the void as *"apparent."* All he wants to show is that nature's abhorrence is limited, as Galileo had done in his *Two New Sciences*.[24] To achieve this, he describes eight experiments that we shall examine in turn.

The Eloquence of Experiments

In the first experiment, a small glass syringe is equipped with a tight-fitting piston. Its advantage over the Torricellian tube is that it is easier to operate. The principle is the same as that of the large suction pumps employed to raise water from wells in Pascal's days, but no such instrument had ever been used for the express purpose of creating a vacuum. The piston is pushed to the bottom, and the opening at the end is then closed by placing a finger on it. When the syringe is submerged entirely in a bowl of water, the piston can be raised without difficulty, and an apparent void is left between the piston and the finger. Pascal notes that when the piston is drawn further, the finger is not *attracted* or pulled with a greater force than before. If the syringe itself is raised so that only its mouth remains under the water when the finger is withdrawn, the water rushes up into the syringe and rises above the level of the water in the bowl.

This experiment is designed to make three things clear: (1) the force required to draw the piston is the same whether the vacuum is large or small, (2) the water rises to fill the empty space, "against its nature as it were," and (3) the resistance to separation that is experienced by the finger is not absolute. In this way, the reader is made to discover by himself that the belief that nature abhors a void is not supported by experiments. In due course, Pascal will go further and argue that it is spurious.

The second experiment, which is performed with bellows instead of a syringe, confirms that only a moderate force is required to create a vacuum. The third experiment is the one that originated with Torricelli but is now carried out in a 46-foot (15 meters) tube filled with red wine (in-

---

[24]In this respect it is interesting to note that Marc-Antoine Dominicy, the editor of Petit's letter to Chanut (Paris, 1647), quotes a passage from the First Day of Galileo's *Two New Sciences*, and ascribes "the honour of the discovery" to Galileo, not Torricelli (*Œuvres de Pascal*, II, p. 544).

stead of mercury) and inverted in a vessel that contains about a foot of water. The wine falls to some 31 feet above the level of the water and leaves a space, apparently void, of about 13 feet. When the tube is slowly tilted, the water rises and completely fills the tube when the top reaches about 31 feet above the level of the water in the vessel.[25] Pascal does not describe the machinery required to lift a 46-foot tube full of water without breakage, but we know that the tube was bound to a mast that was raised and rotated by an ingenious system of pulleys.

The fourth experiment is performed with a gigantic siphon whose legs are 50 feet and 45 feet long. The siphon is filled with water, and the openings of the legs are dipped in two vessels of water, one 5 feet higher than the other. When the siphon is held upright, the water inside each of the legs falls to about 32 feet above the water in the vessels. The longer leg ceases to draw water from the shorter one, "against the expectation of all philosophers and craftsmen," writes Pascal, and the water in both vessels remains at the same height as in the previous experiment.[26] Pascal explains more fully elsewhere why "philosophers and craftsmen" expected the siphon to operate even if their legs extended above 32 feet:

> Take a tube that is bent in the shape of an inverted horseshoe (which is commonly called a siphon), fill it with water and place its legs so that they dip in separate vessels full of water. Then, however small the difference of level between the two vessels, the water from the higher one will rise up the leg to the top of the siphon and will pass down by the other leg into the lower vessel. If water is constantly supplied to the higher vessel, the flow will be continuous. *It is claimed that the water rises on account of nature's horror of the vacuum that would be left in the siphon if the water from these two legs were to fall from each into its own vessel, as it does indeed, if an opening is made at the top of the siphon, and the air allowed to enter.*[27]

---

[25]*New Experiments, Œuvres de Pascal*, II, pp. 502–503; Popkin trans., p. 36. For a good account of Pascal's experiments, see Isabel Leavenworth, *A Methodological Analysis of the Physics of Pascal*. New York: Publications of the French Institute, 1930, pp. 52–55.

[26]Roberval refers to Pascal's experiment with siphons of unequal legs as having been performed with mercury not water (*First Account*, 20 September 1647, *Œuvres de Pascal*, II, pp. 473–475). Did Pascal extend to water results arrived at with a denser (and more manageable) liquid such as mercury?

[27]*On the Weight of the Mass of Air, Œuvres de Pascal*, II, p. 1068; this work is translated in *The Physical Treatises of Pascal*, edited by I. H. B. and A. G. H. Spiers. New York: Columbia University Press, 1937, p. 35.

Engineers knew that water cannot be pumped above thirty-two feet but they failed to realize the implications of their observation. It was only when Baliani, Galileo and, with greater determination, Torricelli and Pascal, began entertaining the possibility of a void that the evidence became significant. The systematic experiments that followed eventually upset the everyday experience that artisans shared with philosophers about what is possible and what is impossible. It had been too easily taken for granted that what had never been done or observed necessarily lay outside the realm of technological possibilities. Pascal exposed their conceit as early as 1645 in his *Notice concerning the Adding Machine*:

> The more they excel in their craft, the more easily do workers become arrogant and assume that they can understand and execute new projects about which they grasp neither the principles nor the application. Intoxicated by a false notion of themselves, they grope about, that is, they work without careful measurements and established guidelines . . . It is important that they should be made aware of their shortcomings and taught that skill is not enough when dealing with new inventions. A grasp of the theory is necessary until practice renders it familiar, and continuous exercise ensures that it is followed as a matter of habit. Now, it would not be in my power, with all the knowledge in the world, to implement my own design alone. I need the help of a worker who masters the hammer, the file and the lathe, and can machine the parts according to the calculations that I made in the light of my theory. Likewise, simple craftsmen, however skilful, cannot satisfactorily put together something that has complicated movements without the help of someone who is able, in the light of the theory, to make the calculations that determine how the pieces fit.[28]

We now return to Pascal's experiments and specifically to the fifth that consists in soaking a 15-foot-long rope to drive out the air in the fibers before placing it in a 15-foot tube, open at one end and closed at the other. The rope is attached to a thread that hangs out of the tube that is filled with water and inverted in a bowl of mercury. Pascal writes that as the rope is slowly pulled out by means of the thread, the mercury rises *proportion-*

---

[28]*Notice Concerning the Adding Machine*, *Œuvres de Pascal*, II, pp. 338–339, English trans. in Pascal, *Selections*, edited by Richard H. Popkin. London: Collier Macmillan, 1989, pp. 27–28.

ally until the height of the quicksilver, plus the fourteenth part of the height of the remaining water, reaches 27 inches. After which, if the rope is pulled further out, the water drops and a space, apparently void, is left at the top. This description of the experiment by Pascal betrays some haste, and the data itself is puzzling, for how can we know what is the fourteenth part of the remaining water unless we are told? If Pascal intends the fourteenth part of the water originally contained in the 15-foot tube, this means a height of almost 13 inches of water, thereby leaving only 14 inches for the mercury in the column whose total height is 27 inches. But this would in no way be near the point where equilibrium with the outside pressure is threatened. Since the specific gravity of mercury, referred to water, is about 14, a column of water at a given height is equivalent to a column of mercury that is one-fourteenth of that height. Hence, the water that must be added to 14 inches of mercury to produce the same pressure as a column of 27 inches of mercury is 16 feet of water!

In the sixth experiment, the siphon that had been used in the first experiment is immersed in one inch of mercury, and the piston is slowly withdrawn. The mercury follows it to the height of 27 inches, but when the piston is pulled further back, the mercury ceases to rise and the apparent void, which is created, grows larger as the piston is drawn higher. From this experiment, Pascal makes an important extrapolation that is printed in italics: *"It is likely that the same thing happens in a suction-pump, for the water only rises there to the height of thirty-one feet, which correspond to two feet three inches of quicksilver."*[29] Pascal also remarks on the fact that the syringe weighs exactly the same whether the piston just leaves the mercury or is raised much higher and creates a greater void.

Pascal does not tell us how he determined the weight of the syringe without removing it from the mercury, but this could have been done by attaching the syringe to a string and hanging it from a balance. Of course, he may also have been relying on the sensation of weight that does not vary as the piston is further drawn back. The significance of the experiment for Pascal lies essentially in its application to suction pumps, and since the syringe weighs the same regardless of how high the piston is raised, there is now a strong case for saying that nothing having weight, hence no physical body, rose to the top of the tube. The snag (clear to us but not to Pascal) is that since the weight of the air contained in the syringe is too small

---

[29]*New Experiments, Œuvres de Pascal*, II, p. 504; Popkin trans., p. 37.

to be determined, the experiment does not preclude the possibility that some matter, as light as air, or even lighter, entered the tube.

The seventh experiment is similar to the fourth, but mercury is used instead of water. The legs of the siphon, which are now 10 and 9½ feet, are immersed to a depth of about an inch, so that the surface of the mercury in one vessel is half a foot higher than in the other. When the siphon is tilted, the mercury flows from the shorter leg into the longer, but when the siphon is made straight, the longer leg sucks up the mercury from the shorter, and the quicksilver falls in both legs to 27 inches above the surface of the mercury in each of their respective vessels. Pascal explored whether mere tilting produced the phenomenon described as *attraction*, and we can follow the steps he took thanks to what Roberval wrote in his *First Account* to Desnoyers in September 1647. First, the siphon was filled and the legs immersed in two vessels of mercury so that the mercury in the siphon stood at 27 inches above the level of the mercury in the vessels. This was achieved by slanting the siphon. In this position, the mercury did not flow. Second, one of the two vessels was lowered, and the mercury flowed from the higher to the lower until the mercury was at the same height in each vessel. Thirdly, the siphon was raised above the critical value of 27 inches, and the mercury fell to 27 inches from the bottom in each vessel while the top became empty, something that appeared "miraculous" to Roberval.[30] In the experiment that we are now considering in the *New Experiments*, Pascal presents the same results in a different order. He skips the case where the two vessels are at the same height, and he inverts the second and the third step because he is interested in the production of a vacuum, whereas Roberval was concerned with the flow of mercury. "The tilting of the tube," stresses Pascal, "always attracts the liquids in the vessels if the openings are not stoppered, and, if they are, then it attracts the finger that seals these openings."[31]

In the eighth, and concluding, experiment, the same siphon as in the foregoing one is used but with a rope, as in the fifth. When the rope is pulled, the mercury rises in both legs so that the fourteenth part of the height of the water in one leg, plus the height of the mercury that has risen in it, is equal to the fourteenth part of the height of the water in the other,

---

[30]We know this from Roberval's, *First Account*, 20 September 1647, *Œuvres de Pascal*, II, pp. 473–474. See the excellent discussion in Dominique Descotes, *L'argumentation chez Pascal*. Paris: PUF, 1993, pp. 268–272.

[31]*New Experiments, Œuvres de Pascal*, II, p. 505; Popkin trans., p. 38.

plus the height of the mercury that ascended in the second leg. This happens as long as the water and the quicksilver do not exceed 27 inches; thereafter an apparent void is produced. In the fifth experiment, a rope had been placed in the tube and a thread fastened at the end. In this experiment, the rope goes round inside the siphon, but Pascal does not tell us whether the thread is fastened to the end of the shorter or the longer leg. He probably felt that it was irrelevant whether the traction was exerted on one end rather than on the other. The serious problem lies indeed elsewhere: if the ratio of water is the same in both legs, and the legs are of unequal length, then one fourteenth of the water in one leg cannot have the same value as one fourteenth of the water in the other. Hence, the height of mercury in each leg must also be different. Pascal is using the ratio that we found puzzling in the fifth experiment. The matter is even more perplexing since the tube in the fifth experiment was fifteen feet, whereas in this experiment the legs of the siphon are ten feet, and nine and a half. We have here yet another instance of haste due, in all likelihood, to Pascal's desire to ensure priority.

Maxims of the First Part

From these experiments, and from others that Pascal does not describe but that he says he performed in vessels of different shapes with various liquids, the following seven maxims are "deduced."[32] They can be divided into two groups of three, with a concluding maxim. The first group concerns Nature's *abhorrence* of a vacuum, the second the *inclination* of bodies to fill the void, and the concluding maxim the force required to produce a vacuum.

Repugnances	*Inclinations*
1. All bodies have a repugnance to be separated or to allow a void in the interval between them; in other words, nature abhors a void.	4. The bodies that limit this void have an inclination to fill it.

---

[32]Pascal writes "we can clearly deduce (*déduire*) these maxims" (*Œuvres de Pascal*, II, p. 505), but the verb *déduire* does not always have the strong sense of "to deduce" in the seventeenth century. For instance, in his letter to Chanut of 26 November 1646, Pierre Petit, when referring to two experiments he performed, writes, "*je vous les déduirai*," and then proceeds to describe them (*ibid.*, II, p. 355).

Repugnances		Inclinations	
2.	This abhorrence or repugnance that all bodies have is not stronger for a larger than for a smaller void, that is to say, it is not greater in a larger space.	5.	This inclination to fill a large void is not greater than for filling a small one.
3.	The force of this abhorrence is limited and is equal to that with which water of a given height (which is about thirty-one feet) tends to flow down.	6.	The force of this inclination is limited and always equal to that with which water of a given height (which is about thirty-one feet) tends to flow down.

Concluding Maxim	
7.	A force, slightly greater than that with which water tends to flow downward from a height of thirty-one feet, is sufficient to bring about a void of any size whatever, that is to say, to separate bodies by an interval of any size, provided that there is no other obstacle to their separation than the abhorrence that nature has for the void.

Anthropomorphic Survivals

Traditional ways of speaking die slowly: bodies are said by Pascal to have a *repugnance* to allow a vacuum, and a *tendency* to fill empty space. Nature still *abhors* a vacuum. But why should repugnance cease at 31 feet in the case of water, at 27 inches in the case of mercury and, for other liquids, at a height proportional to their specific gravity? Why should it disappear when a tube is tilted below a certain level? *Horror* is too melodramatic for a force that is equal to the weight of a column of mercury 27 inches high! *Abhorrence* or *repugnance* is not much better. The correct terminology is simply *force*, and if we read "*the force of this horror*" in the third maxim, and "*the force of this inclination*" in the sixth, we find just "*a force*" in the seventh. Pascal is preaching experimental open-mindedness, and his conservative terminology is meant to avoid the impression that he is rushing ahead of the evidence. He borrows anthropomorphic language in order to mine the semantic field from within.

The void described in the *New Experiments* is always an "apparent void," a point that is stressed by his brother-in-law, Florin Périer, the editor of his posthumous treatises, *On the Equilibrium of Liquids* and *On the*

Weight of the Mass of Air. This is why the maxims are so tame and reach four relatively uncontroversial conclusions: first, bodies have a repugnance to separate and an inclination to fill a vacuum (maxims 1 and 4); second, the repugnance and the inclination do not vary with the size of the apparent vacuum (maxims 2 and 5); third, their force can be measured by a column of water 31 feet high (maxims 3 and 6); and fourth, a slightly greater force produces an apparent void (maxim 7).

Pascal does not mention the difficulty in handling the instruments or the care to be taken in drying the tubes. Why did he skip this information? Perhaps in order to protect his priority claim by concealing the pitfalls of actual experimentation that his rivals had not foreseen. Nonetheless, he provides some concrete details: the vessel of mercury must be large enough for the experimenter to bend his arm (first experiment); the rope must be soaked in water (fifth experiment); small wooden cylinders, which can be used instead of rope, should be tied with brass wire to avoid air slipping in (fifth experiment); and the weight of the syringe should be kept constant (sixth experiment).

The maxims are followed by eight propositions that rule out that the empty space could be filled with outside air, spirits of mercury, or subtle air. Pascal concludes: "My opinion is that this apparently empty space is really empty,"[33] but he makes no attempt to determine its cause, and there is no word about atmospheric pressure.

Pascal's Caution and Descartes' Boldness

It may be interesting at this point to contrast Pascal's cautious approach with Descartes' much rasher procedure. Descartes attempted to show that there is no real difference between empty space and physical extension, and hence no vacuum, by stripping away the properties of bodies:

> It is easy for us to recognise that the extension that constitutes the nature of a body is exactly the same as that of space. Suppose we attend to the idea we have of a body, for example a stone, and leave out everything we know to be non-essential to the nature of body. We will first of all exclude hardness, since if the stone is melted or pulverised it will lose its hardness without thereby ceasing to be a body; next we

---

[33]*Ibid.*, p. 507; Popkin trans., p. 40.

will exclude colour, since we have often seen stones so transparent as to lack colour; next we will exclude heaviness, since although fire is extremely light it is still thought of as being corporeal; and finally we will exclude cold and heat and all other such qualities, either because they are not thought of as being in the stone, or because if they change, the stone is not on that account reckoned to have lost its bodily nature. After all this, we will see that nothing remains in our idea of a stone except that it is something extended in length, breadth and depth. Yet this is just what is comprised in the idea of a space—not merely a space which is full of bodies, but even a space which is called "empty."[34]

Descartes has satisfied himself that all the properties he has enumerated (hardness, weight, color, temperature, and so on) are not essential characteristics of a material body because they can change. But his argument is unconvincing, for although the combined perception of these properties tells us what a body is, the fact that a body always has *some* shape does not entail, as Descartes assumes, that where there is shape there is matter also. In other words, Descartes' definition of body in terms of extension alone is too broad because it does not allow for the crucial distinction between physical magnitudes and magnitudes that are purely mathematical, that is, mental. Roberval was to attack him on just this point.

Descartes and Pascal both insisted on the importance of seeing things clearly, but they did not share the same notion of clarity. For Descartes *intuitus* or *insight* depends on the focused *attention* that results from freeing the intellect of bias; for Pascal, *seeing* is closer to visual perception. This does not mean that Pascal's experiments are mere spontaneous observations. They are arrived at through a process of selection whereby only significant aspects are retained. In other words, they are contrived or even imagined.[35] New conceptual tools had to be forged to look at the world

---

[34]Descartes, *Principles of Philosophy*, Part 2, art. 11 in Descartes, *Œuvres*, edited by C. Adam and P. Tannery, vol. VIII-I, p. 46. See also the second of his *Meditations*, ibid., vol. VII, p. 30. This is discussed in William R. Shea, *The Magic of Numbers and Motion*. Canton. MA: Science History Publications, 1991, pp. 175–177.

[35]Pascal writes, "Experiments are the only principles in physics" in his *Preface to the Treatise on the Vacuum, Œuvres de Pascal*, II, p. 781, trans. in Pascal, *Selections*, edited by Richard H. Popkin, p. 64. See similar statements by Pascal in *Œuvres*, II, pp. 1065, 1101, trans. in *The Physical Treatises of Pascal*, pp. 31–75. Descartes seems content with less, "At the beginning it is better to resort to those experiences that commonly occur rather than to

in this way and Pascal cannot be expected to have developed them fully, but he took a decisive step towards their formulation. He replaced an all-encompassing and dogmatic explanatory model with an empirical form of reasoning that rests on the production of new data. This marks the emergence of a new kind of intelligibility in which exhaustive knowledge is displaced by selective, but empirically verified, information.

Descartes was driven by ontological imperatives to fill empty space with matter, whereas Pascal saw no compelling reason to assert a material substrate where there is an apparent void. He did away with *unobservables*, an exclusion as daring and novel as would be the rejection of the mandragora or sirens in popular sixteenth-century natural histories like those of Conrad Gesner or Ulisse Aldovrandi. Philosophical issues, *as such*, are only raised at the end of the *New Experiments*, where Pascal says that in a forthcoming treatise he will address and dispel the five following objections: (1) empty space is repugnant to common sense; (2) we cannot claim that nature abhors a vacuum and then allow it; (3) everyday experiments reveal that nature cannot tolerate a vacuum; (4) apparently empty space is really filled with a material that is undetected by our senses (for instance, the "subtle matter" of Descartes); (5) light has to be either an accident or a substance: if it is an accident, it cannot be said to travel without a supporting medium; if it is a substance, it must fill the apparently empty space.[36]

Tentative Conclusions

We can now take stock of what Pascal has achieved. He did not prove to his own satisfaction what he had set out to determine, namely, that the apparent void is, unquestionably, a real vacuum. In his conclusion to the

---

seek out those that are rare and more elaborate" (*Discourse on Method*, part 6, A. T., VI, p. 63). In his *Principles of Philosophy*, he provides a brief account of the principal phenomena of nature, "but my purpose," he writes, "is not to offer them as arguments to prove something" (part III, art. 4, A. T. VIII-I, p. 81).

[36]Light will remain an issue for a long time to come. As late as 1674, François Bernier re-examined the experiment of Torricelli and more recent ones to conclude that the resulting vacuum must contain "particles of light," "particles of heat and cold," and "weight-particles," the last to explain why heavy bodies are attracted downwards (François Bernier, *Abrégé de la philosophie de Gassendi*, Paris, 1674, traité I, chap. 10, pp. 98–99). See Jean-Pierre Fanton d'Andon, *L'Horreur du vide*. Paris: Editions du C.N.R.S., 1978, pp. 30–33.

New Experiments, he says only that it is his opinion that the apparent void is a genuine one. What he makes perfectly clear is that the force that resists the formation of a void—be it real or apparent—is measurable and constant within narrow limits. This is equal to the tendency of a column of mercury 27 inches high to run down, and when the mercury is in a tube of uniform diameter, this is equal to the weight of such a column. Any force greater than this will produce a void. If another liquid is used, its height will be to that of mercury in inverse proportion to their respective weights. Pascal studied this force with a variety of instruments including syringes, siphons, bellows, and glass tubes, and he showed, by drawing up the piston of a syringe, that no greater force is needed to produce a large void than a smaller one. Although he speaks of the "horror of the void" in the *New Experiments*, this traditional phrase has shifted meaning, and we have Pascal's own word that he was already thinking of some force outside the tube to counterbalance the weight of the column of mercury or water. Less than five weeks after the publication of the *New Experiments*, he wrote to Périer that he suspected that it was "the weight and pressure of the air that balances the weight of quicksilver."[37]

Much of the work described in the *New Experiments* had been anticipated by Torricelli, who conceived the idea of performing the experiment with tubes of different shapes to test whether the mercury remained at the same level when vacua of different sizes were produced. Torricelli also thought of placing water above the mercury in a vessel and then raising the tube slowly to allow the water to rush in to take the place of the mercury. This constituted strong evidence in favor of the hypothesis that there was a vacuum at the top of the tube.

Torricelli clearly provided Pascal with the incentive he needed, but Pascal deserves credit for the variety and ingenuity with which he proceeded. He produced fresh evidence for the void by tipping the tube and showing that, as the tube was inclined, the mercury gradually rose to the top, but in such a way that it continued to stand at its usual height. This confirmed that it takes the same force to bring about a greater or a lesser degree of rarefaction. He also carried out the experiment not only with water but also with wine, and sometimes with a combination of both. He can also be credited with two new methods of obtaining a vacuum. The first was to fill almost entirely a tube with a solid body, such as a rope, before pouring some liquid in the remaining space. When the rope was gen-

---

[37]Letter of Pascal to Périer, 15 November 1647, *Œuvres de Pascal*, II, p. 679.

tly pulled out, a vacuum was left in the tube. The second was to use large syringes to produce vacua of greater dimensions than had been possible with tubes filled with mercury.

Pascal and Torricelli had two traits in common. They were relatively free from metaphysical presuppositions and they were modest about their objectives. "If Pascal had attempted to give a complete picture of the world in which we live," writes Isabel Leavenworth, "he would have needed many more presuppositions than he needed for the solution of one specific and not very complex problem. But to give such a picture was a thing which Pascal as a scientist would never have attempted."[38] The presuppositions that Pascal makes are minimal, and their justification does not depend on the acceptance of a particular theory. He is careful to describe what he sees, and to provide simple measurements of the "force" at work.

---

[38]Isabel Leavenworth, *A Methodological Analysis of the Physics of Pascal* , p. 78

CHAPTER 4

The Battle of the Void

The Jesuits Against Pascal

The *New Experiments Concerning the Vacuum* created the stir for which Pascal had hoped, but it was not all favorable. The parting shot was fired by Etienne Noël, a Jesuit who had taught Descartes at the College of La Flèche before becoming Rector of the College of Clermont (now the Lycée Louis-le-Grand) in Paris, just across the street from the Sorbonne. At sixty-six, Noël was an influential member of the *Establishment*, and the long letter he wrote to Pascal is a recognition of the importance of the young man's work, but it may also be an indication of the social prominence of his father, Etienne, to whom the *New Experiments* were dedicated.

Noël praises Pascal's "clever and beautiful" experiments, but he will have no truck with the idea that a genuine vacuum had been produced in the inverted tube of mercury. His main argument is that light is refracted in the apparently empty space, which would be impossible unless it encountered some material substance. His second argument is that the mercury falls at a perceptible rate when the tube is inverted, whereas it should fall at an incredible speed if resistance to motion had been completely removed, as would be the case in a vacuum.[1]

In order to understand Father Noël's cast of mind, it is important to briefly recall the Aristotelian natural philosophy that was still taught by the

---

[1]*First Letter of Noël to Pascal*, written between 20 and 25 October 1647 (*Œuvres de Pascal*, II, p. 514).

Jesuits. According to this model, the three states of matter—solid, liquid, and gas—were confused with elements, so that ice, liquid water, and steam represented earth, water, and air. Condensation leads from air to water, and then to earth. Rarefaction is the inverse process: ice melts to form water, which boils to form steam that was considered air. Because flames go up, it seemed that fire was a further rarefaction of air. Purified air was thus held to be igneous air, and we shall see what use Noël made of this idea. The Jesuit also assumed that there is a fundamental *correspondence* between the *microcosm* (the little world of man) and the *macrocosm* (the great world of nature), such that we can make inferences from the similarity between the humors and the elements, more specifically between blood and air. As the blood that courses through our veins is a mixture of the four humors recognized by medicine (blood, phlegm, choler, and bile), so the air we breathe is a mixture of the four elements (fire, water, earth, and air):

> The fluids called humours, which are mixed in the veins, arteries and other passages in our body, enable us to understand, by analogy, the mixture of elements in the *great world* where the actions and movements of the celestial vault, the stars, the planets, and especially the Sun disclose that the elements must be mixed, such that you cannot have one without finding the others present to a greater or lesser degree. *The Sun* continually sends out throughout the world solar spirits that imperceptibly and ceaselessly move and mix everything for the welfare of the world, *just as the heart* sends through the body vital spirits that constantly push and mix everything for the welfare of the body.[2]

Ideology and Physics

Although Noël adduced his comparison between air and blood as an instance of the grand analogy between the macrocosm and the microcosm,

---

[2]*Second Letter of Noël to Pascal, Œuvres de Pascal*, II, p. 536. Noël's position was not unheralded. For instance, in his influential textbook, Abra de Raconis (Charles François d'Abra) compares the humors to the elements and the four ages of life: "Bile corresponds to fire, because it is hot and dry, and thus indicates adolescence; blood corresponds to air, because it is humid and hot, and thus indicates youth" (quoted in Jean-Pierre Fanton d'Andon, *L'horreur du vide*, Paris: Editions du C.N.R.S., 1978, p. 61, n. 5. The first edition of Raconis' *Totius Philosophiae Compendia* appeared in 1617, the sixth in 1637).

he offered the following "experimental" evidence for claiming that glass is always porous: (1) light penetrates glass more easily than any other body, and hence it must contain more pores, and (2) an hermetically sealed bottle does not break when placed on warm ashes, so heat must escape through its pores. Both arguments beg the question: the first by assuming that porosity is necessary for the transmission of light, and the second that heat can only escape through pores in the glass. Noël sensed the difficulty, but he thought it could be circumvented by distinguishing between *ordinary* air, which does not pass through glass, and *purified* or *igneous* air, which does. The plausibility of this hypothesis rests entirely on further comparisons with muddy water that can be made to pass through a thick piece of cloth once it is filtered, or an iron wire that can be made to go through a narrow hole once it is stretched.

Since glass is blown over fire, Noël believed that *purified or igneous air* entered its composition. The mercury in the tube tends to stick to this kind of air, and when the tube is inverted, the weight of the falling mercury pulls in additional *igneous* air from the outside through the pores of the glass until the column of mercury has fallen to a height of 27 inches. At this height mercury is no longer heavy enough to drag in more air. Why this limitation occurs is not further explained, but the result is declared *unnatural*, and the *igneous* air in the tube is said to attempt to escape. "As soon as the violence is removed," writes Noël, meaning as soon as the weight is removed, "the purified air returns to the mixture to which it naturally belongs."[3] This is buttressed by yet another analogy: when a bow is bent some of its *natural spirits* are pushed out on the concave side, which is being compressed, while *unnatural spirits* enter on the convex side, which is being dilated. Both kinds of *spirits* continue to seek their natural place, and once the *external force* that bent the bow is removed, the *natural spirits* re-enter the wood, the *unnatural* ones depart, and the bow straightens itself out! The same process is said to be at work when a sponge, which was squeezed dry, reabsorbs the water that had been expelled.

Pascal's Reply to Noël, 29 October 1647

Pascal was ill and bedridden when he received Noël's letter, but he managed to dash off a reply in less than a week. It contains some of his most

---

[3]*First Letter of Noël to Pascal, ibid.*, p. 516.

significant remarks on scientific reasoning, and it begins with a short lecture on the conditions of rational assent to a given proposition. The first is that it must be so clear and distinct that it cannot be doubted, like the axiom, "when equals are added to equals, the sums are equal." The second condition is strict entailment from such self-evident propositions. For example, the theorem, "the three angles of a triangle are equal to two right angles," follows directly from Euclid's postulate of parallel lines, which was considered obvious. "When one of these two conditions is fulfilled," writes Pascal, "the proposition is certain and true; otherwise it is doubtful and uncertain."[4] Astronomical systems are doubtful because the position of the planets can be worked out equally well in any of the three current models: the Earth-centred system of Ptolemy, the Sun-centred hypothesis of Copernicus, or the compromise system of Tycho Brahe in which planets turn around the Sun and the Sun turns around the Earth. We simply cannot know which system is correct, says Pascal, and it would be rash to favor one model over the others. Galileo, who was condemned while Pascal was still a child, fell foul of the Inquisition for advocating the Copernican model, and Pascal may be slyly suggesting that the rejection of a vacuum could have the same inhibiting effect on science.

After this crash course in logic, Pascal proceeds to examine Noël's two arguments against the possibility of a vacuum. The first was that light is bent or refracted when it goes through the apparently empty space at the top of the tube, a clear indication that it encounters a material body, presumably rarefied air. Had Noël looked more closely, declares Pascal, he would have noticed that refraction does not occur between the walls of the tube but only where the light strikes the glass. He should also have recorded that light shines more brightly as the medium through which it passes becomes rarer, namely closer and closer to a vacuum. Noël's second argument does not fare better. His claim that there must be something in the apparently empty space, otherwise bodies would drop at an incredible speed, is met by recalling that when Galileo dropped balls of different weights from a high tower they fell at the same speed. A vacuum would only shown this more clearly.

---

[4] *Reply to Father Noël*, 29 October 1647, *Œuvres de Pascal*, II, p. 519, trans. in Pascal, *Selections*, edited by Richard H. Popkin, London: Collier Macmillan, 1989, pp. 49–50. This is reminiscent of the First Rule of Reasoning in Descartes' *Discourse on Method*, where we are enjoined to include nothing more in our judgements than what presents itself to our mind so clearly and distinctly that we have no occasion to doubt it.

Pascal stresses the importance of drawing the implications of our theories and putting them to the test. As he later put it, "experiments are the true masters we should follow in physics."[5] Now Noël suggested that mercury pulls down *igneous* air which, in turn, exerts an attractive force on the mercury. If this theory is correct, says Pascal, the attraction should vary with the quantity of air, and mercury should therefore rise higher in taller tubes where there is more empty space and, hence, more *igneous* air. But experiments show that this is not the case: a column of mercury does not rise higher when the empty space above it is increased. Therefore, Noël's theory must be abandoned.

Noël had referred to experts who shared his views but had no better arguments than his own. This earns him a lecture on the role of authority in science. "When dealing with questions of this kind," writes Pascal,

> we do not base ourselves on authorities, and if we refer to them it is to give their arguments not their names, unless we are dealing with historical evidence. Now if your authors had said that they saw small igneous bodies mixed with the air, I would have enough faith in their honesty and integrity to believe that these bodies exist, but I would believe them as witnesses. But since they only say that they *think* that air contains them, you will allow me to go on doubting.[6]

Although we cannot prove that absolutely *nothing* takes the place of the mercury that falls in the tube, this is not enough to affirm that *something* must have replaced it. Pascal compares Noël to a novelist who fashions a world of his choice. "If such a method were granted," he quips, "it would become easy to understand magnetism and the tides," namely, two of the major problems of seventeenth-century physics.[7]

The Ontological Status of the Void: A Long-standing Controversy

Noël's arguments were mainly based on vague comparisons that rest on little more than everyday experience. But there is one serious philosophical

---

[5] Pascal, *Conclusion* of the treatises *On the Equilibrium of Liquids* and *On the Weight of the Mass of Air*, *Œuvres de Pascal*, II, p. 1101, in the trans. by Spiers, p. 75, and in the trans. by Richard H. Popkin, p. 61.
[6] *Ibid.*, p. 523; trans., p. 52.
[7] *Ibid.*, p. 522; trans., p. 51.

issue that he mentions in his first letter when he objects that space cannot be completely empty because it is extended. The point is that what is extended is necessarily material, something that was dear to Descartes but left Pascal cold. Nonetheless, Pascal did not wish to evade the issue, and in his reply to Noël he identifies empty space with what is called a *solid* in geometry, and proceeds to make the startling declaration: "There is as much difference between nothingness and empty space as between empty space and a physical body. Thus empty space is halfway between matter and nothingness."[8] But can a vacuum occupy a halfway house between material existence and nothing? Is Pascal, the logician, creating a logical void? The paradox is real, and in order to appreciate its relevance we must say a few words about the long and heated discussion over the possibility of an effective vacuum that began in Antiquity.

Atomists in Ancient Greece had argued that there must be void space between atoms otherwise they would not be able to move. In his *Physics*, Aristotle countered this argument by claiming just the opposite: a vacuum would render motion unintelligible. His reasoning rests on the assumption that the speed of a body falling through space varies inversely as the density of the medium through which it falls. In a vacuum, a body would encounter no resistance and would move with infinite speed; in other words, it would take no time at all to travel from one point to another. But this is clearly absurd. Furthermore, according to Aristotle, the speed of a falling body varies directly as its weight because the heavier the body the greater its power to overcome the resistance of the medium. In a vacuum, where there would be no resistance to overcome, bodies of different weights would have to fall at the same speed. This is obviously impossible, he contends, and therefore there is always some resistance to motion and no vacuum. The irony will not escape the modern reader who knows that bodies *do fall* at the same speed regardless of their weight!

Because heavy bodies, like stones, naturally fall, while light ones, like fire, naturally rise, Aristotle also thought that this indicates that there are directions (i.e., up and down) in real space. In a vacuum, however, every part of space would be identical with every other part, and so heavy bodies could not fall, and light ones could not rise because up and down would be indistinguishable!

Aristotle's views had not gone unchallenged in Antiquity. Hero of Alexandria argued that matter is composed of minute particles that are

---

[8]*Reply to Father Noël, Œuvres de Pascal*, II, p. 526; trans., pp. 54–55.

separated by small empty spaces or *interstitial vacua*, which he distinguished from the larger and *continuous vacuum* that is occasionally produced by the application of a pump.[9] In the sixteenth century, Giordano Bruno envisaged an infinite material that offered no resistance to motion of bodies or completely penetrated them.[10] But the effective critic of Aristotle was Galileo, who was mainly concerned with problems of motion and not with philosophical considerations about the nature of a vacuum. He was content to assume the hypothetical existence of a vacuum and explore the consequences of motion in the absence of resistance. A vacuum was simply an entity devoid of matter and therefore unable to oppose resistance to any body that might be placed in it.

Galileo on the Vacuum

Galileo's most extensive discussion of motion in a void appeared in his *De motu*, written about 1590, when he was a young lecturer at the University of Pisa. He argues, in opposition to Aristotle, that motion in a void would not take place instantaneously because "natural speed," which is a function of "natural weight," is completely effective only in a vacuum where no material medium can retard it. "My whole point is this," writes Galileo,

---

[9]"Every body is composed of minute particles, between which are empty spaces less than the particles of the body, so that we are wrong when we say that there is no vacuum except by the application of a force, and that every place is full either of air, or water, or some other substance. In proportion as some of these particles recede, others follow and fill the vacant space. There is no continuous vacuum except by the application of some force, and an absolute vacuum is never found, but is produced artificially" (Hero of Alexandria, *Pneumatics*, translated by Bennet Woodcroft. London: Taylor Walton and Maberly, 1851, facsimile London: Macdonald, 1971, p. 10). Pierre Petit wrote that Pascal's father had "long been convinced" of Hero's views (letter to Pierre Chanut, 26 November 1646, *Œuvres de Pascal*, II, p. 354). Hero's *Pneumatics* were published in 1575 in a Latin translation by Federico Commandino (reprinted in 1583 and 1680); in 1589 in an Italian translation by G.B. Aleotti (reprinted in 1647 and 1693), and in 1595 in another Italian translation by Alessandro Giorgi; in 1687 in German; and finally in the original Greek in 1693. An early source of a three-dimensional void is John Philoponus' *De Aeternitate Mundi Contra Proclum* of 529, which was only translated into Latin in the sixteenth century.

[10]See Bruno's *De Immenso et Innumerabilibus*, first published in 1591 (in *Opera Latine Conscripta*, ed. F. Fiorentino, Naples: Morano, 1879–1884, vol. I, pp. 231–233. See p. 141 for Bruno's declaration that a vacuum is neither a substance nor an accident). Bruno assigns the further attribute of impenetrability to space, presumably to preserve its continuity and indivisibility, but for Pascal it is self-evident that space has the property of yielding to bodies without resistance.

Suppose there is a heavy body a, whose proper and natural weight is 1000. Its weight in any plenum will be less than 1000, and therefore its speed will be less than 1000. Thus if the weight of a medium compared to an equal volume of a is only 1, it follows that the weight of a in this medium will be 999. Therefore its speed will also be 999. And the speed of a will be 1000 only in a medium in which its weight is 1000, and that will be nowhere except in a void.[11]

In this example, the determinant of velocity is the effective, rather than the gross weight of a body. Bodies of the same kind of material (say two pieces of iron) will therefore fall at the same speed in the same medium even if they are unequal in size and have different weights. The thrust of Galileo's argument can be seen in his refutation of the Aristotelian claim that doubling a body's weight would result in doubling its speed. Galileo reasons as follows: Drop side by side two bricks of the same size and the same weight. They fall at the same speed. Now imagine that they are joined while falling, and ask yourself whether this will double their speed, as Aristotle maintains. Clearly there is no reason why they should move faster.[12]

If we want to express this in quantitative terms, we can say that the effective weight of a body in a medium, according to Galileo's view, depends on the difference between the specific weight of the body and that of the medium through which it falls. This means that a difference in specific weights determines the velocity of a body, so that the speed of a freely falling body can be represented as:

> Velocity of falling body = specific weight of body − specific weight of medium

But a vacuum has no specific weight. Hence bodies that fall in empty space will do so with speeds that are directly proportional to their specific weights. If they have different specific weights, they will fall with different speeds, but if their specific weights are identical, they will fall with equal speeds. Hence a small and a big cannon ball must necessarily fall with the same speed in a vacuum.

---

[11] Galileo, *De Motu*, chapter 10, in Galileo Galilei, *Opere*, edited by A. Favaro, vol. I, p. 282; trans. in Galileo Galilei, *On Motion and On Mechanics*, edited by I.E. Drabkin and S. Drake. Madison: University of Wisconsin Press, 1960, p. 47.

[12] Galileo, *De Motu*, chapter 8, *Opere*, vol. I, p. 266, Drabkin and Drake, trans., p. 30.

In his survey of early ideas on the void, Edward Grant points out that in reaching the conclusion that bodies of the same kind fall in a void with the same speed regardless of their weight, Galileo made the same point as Thomas Bradwardine and Albert of Saxony two centuries earlier, but with a difference that is crucial. Whereas his medieval predecessors assumed that heaviness and lightness were absolute properties of bodies, Galileo completely abandoned the concept of absolute heaviness and lightness. He insisted that everything has weight except the void. Gone are absolute heaviness and lightness, and the distinction between simple elemental bodies and mixed bodies. All bodies are now alike. In the medieval conception, mixed bodies could move naturally up or down in a vacuum, and the direction of motion was determined by the dominance of lightness or heaviness. A material mixed body that contained more air and fire than earth and water would, despite the possession of a certain determinate weight, move upward in a vacuum. The mixture of light and heavy elements in bodies functioned therefore both as motive force and internal resistance. For Galileo this is no longer the case. Void space is truly homogeneous and without any capacity to influence the behavior of material bodies.[13]

Pascal vs. Descartes

Pascal had learned his lesson well, and he agreed with Galileo's analysis of free fall. But in France the debate turned more easily to metaphysics, and Pascal was probably influenced by Roberval, who held that void space is prior to creation and somehow grounds the eternal truths of geometry. This was discountenanced by Descartes, who identified the extension of space with the extension of a material body and ruled out the possibility of a vacuum.[14] Now Pascal's *New Experiments* were published in 1647, the year the French translation of Descartes' *Principles of Philosophy* appeared, and this may explain the caution with which Pascal declares:

---

[13] See the excellent discussion in Edward Grant, *Much Ado About Nothing: Theories of Space and Vacuum from the Middle Ages to the Scientific Revolution*. Cambridge: Cambridge University Press, 1981, pp. 60–66.

[14] Descartes, *Principles of Philosophy*, part 2, art. 16. Descartes, *Œuvres*, edited by C. Adam and P. Tannery. Paris: Vrin, vol. VIII-A, p. 49. In the *Philosophical Writings* translated by J. Cottingham, R. Stoothoff and D. Murdoch. Cambridge: Cambridge University Press, 1985, vol. I, pp. 229–230.

"Having demonstrated that none of the materials that are perceived by our senses, and of which we have any knowledge, fills this apparently empty space, my view, until someone demonstrates the presence of some material that fills it, is that it is truly empty and devoid of all matter." But he promised to reply at a later date to the objection that "an imperceptible material, unheard of and unknown to all the senses, fills that space."[15] In his letter to Father Noël, it is clear that he has Descartes in mind:

> For things of this nature, whose existence are not revealed to any of the senses, are as difficult to believe as they are easy to invent. Many persons, including some of the most learned alive today, have proposed the same material before you (but just as a simple idea, and not as an unalterable truth), and this is why I mentioned it. Others, in order to fill the empty space with some matter, have imagined one with which they fill the whole universe, because our imagination can create with as little time and trouble the greatest as well as the smallest thing. Some have made this material out of the same substance as the sky and the elements, others, out of a different substance, according to their fancy, since they treated it like a creation of their own. When they are requested, as you are, to show us this material, they answer that it is not visible. If they are asked whether it makes some sound, they say that it cannot be heard, and so on with all the other senses. They think they have accomplished a great deal when they have placed others in the position of not being able to prove that this material does not exist, when they have deprived themselves of the power of proving that it does.[16]

The letter ends with an even more explicit reference to Descartes. "You do not believe," Pascal adds,

> that a material unheard of and unknown to all the senses was proposed by *any physicist*. I can assure you of the contrary, since it comes from one of the most famous of our time. You can see in what he writes that he fills the entire universe with a universal matter that is imperceptible and that no one has heard of but which he claims is similar to that of the sky and the elements. Further, upon examining

---

[15]Pascal, *New Experiments*, Œuvres de Pascal, II, p. 508.
[16]*Reply to Father Noël*, Œuvres de Pascal, II, p. 522; Popkin trans., pp. 51–52.

yours, I find it so imperceptible, and its qualities so unheard of, that it must be of the same kind.[17]

Descartes had argued that the term *empty* does not imply the total absence of bodies but simply the absence of things that we expect to find there: a pitcher is called *empty* when it is only full of air, a fishpond when it only contains water, and a merchant ship when it is only loaded with sand ballast. If we supposed that space contains nothing whatsoever, we would be just as mistaken as thinking that air in a jug is nothing. But this invites the question, What if God removed the air from a vessel and prevented any other body from taking its place? Descartes replied that if God were to take away every single body contained in a vessel, without allowing any other body to take the place of what has been removed, the sides of the vessel would have to be in contact. For when there is nothing between two bodies they necessarily touch each other.

Empty Philosophical Categories

Noël's target was Pascal's notion of a vacuum as lying between being and nothingness, a halfway house between mind and matter, as he put it. Such a ghostly entity could neither transmit light nor slow down falling bodies. Noël made much of the philosophical tenet that whatever exists must be either a *substance* or an *accidental property*. Pascal flatly denied the dichotomy and had added that some Jesuits were on his side. One of these was Pedro Fonseca (1528–1599), who inspired an influential series of commentaries on Aristotle published by the University of Coimbra under the collective title of *Conimbricenses*. Fonseca denied that space was a substance or an accident, and he called it an imaginary rather than a real being, although he insisted that it was not a mental fiction.

Another natural philosopher who was closer to Pascal was the Italian Francesco Patrizzi (1529–1597), who considered space as the first of God's creation because all other things require it for their existence. For Patrizzi, space offers no resistance to the reception of bodies, and it continues to exist even as it coincides with the body that fills and occupies it. The

---

[17]*Ibid.*, pp. 526–527; Popkin trans., p. 55.

French philosopher, Pierre Gassendi (1592–1655), with whom Pascal was acquainted, took the further step of reviving the ancient atomism of Democritus and Epicurus but with a Christian twist by filling the void with God, the uncreated Spirit. Although God is not an extended magnitude like a body, we imagine Him *as if* He were infinitely extended. Before the creation of the world, God was everywhere, that is, in every place, not only the place in which the future world would be, but also in an infinity of other places. To characterize this space, Gassendi resorted to a radical device: the imaginary annihilation of all or part of the world. If we imagine that the infinite world was destroyed, what would be left behind would be a three-dimensional, endless space which, Gassendi concluded, must have existed before the creation of the world. This space would be absolutely immobile, for if God were to move the world through it, it would remain motionless. It could offer no resistance to bodies that move in and out of it, and would not be something positive like a substance or an accident. Now God creates only positive things that are either substances or accidents; therefore God did not create space. We are thus left with two eternal, uncreated beings: God and space, each independent of the other. Gassendi could have identified space with God's immensity but he shunned this, probably out of fear that if space were God's immensity, we would have to assume that God Himself is an extended being!

Pascal's reference to space as a homogeneous, infinite void that exists by itself whether or not bodies occupy all or part of it belongs, therefore, to a distinguished but highly controversial tradition. It may have seemed paradoxical to Father Noël, but it greatly contributed to the demise of the categories of substance and accident. Almost half a century later, John Locke, in his *Essay on Human Understanding*, tackled the problem in his mock-modest way: "If it be demanded (as usually it is) whether this space, void or body, be *substance* or *accident*, I shall readily answer I know not; nor shall be ashamed to own my ignorance, till they that ask show me a clear and distinct idea of substance." Locke saw that what was at stake was the principle of plenitude, and that against such metaphysical assumptions, experiments are of little avail. What follows could have been written by Pascal himself:

> I endeavour, as much as I can, to deliver myself from those fallacies that we are apt to create for ourselves, by taking words for things. It helps not our ignorance to feign a knowledge where we have none, by making a noise with sounds, without clear and distinct significations.

Names made at pleasure neither alter the nature of things, nor make us understand them, but as they are signs of and stand for determined [this in the 4th edition; the first three had "clear and distinct"] ideas.[18]

Pascal may have anticipated Locke in his insistence on experiments, but Noël had no intention of allowing him to neglect metaphysics.

Noël's Second Volley

Pascal's reply to his letter had reached Noël on Thursday evening, 31 October 1647, and within a week he strode out lustily into battle. His second letter accuses Pascal of failing to realize that if a vacuum was real and not merely imaginary space, then it could only be the immensity of God. The very idea made him shudder, for God cannot have "parts outside parts." Pascal refused to be drawn onto this slippery path and preferred to declare that the "the mysteries that concern the Godhead are too holy to be profaned by our disputations. They are objects of adoration, not subjects of disputation."[19] That a Christian physicist should combine reverence for the mysteries of Faith with indifference towards theological wrangling came as a surprise to Noël, who was not genuinely interested in experimental physics but in the way science could be used to illustrate philosophical concepts. Noël distinguished between mathematical space, an abstraction that exists in the mind of geometers, and "material matter, which alone can be set in motion."[20] The issue was a live one among philosophers since Descartes had ruled out the distinction between a real geometrical object and a material body, as a case of *false abstraction*.[21]

---

[18]John Locke, *Essay on Human Understanding*, edited by Frazer (1894). New York: Dover reprint, 1959, book II, chapter 13, §17–18, Vol. I, pp. 228–229. The work was first published in 1690.

[19]*Letter to Le Pailleur, Œuvres de Pascal*, II, p. 564. The letter is undated but was probably written in February 1648.

[20]*Noël's Second Letter, Œuvres de Pascal*, II, pp. 530–531.

[21]The main characteristic of a body for Descartes is its endless divisibility: "I say, in the first place, that there is a great difference between mind and body, inasmuch as body is by its nature always divisible while the mind is utterly indivisible" (Descartes, *Œuvres, Sixth Meditation*, vol. VII, pp. 85–86). In the *Philosophical Writings* translated by J. Cottingham, R. Stoothoff and D. Murdoch. Cambridge: Cambridge University Press, 1985, vol. II, p. 59.

Pascal was later to address the problem of the explanatory power of mathematics in his essay, *On the Geometrical Mind*, where we find the strong declaration, "What is beyond geometry is beyond our reach."[22] For Pascal, the best demonstrations are found where the whole philosophical tradition located them, in mathematics. Descartes alone believed he had found a more rigorous method that enabled him to provide metaphysical demonstrations "as certain and evident as proofs in geometry, if not more so."[23]

Noël as Gatekeeper

Noël saw himself as the custodian of sound philosophy, and made much of what he considered basic "experiments" that include: (1) lighting a piece of paper with a magnifying glass to show how fire particles can be taken out of the air, (2) having them shoot out as sparks by pulling a handkerchief out of a coat pocket in winter, (3) making them fly off when a piece of metal is being drilled, (4) clapping our hands to experience the prickling sensation that these fire particles cause when they are pushed out of the surrounding air, and (5) noticing how they rise in the form of bubbles from the bottom of a kettle when water approaches the boiling point.[24]

These particles of fire or igneous spirits, as Noël also calls them, are used to explain why a ball that is dropped from a moving vehicle sometimes moves ahead upon striking the ground:

> A ball dropped from a moving carriage does not fall straight down but a little ahead the faster the carriage is going. This is because the air in front and on the sides of the carriage is squeezed in such a way that igneous spirits come out, roll in the direction of the carriage and push the ball.[25]

---

[22]Pascal, *On the Geometrical Mind*, Œuvres de Pascal, III, p. 393; trans. by Popkin, p. 174.
[23]Descartes, Dedicatory letter to the *Meditations*, Descartes, Œuvres, vol. VII, p. 4. In the *Philosophical Writings* translated by J. Cottingham, R. Stoothoff and D. Murdoch. Cambridge: Cambridge University Press, 1985, vol. II, p. 5.
[24]*Noël's Second Letter*, Œuvres de Pascal, II, p. 533. No one had thought of static electricity at the time. Pascal conjectured that the sparks that appear when a handkerchief or a piece of fur is rubbed are caused by friction that makes fatty or oily parts catch fire (*Letter to Le Pailleur*, Œuvres de Pascal, II, p. 570).
[25]Etienne Noël, *Le Plein du Vide*, p. 115.

Noël is thinking of the ancient theory of *antiperistasis*, according to which projectiles are propelled by a flow of air, and his suggestion that the motion is caused by "igneous spirits" rather than the air itself can hardly be called an improvement. By contrast, here is Galileo's analysis of the same phenomenon:

> Fasten a tilted board to the side of a carriage with the lower part of the board toward the horses and the higher part toward the back wheels. When the carriage is going full speed, drop a ball down the board. The ball will acquire a spin by rolling, and this motion, added to the one received from the carriage, will make it move somewhat faster than the wagon when it strikes the ground. If another board were tilted in the opposite direction, we could adjust the motion imparted by the carriage in such a way that the ball falling down this board would stop or even run in the opposite direction when it hits the ground.[26]

We shall never know whether Noël bothered to read Galileo. In any case, he had a more urgent task: that of vindicating his definition of light as "a luminary movement of rays composed of lucid bodies, that is to say, luminous ones" that Pascal had ridiculed as a tautology. A *luminous* object, says Nöel, becomes visible because it sends out minute *lucid bodies* that impinge upon the eye and are transmitted by the optical nerves to the brain. These *lucid bodies* are present in all transparent materials but they can only travel if a *luminous motion* is conferred to them by an external impulse. This, he adds, should not appear more mysterious than current explanations of how a magnet moves iron filings.

Noël worked hard to relate his *igneous air* to the way mercury dropped in an inverted tube. The weight of the mercury seemed to play a role, and Noël wondered whether the mercury in the tube and the mercury in the vessel could be compared to weights that are balanced at the ends of a pair of scales. There was a snag, however, since the mercury inside the tube remains at the same level whether the weight of the mercury in the vessel is increased or decreased. Noël recognized the difficulty and suggested the following way out: might not the outside, common air exert more pressure on the mercury in the vessel than the *igneous* air on the

---

[26]Galileo Galilei, *Dialogue on the Two Chief World Systems*, Second Day, *Opere*, VII, p. 188; Drake trans., p. 162.

mercury inside the tube?²⁷ The question was not addressed in precise quantitative terms, and Noël felt no need to proceed to actual weighing, yet he was not far from realizing that the atmospheric pressure of common air is all that is needed. This is the step that Pascal took. Stimulated as well as irked by Noël, Pascal wrote to his cousin Florin Périer on 15 November 1647, within a fortnight of receiving the Jesuit's second letter, to say that he now thought he could explain "by the weight and the pressure of air" all the vacuum experiments, which were "particular cases of a universal proposition concerning the equilibrium of liquids."²⁸ But although Noël and Pascal might appear to be sharing variants of the same hypothesis, they were actually working within different frameworks. Noël's mentor was not Torricelli but Descartes, as Pascal clearly saw:

> Noël puts into the tube a subtle matter disseminated throughout the universe and he gives the external air the force required to sustain the liquid suspended. It can easily be seen that this *fine thought* is in no way different from that of Descartes, and that he agrees with him about both the cause of the suspension of the mercury and the material that fills this space, as you can see from his own words on page 6 where he says that this material, which he calls *subtle air*, is the same as Descartes' *subtle matter*.²⁹

A Letter Full of Indignation

When Pascal did not reply to Noël's second letter, Jacques Le Pailleur, a close friend of the Pascals, expressed surprise. In a letter meant for wider circulation, Pascal explained to Le Pailleur that Noël had requested that his communication be kept confidential. Pascal had agreed not to engage in further epistolary warfare, although he feared that his silence might be interpreted as a capitulation. Noël did not respect the truce and, in January 1648, published a book entitled *The Fullness of the Void*, in which he used material (sometimes *verbatim*) from his two letters to Pascal. The Latin version that followed had the less bumptious title of *The Plenum*

---

²⁷*Noël's Second Reply*, Œuvres de Pascal, II, p. 537.
²⁸*Letter to Florin Périer*, Œuvres de Pascal, II, pp. 678–679; Popkin trans., p. 43.
²⁹*Letter to Le Pailleur*, Œuvres de Pascal, II, pp. 571–572.

Confirmed by Experiments, but Pascal still felt sorely abused. He had acted out of consideration for Noël's age and status, and he had kept his word religiously. "I had resolved," he tells Le Pailleur, "to postpone my reply until after the publication of the treatise in which I intend to answer all the objections raised against my summary [the *New Experiments*], and specifically *that this space does not contain anything that is known in nature or can be apprehended by our senses*."[30]

Pascal justifies his conception of empty space as something that lies halfway between matter and nothingness on the grounds that it is neither a material body, because it is fixed and immovable; nor a mere nothing because it is extended. Pascal does not make a very convincing case, and he is even driven to appeal to authority in a way that belies his own strictures against Noël's style: "I would remind Father Noël that when a position is embraced by many learned persons, no attention should be paid to objections that seem cogent if these objections could easily be foreseen. We should rather assume that they have already been solved."[31] Had Noël not been blinded and his imagination bewitched, he would have noticed, says Pascal, that a vacuum cannot be transported. When an evacuated tube is carried from one place to another, the empty space above the mercury does not follow it but remains where it is. At its new location, the tube occupies a *new part of immobile space*. This way of looking at the problem may strike the modern reader as odd, but it adumbrates the distinction between absolute and relative space that Newton embraced in his *Principia Mathematica* at the end of the century:

> Absolute space, of its own nature and without reference to anything external, always remains homogeneous and immovable. Relative space is any movable measure or dimension of this space . . . . For example, if the earth moves, the space of our air, which in a relative sense and with respect to the earth always remains the same, will now be one part of the absolute space into which the air passes, now another part of it, and thus will be changing continually in a absolute sense.[32]

---

[30]*Letter to Le Pailleur*, Œuvres de Pascal, II, p. 570.

[31]*Letter to Le Pailleur*, II, p. 565.

[32]Isaac Newton, *Principia* (1687), translated by I. Bernard Cohen and Anne Whitman. Berkeley: University of California Press, 19, pp. 408–409.

Pascal and the Rhetoric of Science

As we have seen, Noël did not waver on the notion of space, but he was willing to make some concessions about the weight of air. Pascal had no intention of letting him off so lightly, however, and he poked fun at him for weighing down his igneous matter and letting it "slide by degrees into nothingness from which it is only held back by a small residual bias."[33] But by the time Pascal had written to Le Pailleur, Noël had shifted his ground once more and had reverted to the notion that air has no weight. In his *Fullness of the Void*, mercury no longer drops because air is heavy but on account of the presence of *subtle* air, which is endowed with a "moving lightness." Mercury is now said to fall until equilibrium is established between the downward action of its weight and the upward striving of the subtle air. But, if this is the case, why does the mercury not rise higher in a tall tube where there is more *subtle* air? Noël answers that this is because a larger amount of subtle air does not possess a greater *moving lightness* when it has no place to expand. And how do we know about this remarkable property of *subtle* air? Why, we only have to look at the action of cupping glasses, which are commonly placed on sores and make the flesh rise, says Noël.[34]

Physics, for Noël, was a question to be debated rather than tested by experiment, and within a few days he made yet another somersault, abandoned the *moving lightness* of subtle air and reverted to the weight of atmospheric air. "As I was writing these words," writes Pascal, "the Father sent me a printed sheet that overturns most of what he had written in his book. The *moving lightness* of the ether is revoked, and the weight of the air is summoned back to sustain the mercury. . . . The ease with which he disposes of this *moving lightness* shows that he made it up himself and that it exists only in his imagination."[35]

Noël's reasoning seems absurd when it is severed from its dogmatic roots, and Pascal reduced it to a purely subjective opinion, but it was the logical consequence of a commitment to a "plenist" philosophy of the kind that Descartes advocated. Pascal presented his own method as the reverse of dogmatism, but he knew full well that he could not prove that

---

[33]*Letter to Le Pailleur, Œuvres de Pascal*, II, p. 571.
[34]Etienne Noël, *Le Plein du Vide*, pp. 126–127.
[35]*Letter to Le Pailleur, Œuvres de Pascal*, II, p. 575.

there is, strictly speaking, a perfect vacuum in the empty space above the mercury. Nonetheless, when faced with the status of a vacuum, Pascal ventured onto the wind-swept plains of metaphysics and located it halfway between matter and nothingness. This was going well beyond the prudential "This space does not contain any material that is known in nature or can be apprehended by our senses" of his *New Experiments*.

What we witness in the correspondence between Pascal and Noël is a clash of two styles of reasoning. The fact that mercury falls to a certain height in the tube is not queried by either side. But whereas Pascal accepts the void and looks for the way it could have been produced (an approach that will lead him to Torricelli's hypothesis of the weight of air), Noël expends all his energy in denouncing the vacuum as a vacuous notion. Like Descartes, but on somewhat different grounds, Noël believed that the world is full of matter. The great chain of being knows no gaps. Everything is connected in the macrocosm and the microcosm, and the best analogy with which to describe the world is a living organism rather than a lifeless machine. From the outset, Pascal was sceptical—we might even say deeply suspicious—of this organic analogy. He believed that outside the realms of theology, where we find divine truths, and mathematics, where we have rigorous demonstrations, nature can only be known through experiment. But experimentation was still in its infancy and few realized that nature would only answer questions that were framed in the new quantitative language.

The Pride of a Physicist

Pascal was so annoyed with Noël that he asked his father to intervene. A high-ranking tax official with little time or patience for this kind of wrangling, Etienne Pascal drafted a letter to Noël in April 1648 in which he complained that he was overworked and had to cut down on his sleep in order to write to him, thereby implying that the priest, who was the Rector of the College de Clairmont, had all the time in the world! The letter is querulous and unworthy of a man of his standing. Although his own son had not referred to Torricelli in his *New Experiments*, Pascal senior found fault with Noël for not stating explicitly that Torricelli was the first to declare that mercury is supported by a column of air.[36] This rankled with the

---

[36]Letter of Etienne Pascal to Noël, *Œuvres de Pascal*, II, pp. 588–589.

Jesuit colleagues of Noël and they waited for a convenient moment to teach Pascal a lesson. An opportunity presented itself in June 1651, when one of their members who taught at the College of Montferrand (the twin city of Pascal's native Clermont) was invited to deliver a public lecture before Antoine de Ribeyre, the President of the Administrative Court to which Pascal's father had belonged. The Jesuit asserted that certain persons had tried to pass themselves off as the authors of an experiment designed by Torricelli, and carried out in Poland. Although Pascal was not mentioned by name, he was clearly intended, and someone in the audience promptly informed him. Pascal penned a letter of protest to de Ribeyre and published it in all haste. The 11-page pamphlet, which bears no name of place or publisher, claimed that Magni performed his experiment after reading Pascal's *New Experiments*. When Antoine de Ribeyre declared that it was all a storm in a teacup and that the letter should not have been published without his consent, Pascal answered that he had rushed into print because it was a grave matter of honor. Had he failed to reply, he would have confirmed the impression that he had stolen intellectual property, something that would make him a common thief!

There is more than a hint of haughtiness in Pascal's vindication of plagiarism, and he is unfair to Magni whom he accuses of the very same offense. Magni had performed his experiment in Poland in July 1647, several months before the publication of Pascal's *New Experiments*.[37] Roberval was violently antagonistic to Magni, and Pascal relied uncritically on his assessment when he stated that Magni's priority was rejected "by all of us and particularly by Roberval, who used my publication as proof."[38] Pascal was also grossly wrong when he added that Magni did not reply to Roberval, for Magni had sent him a vindication of the independent character of his research in November 1647.

The fact of the matter is that Magni had become interested in the vacuum after reading in Galileo that pumps cannot raise water much above 18 *braccia*. In 1644, he had tried to find a suitable glass tube to use mer-

---

[37] Pierre Desnoyers, the secretary of the Queen of Poland, wrote to Roberval and Mersenne on 17 and 24 July 1647, and gave a full account of Magni's experiment with mercury (*Correspondance de Mersenne*, XV, pp. 311–314, 319–321). On 31 July 1647, he forwarded a copy of Magni's *Demonstratio ocularis* (*ibid.*, p. 335), and on 25 September 1647, he informed Bruslart de Saint-Martin that he was sending the second part of this work. All this occurred before the publication of Pascal's *New Experiments*.

[38] Letter of Pascal to Antoine de Ribeyre, 12 July 1651, *Œuvres de Pascal*, II, p. 811.

cury, but those from the glassworks near Cracow were too fragile, and attempts to use a wooden tube gave unsatisfactory results because the mercury did not always fall to the same level. Magni did not realize that he might be facing a fluctuation in atmospheric pressure. Indeed, he did not think of the weight of the air at all but assumed that the level of the mercury was related to the "amount of air attracted by the pores of the wood."[39] Magni eventually managed to get good tubes from a Venetian glass blower whom the King of Poland had called to Warsaw, but he was too eager to use the vacuum as a weapon in the battle he was waging against Aristotle to think of anything else. His obsession with a rival school of thought precluded an objective and dispassionate appraisal of the phenomenon at hand.

Conflicting Viewpoints

What we have been able to learn of Pascal's method can be summarized by contrasting it with the four types of argument that Noël uses in defending his own position. First, Noël relies on authority, and if a proposition was held by Aristotle then it may be accepted without further question. Pascal, by contrast, stresses logical reasoning and experimentation as the only sources of knowledge, and if he quotes other physicists it is not to give their names but their arguments. He acknowledges that authority reigns supreme in history and theology, but dismisses it in the natural sciences. Whereas Noël considers it necessary to show how physics is compatible with theology, Pascal avoids any reference to religion when tackling a physical problem.

Second, Noël erroneously assumes that all that is required to prove that a hypothesis is true is to show that the observed phenomena would follow from it. Pascal points out that all logicians know that a true conclusion can follow from a false premise. It is easy enough to invent hy-

---

[39]Magni's *Apology* was included in his *Admiranda de vacuo et Aristotelis philosophia* sent by Desnoyers to Roberval on 4 December 1647 and to Mersenne on 9 December of the same year (*Œuvres de Pascal*, II, pp. 448–450). The text of Magni's reply is printed in Cornelis de Waard, *L'expérience barométrique*, pp. 173–176, and in the *Correspondance de Mersenne*, pp. 527–531). On Magni's frame of mind, see Massimo Bucciantini, "Valeriano Magni et la discussion sur le vide en Italie" in E. Festa, V. Jullien and M. Torini, *Géométrie, atomisme et vide dans l'école de Galilée*. Florence: Istituto e Museo di Storia della Scienza, 1999, pp. 129–152.

potheses that entail what we already know to be true, especially if we are willing to entertain the existence of properties that can in no way be observed. Pascal will have none of this and considers such hypotheses unworthy of examination

Third, Noël favors the practice of providing a definition and then proceeding as though it was an ascertained fact. For instance, he defines the word *body* in such a way that it means what is ordinarily called *space*, and then argues that wherever there is space there must also be body. Pascal, in his letter to Le Pailleur, offers an excellent criticism of the arbitrariness of this procedure. Finally, a fourth type of argument invoked by Noël is the belief that everything in nature has a purpose and obeys a design. This was characteristic of Aristotelian natural philosophy, but Pascal was distrustful of such explanations and particularly of the idea that nature could experience horror of the void.

Taking Stock

We can now review the development of Pascal's views up to the time of his *Letter to Le Pailleur*. In the first place, Pascal has not yet been able to determine to his own satisfaction what he set out to find, namely, whether the apparent void is a real one. In the conclusion to his *New Experiments* and in his letters to Noël and Le Pailleur, he does not go beyond giving as his considered opinion that the empty space above the mercury in the tube is a genuine vacuum. He makes abundantly clear, however, that the force that resists the formation of a void, be it apparent or real, can be measured and is equal to the tendency of a column of mercury 27 inches high to run down. In a tube of uniform diameter, this is equal to the weight of such a column, and any greater force will produce a void. If a column of some other fluid is used, its height will be to the height of the column of mercury inversely as the density of the fluid is to that of mercury.

Pascal experimented with glass tubes, syringes, and siphons of varying sizes and lengths. He filled the tubes with mercury, water, or wine, but also with cords that were slowly pulled out to create a vacuum. Always the result was the same. Next he showed conclusively that it takes no greater force to produce a large than a small vacuum, and he used this fact against those who claimed that the suspension of mercury depended on *igneous* air, or some kind of vapor above the column of mercury. If he could not rule out that one of these substances might be found inside the tube, he at least demonstrated that its presence did not appreciably affect the height of the mercury.

Thus, if Pascal did not give the correct explanation of the suspension of the mercury, he disproved current accounts. He still used the phrase, *horror of the void*, but by the time he published his *New Experiments* he had begun to look outside the tube for the cause of the suspension of the liquid inside the tube. We have his own word for this and it is clear from the trend of his argument. He suspected that this was atmospheric pressure but he practiced the restraint that he preached to Noël, and he waited for further confirmation.

As we have mentioned earlier on, much of Pascal's work had been anticipated by Torricelli, such as the idea of placing water above the mercury in the vessel and raising the tube slowly to allow the water to rush in to take the place of the mercury. This constituted strong evidence in favor of the hypothesis that the empty space at the top of the tube was really a vacuum. Torricelli had also performed the experiment with two tubes of different shapes so that voids of different volumes were created, although the mercury remained at the same height in both, a powerful argument against the hypothesis that the cause of the suspension of mercury was inside the tube.

If Torricelli provided Pascal with the central idea in the light of which he carried out his experiments, Pascal deserves credit for the ingenuity and the care with which he drew his conclusions. He produced two new ways of producing a vacuum in his *New Experiments*. The first consists in filling a tube almost entirely with a solid body, such as a cord or a rope, and pouring water on top. When the solid is gently pulled out, a void is left in the tube. The second method, which is simpler and even more effective, consists in using a large syringe to produce a vacuum of considerable dimensions. This made it possible to perform experiments that could not be carried out in a narrow glass tube.

At a more general level, Pascal's scientific approach can be characterized by its relative freeness from metaphysical presuppositions and the modesty of his aim. The two are undoubtedly linked. Had he attempted to give a complete account of physics, he would have needed more principles than he could muster for their solution. Pascal's arguments and his interpretation of the experiments he describes do not depend on a particular philosophical theory. There is one point, however, where Pascal makes a cosmological assumption, that is, when he argues for a vacuum from condensation and rarefaction, an argument that only holds if matter is composed of atoms. For the rest, Pascal is content to show how things work and to make simple measurements of the force that produces the vacuum. His distrust of speculative flights can be seen in the matter-of-fact way with which he dismisses hypotheses that entail conclusions that are at variance with the results of experiments.

The Hesitations of Roberval

We have seen how Pascal's ideas fared with the Jesuits. We now have to consider how convincing they appeared to his friend and early mentor, Roberval. In his letter to Desnoyers of 20 September 1647, namely his *First Account* of the experiments on the void, Roberval had agreed to keep his friend in Warsaw informed of further developments. He was slow in complying, and Desnoyers reminded him of his promise in a witty letter, in which he admitted that he had himself been remiss in informing Roberval about an Italian engineer who claimed to have invented a flying machine. "But it is more than a year that you said as much about your *vacuum*," he added, "and you will allow me to use the excuses that you would invent if I were to scold you."[40] Roberval's dilatoriness was well known, but the real reason for his delay was his growing uncertainty about the nature of the void. Desnoyers must have suspected as much since he promised not to print, or even let out of his hands, any communication that he might receive from his friend.

Roberval finally sent off a long letter at the end of 1648.[41] In this *Second Account*, he tells how he put mice and small birds into the empty space in the tube above the mercury. When they did not survive, he tried flies and other insects. These also seemed to die, but they revived when brought out. Earthworms were sturdier and remained unaffected by the vacuum. This intrigued Roberval and he decided to make a series of tests to determine whether the space was really void. He carried out six experiments, and as they are good examples of the kind of research that was possible at the time, we shall briefly examine them in turn.

---

[40] *Œuvres de Pascal*, II, p. 603.

[41] The *Second Account* is dated 15 May 1648, but Roberval added a postscript after Périer's experiment on the Puy-de-Dôme on 19 September 1648. The full text is given in the Brunschvicg edition, *Œuvres de Pascal*, edited by L. Brunschvicg, P. Boutroux and F. Gazier, 11 vols. Paris: Hachette, 1908–1914. Reprinted Vaduz: Kraus, 1965, vol. II, pp. 310–340, with the conclusion on pp. 359–361; an abridged version is printed in the *Œuvres de Pascal*, edited by Jean Mesnard, II, pp. 605–611.

In the first experiment, the upper part of a tube in which mercury was suspended was heated with a warm towel. The apparent void expanded and the mercury dropped. This made Roberval suspect that something inside the tube had been rarefied, but he did not initially think it was air. In a second experiment, mercury was poured into a tube to a height of an inch and a half from the top, and the rest filled with water. The tube was inverted and plunged into a bowl of mercury, and as the mercury fell, *bubbles* of varying sizes were seen rising. The tube was then tilted until the empty space at the top vanished, and the bubbles combined into a very small one. This suggested to Roberval that some kind of attraction was at work, an hypothesis that fitted his own theory that the world is held together by natural attraction, as he had argued in 1644 in a book that he claimed had been written by to the ancient astronomer Aristarchus.[42]

For his third experiment, Roberval took a tube just under three feet long and filled it with mercury, leaving an inch and a half of air at the top. When the tube was inverted, the bubbles that rose through the quicksilver appeared to be less numerous, but Roberval guessed that although as many had been produced as before they could not be seen because mercury is opaque. In order to confirm this, he added a little water and some air. "Innumerable bubbles" immediately rose and depressed the mercury to a lower level. When more air was introduced into the tube, the column of mercury dropped below its usual height, and Roberval found it tempting to conclude that it fell because of the air. There was a difficulty, however. In a taller tube, where more space was available above the mercury, the column was less depressed by a given amount of the air. This seemed to indicate that when the air was rarefied it pressed less strongly against the column of mercury, and this is what made Roberval change his mind and revert to the idea that there was some air in the apparently empty space above the mercury. His recantation is recorded in the following passage in which the words in parentheses were so carefully struck out that some are difficult to decipher:

> I reasoned according to the laws of mechanics and concluded that there is no better explanation of the fact that mercury is depressed when air is introduced than by saying that, according to the laws of

---

[42]*Aristarchi Samii de Mundi Systemate*. Paris: A. Bertier, 1644. The hoax was perpetrated with the connivance of Mersenne. See L. Auger, "Les idées de Roberval sur le système du monde," *Revue d'histoire des sciences* X (1957), pp. 226–234.

nature, air spontaneously (*although I said something else, as many others did, in my First Account*) expands and becomes rarefied in the tube so that it occupies all the apparently empty space, but without completely exhausting its tendency to rarefaction so that it continues to try to expand and thereby exerts pressure.[43]

This a dramatic reversal of the position Roberval had taken a year earlier in his *First Account* where he had vigorously denied that "even an imperceptible drop of air" had been left in the tube. The Aristotelian Pierius was delighted but felt that Roberval had not gone far enough, and he reminded him that air is essentially light and, hence, has no weight whatsoever![44]

Roberval effectively gave air a twofold tendency: one to condense, the other to expand and become rarefied. This dual tendency, he claimed, diminished as the *force* (i.e., pressure) decreased. But how could a natural tendency, such as the air's inclination to rarefaction, be said to have degrees? If the air expands in order to fill the entire space inside the tube, why is the column of mercury not pushed all the way down? Those who maintained that the space above the mercury was filled with rarefied matter claimed that the mercury did not fall all the way down because, below a certain level, its weight was no longer sufficient to bring about further rarefaction. But if, as Roberval contended, *no force* is required to produce rarefaction and air becomes thinner of its own accord, why did the mercury remain suspended? Roberval brazened this out, and declared that it was a trifling objection, easily removed "by anyone who knows anything about mechanics." What had to be considered, he said, was the force that joins all parts of nature and "is commonly called weight," although it is really "a mutual striving of all parts of nature to come together."[45] Roberval would have his reader believe that this is enough to infer that air has a tendency to rarefaction that is proportional to the force that compresses it. When the mercury begins to fall in the tube, the air gradually loses its power of dilatation because the force decreases. This is why the mercury does not fall all the way down.

---

[43]Roberval, *Second Account*, Brunschvicg, vol. II, p. 314. Roberval also observed that mercury and air do not pass through the membrane but that water does. This is a case of osmosis, recorded but not explained.

[44]Pierius, *A reply to a Recent Experiment Concerning the Vacuum*, p. 14, quoted in the Brunschvicg edition, II, p. 294.

[45]Roberval, *Second Account*, Brunschvicg, II, p. 317.

In the fourth experiment, Roberval observed that when mercury was poured into the glass tube, bubbles, which he now presumed to be air, were left clinging to the side. Like Torricelli before him, Roberval had probably failed to dry the tube, and water stayed inside. When he removed his finger from the opening of the tube, bubbles ascended and entered the apparently void space with such "an impetus to rarefy" that they disrupted the surface of the mercury "like nitric power that is ignited." Roberval states that he saw the bubbles being rarefied and that he had them witnessed by others. Now mercury falls very rapidly, and the *rarefaction*, that is the swelling of the bubbles, can only be observed by eyes that move as fast as the mercury drops. Here Roberval displayed real ingenuity in slowing down the quicksilver. He made a small hole in the membrane with which he had sealed the opening and allowed the air in gently, thereby reducing the speed of the mercury as it dropped. Roberval also poured a couple of inches of water over the mercury in the tube and, as he expected, bubbles that had hardly been visible in the mercury swelled enormously because they were free to expand. This finally convinced him that air was present in the apparent void and that no genuine vacuum had been produced. The question was now to determine the size of the expansion or rarefaction. His fifth and most striking experiment was designed to show just how stupendous this is.

The Tell-Tale Membrane

From the bladder of a carp, Roberval formed a small balloon that he flattened out to expel all the air before tying a string about its neck so that no air could enter. He then introduced the bladder into the apparent void above the column of mercury. It immediately swelled and became as full as in the belly of a live carp. When the tube was tilted, the bladder became flabby, and when the tube was straightened again, it swelled once more. Finally, a small hole was perforated in the membrane that sealed the tube and, as the air slowly entered, the bladder decreased in size. Roberval was astute enough to realize that the crucial question was whether some air had been unwittingly left in the bladder, but he was confident that he had succeeded in expelling all the air, although another bladder, which contained a minute amount of air, had burst when placed in the evacuated tube.

The bladder caused a stir not only in France but all over Europe. In a letter to Hevelius in Prussia, Mersenne described the "admirable" experiment and conjectured that the bladder, which had been regularly inflated and deflated when it was in the belly of a live carp, retained a disposition to swell in the absence of pressure.[46] Gassendi provided a similar explanation and argued that organic matter has a natural tendency to return to its prior condition, as is evidenced in compressed fibers that spontaneously regain their original size.[47] Although compression and expansion are understood today as purely mechanical, we must remember that it appeared more natural for Mersenne and Gassendi to broach the subject within a conceptual framework that attended to vital reactions rather than coiled springs.

Robert Boyle heard about the bladder from his correspondent Samuel Hartlib, who wrote to him from Paris on 9 May 1648:

> They prepare a long tube like a weather glass, which is filled with quicksilver; and being stopped as close as may be with one's finger, the tube is inverted, and plunged in a vessel half or more full of quicksilver. The quicksilver in the tube will force the quicksilver in the vessel to rise, by adding more quicksilver to it, and so leaves a space in the top of the tube vacuum, as is supposed. But a bladder being hung in that vacuum was as perfectly seen as could be, so that there must be some body there to convey the action of light to the eye, as I suppose and divers other here. That bladder was made as flat as they could, when they put it in; and when the quicksilver left it, it swelled in that supposed vacuum like a little football.[48]

Boyle repeated Roberval's experiment with the bladder of a lamb, which is bigger than that of a carp, and he describes how the bladder, which was half-empty, started to swell when pumping began and the air in the container decreased. He explains the plumpness of the bladder as "the surmounting of the debilitated spring of the ambient air remaining in the vessel by the stronger spring of the air remaining in the bladder." In order to show this, he allowed air back into the receiver and, as the air came in, the distended air in the bladder was gradually compressed and

---

[46]Letter to Johann Hevelius, 1 June 1648, *Corrrespondance du P. Marin Mersenne*, XVI, p. 337.
[47]Letter of Gassendi to Bernier, 6 August 1652, in *Œuvres de Pascal*, II, p. 933.
[48]Quoted in E.W.F. Middleton, *The History of the Barometer*, p. 56. See René Pintard, "Autour de Pascal: l'académie Bourdelot et le problème du vide" in *Mélanges offerts à Daniel Mornet*. Paris: Librairie Nizet, 1951, pp. 78–79.

the bladder grew more and more flaccid. In another experiment, when the air was pumped out, a bladder that was moderately full of air "gave a great report, almost like a cracker," and was rent apart "as if it had been forcibly torn asunder."[49]

We now return to Roberval and his sixth and concluding experiment, which is the one Pascal had performed so successfully with water and wine in tubes over twelve meters long. Unfortunately, Roberval missed the point of Pascal's demonstration because he focused on the air bubbles that could be seen rising and growing in size, a sign, he believed, of their power of *rarefaction* and therefore evidence against a vacuum. When commenting on this passage, Pierius notes that Roberval would have been even more impressed "had he placed his mouth on the opening of the fifty-foot tube that Pascal used on several occasions to make his experiment in Rouen."[50] We are left to imagine how anyone could have applied his mouth to such a long tube.

Roberval was anxious not to appear dogmatic. Although he had showed to his satisfaction that there was some residual air in the apparently void space, he did not go so far as to declare that a vacuum was impossible. This remained an open question, and he scoffed at his arch-enemy, René Descartes, who claimed that the concept of space without matter—a vacuum—was unthinkable.

Natural Scientists at War

In the summer of 1648, Roberval met Descartes in Paris and they clashed over the nature of rarefaction. Descartes declared that it was to be understood on the analogy of a sponge that absorbs water, not by increasing its quantity of matter but merely by opening its pores wider. Roberval thought that this ease of conception was no more than ease of imagining.[51]

---

[49]Robert Boyle, *New Experiments Physico-Mechanical, Touching the Spring of Air* (Oxford, 1660), in *Works*, edited by Thomas Birch, 6 vols. (London, 1772). Facsimile Hildesheim: Georg Olms, 1965, vol. I, pp. 18–19.

[50]*Second Account*, Brunschvicg edition, II, pp. 328–329; Pierius' remark is quoted in the Brunschvicg edition, II, p. 328, no. 2.

[51]*Second Account*, Brunschvicg edition, II, p. 339. The meeting between Roberval and Descartes took place in June or July 1648 (Adrien Baillet, *La Vie de Monsieur Des-Cartes*, 2 vols. Paris: Daniel Horthemels, 1691. Facsimile Geneva: Slatkine Reprints, 1970, vol. II p. 344).

An alternative to Descartes' view was available in the writings of Galileo, who explained rarefaction and condensation by postulating that matter is composed of atoms separated by indivisible vacua.[52] Galileo applied his theory to the famous paradox known as the "Wheel of Aristotle," which concerns two concentric circles that turn together (see Figure 1). We can think, for instance, of the rim and the hub of a cartwheel. The difficulty lies in explaining how, after one revolution, the larger circle, whose radius is AB, traverses a line equal to the distance covered by the smaller circle, whose radius is AC. Galileo argued that when the small circle is rotated, the larger one describes a straight line that is shorter than its circumference because the infinite number of indivisible and unquantifiable atoms are superimposed, thereby allowing for contraction and hence condensation. On the other hand, when the larger circle is rotated, the smaller circle describes a line greater than its circumference because the infinite number of indivisible vacua between the equally infinite number of indivisible atoms render expansion and thus rarefaction possible. Thus, Galileo believed that "by imagining a line resolved into unquantifiable parts—that is, into its infinitely many indivisibles—we can conceive it as immensely expanded without the interposition of any quantified void spaces, though not without infinitely many indivisible voids."[53]

In the *Two New Sciences*, Galileo has the Aristotelian professor, Simplicio, call this argument fanciful since it implies that an ounce of gold could be made as large as the Earth or the Earth reduced to the size of a walnut! Salviati, Galileo's spokesman, welcomes the comment and replies that the spectacular expansion of gunpowder into heat and light suggests that equally stupendous contractions are possible in the opposite direction. But we see wood turn into fire and light, says Simplicio, we never see fire and light condense to make wood! Right, replies Galileo, "and this is why speculation must step in where observation is lacking."[54] With less speculation but more mathematical insight, the engineer Antonio de Ville solved the conundrum of the two concentric circles by pointing out that the revolution of the smaller circle is the combined outcome of two motions: (a) the rotation of the small circle, and (b) the added motion of

---

[52]The idea is outlined in the First Day of the *Two New Sciences* (*Opere di Galileo*, VIII, pp. 91–96). Galileo wrote to Fulgenzio Micanzio on 19 November 1634 that it flashed upon him when he was attending a church service in 1616 (*ibid.*, XVI, p. 163).
[53]*Ibid.*, VIII, p. 72.
[54]*Opere di Galileo*, VIII, p. 105.

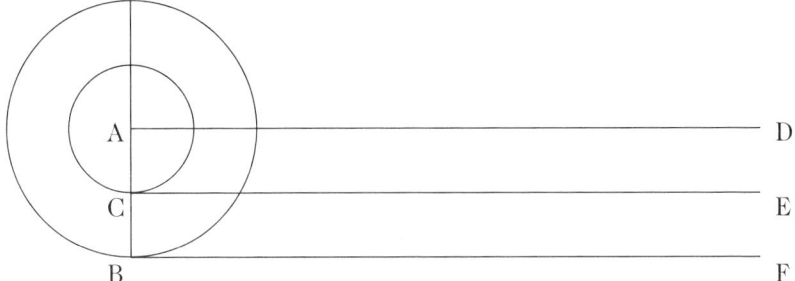

FIGURE 1   *When two concentric circles rotate together, the line traced out by the large one is not longer than the one made by the smaller.*

translation imparted to it by the large circle. Since Roberval and Pascal were both acquainted with Galileo's *Two New Sciences*, they may have been influenced by what he wrote but they do not refer to him. The person who was uppermost in their minds was Descartes.

Descartes' Packed Universe

Descartes had made a point of meeting Pascal when he passed through Paris on his way back to the Netherlands in the autumn of 1647. On Sunday, 22 September, he sent a friend to request an interview for the next day at 9:00 A.M. Pascal's sister, Jacqueline, was embarrassed about how to reply, because her brother was ailing and often found it difficult to carry on a conversation early in the morning, but she nonetheless agreed that Descartes should come at 10:30 A.M. When Descartes arrived, he found, much to his surprise, that Roberval had been invited to give him a demonstration of Pascal's calculating machine. Whether Descartes was impressed by the invention is not recorded, but the conversation soon turned to experiments on the void. When asked whether he believed that something had entered the tube, Descartes replied with great earnestness that subtle matter had. Pascal responded as best he could, but Roberval, fearing that he would tire himself, did most of the talking and expressed himself with his customary vigor. "Descartes was irritated," writes Jacqueline, "and told him he would gladly discuss with my brother for as long as he wished, but not with him because he was so biased." Glancing at his watch and realizing that it was twelve, Descartes, who had a luncheon appointment on the other side of the Seine (the Pascals lived in the rue Brisemiche on the right bank), rose to take his leave. As Roberval was going in the same direction, he offered him a lift in his carriage, and the two left together. We can only hope that the ride was not too bumpy for either of them.

Upon leaving, Descartes promised to return on the morrow, and he arrived promptly at eight o'clock and stayed until eleven. According to Jacqueline, who was not present, Descartes had come to advise her brother about his health, but had little to say beyond recommending a diet of soup and staying in bed as long as possible in the morning, as was his own practice.[55] They also spent some time discussing the vacuum. Descartes did not deny that the mercury dropped to a level where it was balanced by the weight of the atmosphere, but he maintained that the space above the mercury was not a genuine vacuum since it was filled with subtle matter. Such a conversation could have led to the idea of performing the experiment at the top and the bottom of a mountain and Descartes may have been the first to suggest this, as he later claimed, but Pascal was always to insist that he had already thought of it when they met.

Cosmological Implications of the Experiment on the Void

Roberval had voiced the hope that experiments on the vacuum would shed new light on "the system of the world, the Earth and its elements, so that something could be decided in favour of Ptolemy, Copernicus or Tycho."[56] At a more mundane level, he expected that the "continuous experiment," namely the regular observation of a fixed tube (what we now call a barometer), would provide a clue to the cause of the variation in the height of a column of mercury when the weather changes. The *Second Account* originally ended with this hope, but Roberval added, a few months later, a postscript in which he summarized the experiment that Florin Périer, acting under Pascal's instructions, had performed at the base and at the top of the Puy-de-Dôme, and which we shall discuss in the next chapter. Roberval was unconvinced, and maintained that the weight of the

---

[55]Our source of information about the meeting is the letter that Jacqueline Pascal wrote to her sister Gilberte on 25 September 1647, *Œuvres de Descartes*, II, p. 481.

[56]*Second Account*, Brunschvicg edition, II, p. 359.

atmosphere alone could not explain the varying level of the mercury, and that a force "of mutual and reciprocal attraction" had to be added.[57] His fascination with the general problem of universal gravitation is therefore partly to blame for his inability to grasp the cogency of the much more modest claim that the young Pascal was making.

Although Roberval's six experiments seemed to show that there was air in the apparent void and that a vacuum has yet to be realized, Pascal did not react with the outright indignation that he had displayed upon receiving Father Noël's letter. Beyond the fact that Roberval was a friend of the family and his agent for the sale of his calculating machine, there is the more important consideration that Pascal wanted it known that the quality of his experiments did not depend on the acceptance of an a priori theory. His aim had been to refute the claim that the suspension of mercury depended on air or some other substance *inside* the tube, and Roberval had recognized that the *outside* atmospheric pressure was *partly* the cause of the suspension of mercury. Pascal thought he could bring him to see that atmospheric pressure was all that was needed. The Puy-de-Dôme experiment would take care of that.

---

[57]*Ibid.*, p. 361. This passage is cited in *Œuvres de Pascal*, II, p. 611.

CHAPTER 5

The Great Experiment on a Mountain in France

The debate on the vacuum was not settled by the experiments performed in Rouen or the publication of the *New Experiments*. Even Pascal's friend, Roberval, wanted more stringent proof. The problem was that the experiments could be discussed in the light of two very different questions: one is, *What is left* in the apparently empty space above the column of mercury? and the other, *What holds up* the column of mercury in the tube? Each question opened up a different avenue of inquiry. Noël focused exclusively on the first and imagined that the empty space was filled with *igneous air*, *spirituous vapors*, or the *subtle matter* of Descartes. In this context, experiment was only of marginal interest, for how could anyone prove that there was really nothing in the seemingly empty space? How could nothingness be demonstrated? This kind of question was truly metaphysical and, by the autumn of 1647, Pascal knew that he was running the risk of being trapped in a vacuous philosophical discourse. Fortunately he heard, at long last, of Torricelli's hypothesis that we live submerged in a sea of air, and he shifted all his attention to the second question, which can be formulated more sharply as, What is the mechanical cause of the elevation of the column of mercury in the evacuated tube?

Three steps took Pascal to his goal. The first step was straightforward, and consisted in comparing the relative weight of water and mercury and drawing the implications as Torricelli had done. A column of mercury reaches only one fourteenth of the height of a column water, and Torricelli, who knew that mercury weighs about 14 times more than water, had interpreted this as a case of equilibrium produced by one and the same force, namely the weight of the air on the surface of the liquid in the vessel in which the tube is inverted. The second step was to find a convincing proof. Pascal realized that if the air could be removed from the room in which the experiment was performed, the column of mercury would fall all the way down since there would no longer be any weight to sustain it. Unfortunately, neither he nor anyone else could live in such a room long enough to set up the experiment, let alone perform it!

A Vacuum inside a Vacuum

Pascal's experimental ingenuity came to the rescue and made possible what seemed naturally impossible. He saw that he did not have to exclude the air from the whole room but only from the space surrounding the tube of mercury inverted in a vessel full of mercury. In other words, he could produce a vacuum in a vacuum! The illustrations show how he went about it.[1]

One end of a glass tube, 3 feet long, filled with mercury and closed at both ends by membranes (Pascal used the bladder of pigs), is dipped in a basin of mercury and the whole—the tube and the bowl—is fastened on the inside of a larger tube 6 feet long, towards its upper end (see Figure 1, illustration I). This larger tube is left open at the top but the bottom is closed by a membrane and dipped in a basin of mercury. The tube is then filled with mercury and closed at the top with a membrane. The result is an apparatus with a Torricellian tube and basin inside another Torricellian tube, the bottom of which is also in a basin. When the membrane at the lower end of the larger tube is broken, the mercury descends to the usual height of roughly 27 inches, leaving the smaller tube in the vacuum created above the mercury (see Figure 1, illustration II). Next the membrane at the bottom of the smaller tube is broken and the mercury falls entirely into the basin beneath (see Figure 1, illustration III). This shows that the force that ordinarily causes the suspension of the mercury is not op-

---

[1]Pascal's original experiment is illustrated in a series of four diagrams by J. Thirion, who drew them after the description given by Noël in his *Gravitas Comparata* of 1648. See J. Thirion, "Pascal: l'horreur du vide et la pression atmosphérique," *Revue des questions scientifiques*, 3e série, t. 13 (1908), p. 179. The diagrams are reproduced in *Œuvres de Pascal*, II, p. 636.

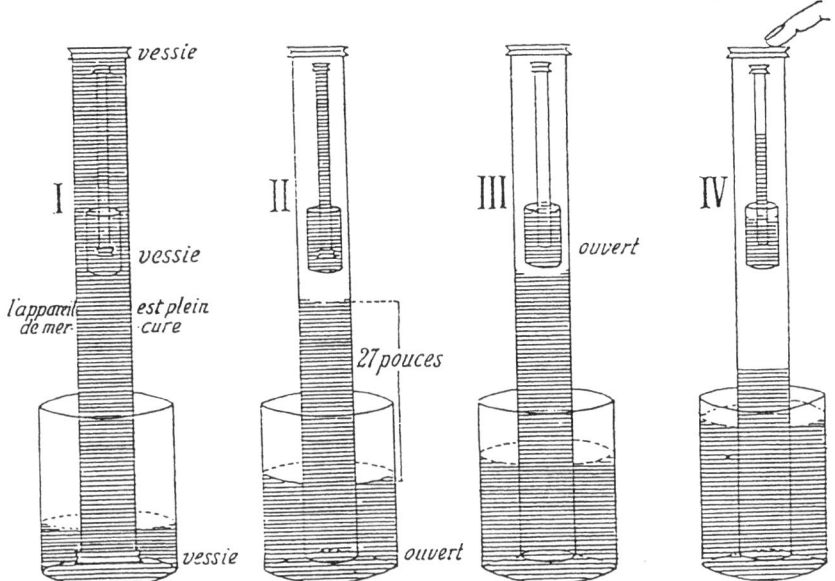

FIGURE 1 *Pascal's Early experiment to create a vacuum in a vacuum.*

erating. Finally, a small hole is perforated in the membrane at the top of the larger tube and, as the air flows in, the mercury in the inner tube rises while the mercury in the larger tube descends (see Figure 1, illustration IV). The height of the mercury, as Pascal notes, "increased or diminished in proportion as the pressure of the air increased or diminished."[2] Using a later terminology, we can say that Pascal's results were reached by using the method of difference and the method of concomitant variation. It is not clear from the diagram (which is not by Pascal) how the membrane at the top of the inner tube was removed, but he could have made a rod pass through the wall or the bottom of the large outer tube. Galileo faced a similar problem when attempting to determine the weight of air, and he inserted a metal rod through a small hole that he had drilled at the bottom of the container.[3]

---

[2] Letter of Pascal to Périer, 15 November 1647, *Œuvres de Pascal*, II, p. 679; Popkin trans., pp. 43–44; Spiers trans., p. 100.
[3] Galileo Galilei, *Two New Sciences, Opere*, VIII, p. 123.

How Pascal could gradually *increase* the pressure by letting in a little air at a time is not difficult to understand. How he could have gradually *decreased* the pressure is less obvious. One way of reducing the pressure in the larger of the two tubes would be to raise it somewhat to create a greater volume of empty space. This would result in a decrease in the pressure, in agreement with the well-known law that pressure, at constant temperature, is inversely proportional to the volume, namely $P = T/V$, although the law was never formulated as such by Pascal but by Boyle in England and Mariotte in France. Pascal's own terminology was still in a state of flux. He spoke of "suspension," which is proper to the theory that considers that the mercury is *pulled* upward, and at other times of "pressure," a word that is more fitting if the mercury is *pushed* up by the outside air.

The apparatus was shown to Roberval, who rapidly improved it by designing a heart-shaped flask with two projecting tubes.[4] In Figure 2, illustration I, the tube on the right extends over 27 inches above the flask, and its lower end almost reaches the bottom of the flask; the one on the left barely extends above the upper part of the flask but plunges through an opening at the bottom and stretches down below for a little over 27 inches. The bottom of this tube is stopped with a membrane, as is the top of the other tube that rises above the flask (see Figure 2, illustration II). The containers are filled with mercury and the end of the lower tube is placed in a vessel of mercury. A membrane also seals the top of the shorter tube. When the seal at the bottom of the tube immersed in the vessel of mercury is broken, the mercury falls to 27 inches in the tube, while the mercury in the flask drops to the level of a small hole in the short pipe that is left empty in the flask, and the longer tube that extends above the flask is emptied to this level (see Figure 2, illustration III). The membrane is then removed from the shorter tube that extends above the flask: the mercury in

---

[4]The four diagrams drawn by Pierre Duhem are reproduced in *Œuvres de Pascal*, II, p. 637. Twenty-seven inches is the value given in Noël's *Gravitas Comparata*. If he used the value for an inch given by Mersenne in *La vérité des sciences* (quoted in *Mersenne ou la naissance du mécanisme*, p. LXIII), namely 27.35 mm, this would give $27 \times 27.35 = 738.45$ mm. In his *Account of the Great Experiment*, Périer uses the slightly lower value of 27.17 mm for one inch (*Œuvres de Pascal*, II, p. 687), which yields $27 \times 27.17 = 733.59$ mm. Taking Mersenne's value, which is the highest, we have, in modern English inches (where 1 inch = 25.4 mm), $738.45 \div 2.54 = 29.07$ inches, which is a lower limit since a column of mercury at sea level varies between 29 and 31 inches. The standard unit (one atmosphere) is the mean pressure at freezing temperature at latitude 45°. This is 760 mm or 29.92 inches of mercury, and about 34 feet of water.

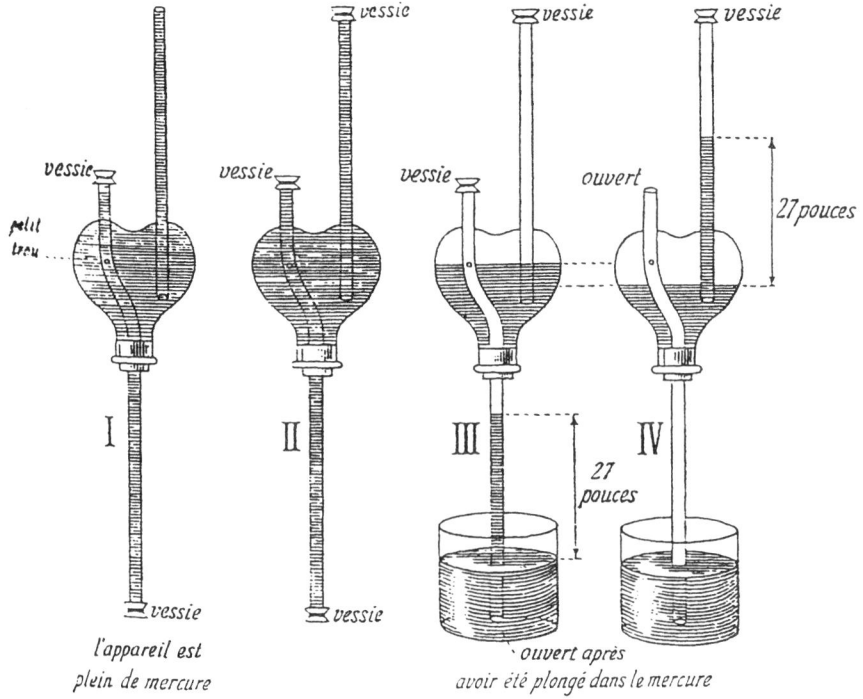

FIGURE 2 Roberval's heart-shaped flask to produce a vacuum in a vacuum.

the flask now falls to a lower level while the mercury in the tube to the right rises to 27 inches, and all the mercury in the long tube that extends below the flask flows into the vessel (see Figure 2, illustration IV).

The next improvement was made by a friend of Pascal, Adrien Auzoult, who used, as we see on the left of Figure 3, a very long-necked flask, AB, with a small tube, G, which is open at one end and can be sealed with a membrane.[5] The drawing on the right shows the flask inverted in a large trough after a tube, CF, has been sealed in the flask at B and made to rest in a small rectangular trough C. The trough D at the bottom is partly filled

---

[5]The source is Jean Pecquet's *Dissertatio anatomica de circulatione sanguinis et chyli motu*. Paris: Sébastien et Gabriel Cramoisy, 1651. In a section entitled *Experimenta physicomathematica de vacuo*, he credits Auzoult with being the first to have successfully performed this experiment (cited in *Œuvres de Pascal*, II, p. 767, with the diagram appearing on p. 769).

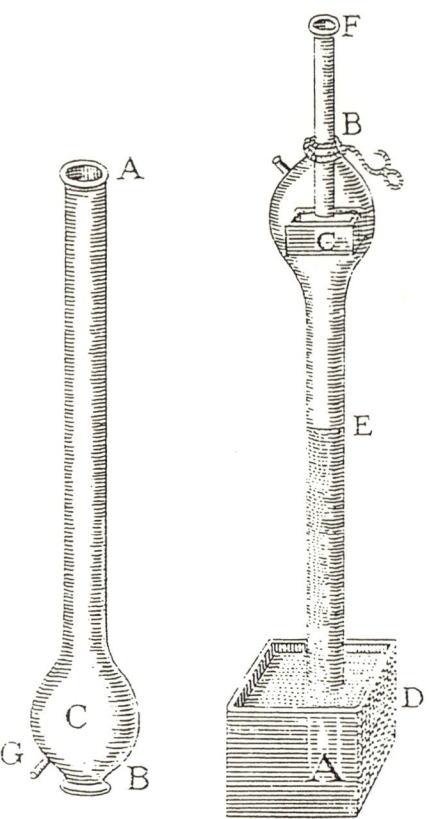

FIGURE 3 *Auzoult's long-necked flask.*

with mercury, and the inverted flask, AB, is stopped at the bottom end, A. The flask is filled with mercury through F, which is then sealed with a membrane. When the seal is removed from A, all the mercury runs out of the tube, FC, and the mercury stops at the height E in the inverted flask. Next, the membrane at G is pierced with a pin, and the mercury in the flask falls all the way down while the mercury (from the trough C) rises to a height of 27 inches in the upper tube CF. The advantage of this set-up is that the rectangular trough of the inner tube rests on the neck of the inverted flask.

But all these experimental devices were cumbersome, and Pascal wanted something less complicated. He pondered possible improvements and eventually came up with a much simpler apparatus that he described a few years later in his *Treatise on the Weight of the Mass of Air*. We shall consider this improvement later when we examine the *Treatise*, but let us note that Pascal does not say he was the first to think of the experiment of a vacuum-in-a-vacuum. In a sense, Torricelli may be called the forerunner of the experiment. We have seen that Michelangelo Ricci had objected to him that the weight of the air on the mercury in the vessel, in which the inverted tube of mercury was plunged, could be rendered inoperative simply by placing a cover over the mercury so that the air weighed on the cover and not on the liquid. Torricelli had replied that if the air left under the cover were somehow rendered thinner, the mercury would fall, and if a vacuum were produced, the mercury would drop all the way down. Could such a vacuum be produced without removing all the air from the room in which the experiment was performed? The first to ask this question was Pascal, and we have seen with what experimental flair and acumen he provided an answer.

The Decisive Experiment

The experiment of a vacuum-in-a-vacuum seems so naturally explained by the pressure and weight of the air that the modern reader may wonder what more could be desired. Pascal, however, had his qualms and he wrote to his brother-in-law, Florin Périer, that he had not entirely ruled out the abhorrence of a vacuum. Pascal does not explain in detail why he regarded his experiment as inconclusive, but he felt that it did not do away completely with the possibility of a *limited* horror of the void. Indeed, the three steps of the experiment that Pascal described could be interpreted as showing just this. In the first step (see Figure 1, illustration II above), the horror exerts itself only on the interior tube; in the second (see Figure 1, illustration III above) it no longer exerts itself on the interior tube because it has exhausted its strength in raising the mercury in the large tube to the usual height. Beyond this height, nature gives in and the void holds sway. Now, in the third step, some air is allowed into the large tube (see Figure 1, illustration IV above) and the horror of the void does not have to exert itself so strongly against the lower part of the apparatus. As a result, the force of the void changes its point of application and, to the extent to which it is liberated, exerts itself upon the interior tube. In other words, tenants of a natural but limited horror of the void were not reduced to silence.

Pascal's goal was a knockdown argument. When he did not find it in the laboratory, he sought elsewhere. On 15 November 1647, he told Périer that a new idea had occurred to him and he asked for his help in testing it. What Pascal had in mind was to repeat the standard Torricellian experiment many times on the same day, in the same tube, and with the same mercury at the bottom and on the top of a mountain no less than five or six hundred fathoms high (roughly between 1000 and 1200 meters), to ascertain whether the height of the mercury suspended in the tube is the same in both cases. "You undoubtedly see," he writes,

> that this experiment will decide the question and, if the height of the mercury is less at the summit than at the bottom of the mountain (as I believe will be the case, although all those who have considered the matter disagree), it will necessarily follow that the weight or pressure of the air is the only cause of the suspension of the mercury, and not the horror of the void. For it is quite certain that much more air weighs on the foot of a mountain than on the summit, and one can hardly hold that nature has a greater horror of a vacuum at the foot of the mountain than at the top.[6]

Florin Périer had witnessed Pascal's experiments in Paris, and he had repeated some of them in Clermont where he lived. It was natural that Pascal should turn to him for assistance, but the young man was busy and Pascal begins his letter with just the right blend of flattery and humor: "I would not dare interrupt you in your official duties (Périer was a lawyer) to raise problems of physics if I did not know that they will provide entertainment in your hours of leisure, and prove as amusing to you as they would be burdensome to others." Pascal then states his conviction that atmospheric pressure accounts for the effects usually attributed to nature's horror of a void: "For, to tell you frankly what I think, I find it hard to believe that nature, which is not animated or sensitive, should be susceptible to horror, since passions presuppose a soul capable of feeling." He insists that he already felt that way when he published his *New Experiments concerning the Void* "but for lack of convincing experiments, I dared not then (and I dare not yet) give up the idea of the horror of a vac-

---

[6]Letter to Périer, 15 November 1647, *Œuvres de Pascal*, II, p. 680; Popkin trans., p. 44; Spiers trans., pp. 100–101. The letter (which may have been revised before appearing in print) was published in 1648 along with Périer's account of the experiment in a 20-page pamphlet entitled *Account of the Great Experiment of the Equilibrium of Liquids*.

uum." His strategy was never to abandon what the Ancients had handed down "unless compelled to do so by indubitable and unanswerable proofs."[7] In the *New Experiments*, the view that a vacuum is absolutely impossible had been destroyed. The next step was to find an experiment that would show not only how but why a vacuum can be produced.

This quest led Pascal, first, to the experiment of a vacuum-in-a-vacuum, and now to the idea of repeating the Torricellian experiment at the base and the top of a high mountain that was close enough to a town to avoid too great a risk of breaking the fragile glass tubes or spilling the mercury on a long journey. Now Clermont is just next to the Puy-de-Dôme, a mountain that soars to 1465 meters, although Pascal could not determine its exact height.[8] Trusting that Périer would carry out the experiment, Pascal told several Parisian friends about it, "among others the Revd. Father Mersenne, who has already pledged himself in letters written to Italy, Poland, Sweden, Holland and elsewhere, to inform the friends that his great merit has won him in those countries."[9] This claim is surprising, for there is no evidence that Mersenne knew about Pascal's plan, let alone advertised it. As early as October 1647, however, he had suggested that residents of Nantes, Nevers and Langres, three towns along the Loire river, compare the level to which mercury drops in their respective localities.[10] Because the plateau of Langres is the watershed between the rivers that flow to the English Channel and those that flow toward the Mediterranean, Mersenne had assumed that it was the highest place in France. Actually it is only 500 meters above sea level. On 4 January 1648, Mersenne informed

---

[7]Letter to Périer, 15 November 1647, *Œuvres de Pascal*, II, p. 678; Popkin trans., p. 43; Spiers trans., pp. 98–99. In the *New experiments concerning the Vacuum*, Pascal's terminology was still strongly colored by analogies with sentient beings. For instance, he wrote, "I plan to show what force nature employs to avoid a vacuum which it nonetheless *allows* and *tolerates*"; "all bodies have a *repugnance* to separate themselves"; "nature *abhors* this apparent vacuum"; "the strength of *this inclination* is limited" (*Œuvres de Pascal*, II, pp. 500–506).

[8]The highest point of the Puy-de-Dôme is more than 1000 meters above Clermont but less than 500 meters from the plain from which it rises. Périer estimated the difference of altitude between the summit of the mountain and the town of Clermont at 500 fathoms (*toises*), or about 978 meters (*Œuvres de Pascal*, II, p. 687).

[9]Letter to Périer, 15 November 1647, *Œuvres de Pascal*, II, pp. 680–681; Popkin trans., pp. 44–45; Spiers trans., pp. 101–102.

[10]Marin Mersenne, *Preface* to the *Novarum Observationum Physico-Mathematicarum . . . Tomus III*. Paris: Antoine Bertier, 1647, a work that came off the press on 1 October of that year (*Œuvres de Pascal*, II, p. 487).

Constantin Huygens that he wanted to climb a mountain one league (4920 meters) high, and four days later he sent a letter to Le Tenneur to ask him to ascend the Puy-de-Dôme, but he apparently made no reference to Pascal.[11]

When Mersenne had written to Le Tenneur, he had assumed that he was still in Clermont, but Le Tenneur had left the region and was therefore unable to climb the Puy-de-Dôme. "In any case," he wrote to Mersenne from Tours, "I have to say that I agree with Roberval that the same thing would happen on the summit as at the base." Although he had witnessed the vacuum experiment performed by Périer, Le Tenneur had trouble grasping what was meant by a "cylinder of air" that was not enclosed in a solid container. He also felt that Mersenne underestimated practical difficulties, such as climbing a high mountain with 20 pounds of mercury in a glass tube.[12] In an earlier letter, he had also asked about the cost, and when Mersenne had suggested that he might apply for financial assistance to a great lady of his acquaintance, Le Tenneur guessed that Mersenne had in mind "the fat Marquise d'Effiat," who gladly danced with anyone but would not consider "spending a penny for the advancement of science."[13] On the strength of what he had just read in Descartes' *Principles of Philosophy*, which had come out in French translation, Le Tenneur doubted that the space above the mercury in the evacuated tube could not be emptied of all matter.

Meanwhile, Mersenne tried repeatedly to get good glass tubes, and he ordered a tube one inch wide and three feet high (27 mm × 872 mm) from the glassworks in Rouen. Raoul Hallé de Monflaines, a friend of Pascal, went to speak to the glass blowers personally but only to find that they could not carry out the order. "They were unable," he reported to Mersenne, "to seal the tube at both ends for, as soon as they took it off the fire, the air inside condensed and shattered the glass."[14] Nonetheless, Monflaines forwarded the damaged tube in the hope that it could be plugged at the broken end and rendered serviceable. This shows how

---

[11]Letter of Mersenne to Huygens, 4 January 1648, in *Correspondance du Père Marin Mersenne*, vol. XVI, pp. 12–13.

[12]Letter of Le Tenneur to Mersenne, 18 January 1648, *Œuvres de Pascal*, II, p. 492. When Périer made the ascent, he carried 8 pounds of mercury, roughly 4 kilograms (the Parisian pound was worth about 490 grams).

[13]Letter of Le Tenneur to Mersenne, 18 January 1648. The passage is omitted by Mesnard but is found in the *Correspondance de Mersenne*, XVI, p. 57.

[14]Letter of Hallé de Montflaines to Mersenne, 1 February 1648, *Œuvres de Pascal*, II, p. 624.

lucky (or how wealthy) the Pascals were, since they managed, without apparent difficulty, to have robust and properly sealed tubes made in Rouen, probably in the same glass factory.

The Great Day Arrives

The request that Pascal had made to his brother-in-law in the letter of 15 November 1647 was only met in September of the next year, which, at first sight, seems an unconscionable delay. It is explained by the fact that Périer was away on business for extended periods of time and that when he returned to Clermont the mountain was covered with snow or lost in fog, as is often the case in winter. Pascal knew the climate well, and in his *Letter to Le Pailleur*, written in February 1648, he had mentioned the forthcoming experiment with the proviso, "I do not expect it soon because, in answer to the letters that I wrote over six months ago, I have invariably been told that snow rendered the summits inaccessible."[15] Finally, weather conditions improved and Périer was able to carry out the experiment on 19 September 1648. His account is one of the most famous in the history of science and I shall give it in full:

> The weather on Saturday last, the nineteenth of this month, was very unsettled. At about five o'clock in the morning, however, it seemed sufficiently clear; and since the summit of the Puy-de-Dôme was then visible, I decided to make the attempt. To that end I notified several people of standing in this city of Clermont who had asked me to let them know when I would make the ascent. Some were members of the clergy, others laymen. Among the clergymen were the Very Reverend Father Bannier, a Minim Friar of this city, who on several oc-

---

[15] The reference to letters written more than six months prior would push back his correspondence with Périer to September, some two months earlier than the first extant letter, which is dated 15 November 1647. Since snow is mentioned, Pascal must be off by a couple of months or else the *Letter to Le Pailleur* was written in April, not February as the editor of Pascal's works conjectures on the grounds that the work makes a reference to Noël's *Fullness of the Void* that appeared at the end of January (*Œuvres de Pascal*, II, p. 570). But Etienne Pascal, who read the draft of Blaise's *Letter to Le Pailleur* told him to await his arrival in Paris to make a clean copy (*Œuvres de Pascal*, II, p. 588). Since Etienne only arrived in May we can assume that the clean copy was only made at that time. Further evidence is the letter Mersenne wrote Huygens on 15 May 1648 in which he promises to send him "the letter of Pascal who is a new Archimedes" (*Correspondance de Mersenne*, XVI, p. 314).

casions has been "Corrector" (that is, Father Superior), and Monsieur Mosnier, Canon of the Cathedral Church of this city. Among the laymen were Messieurs La Ville and Begon, councillors to the Court of Aids, and Monsieur La Porte, a doctor of medicine who practises here. All these men are distinguished, not only in their profession, but also for their broad intellectual interests. It was a delight to have them with me in this fine work.

On that day, therefore, at eight o'clock in the morning, we met in the garden of the Minim Fathers, which is practically the lowest spot in town, and there the experiment began in this manner. First, I poured into a vessel sixteen pounds of quicksilver which I had rectified [i.e., re-distilled] during the three days preceding; and having taken glass tubes of the same size, each four feet long and hermetically sealed at one end but open at the other, I placed them in the same vessel and performed with each of them in turn the usual vacuum experiment. Then, having set them up side by side without lifting them out of the vessel, I found that the quicksilver left in each stood at the same level, which was twenty-six inches and three and a half lines above the surface of the quicksilver in the vessel. I repeated this experiment twice at this same spot, in the same tubes, with the same quicksilver, and in the same vessel; and found in each case that the quicksilver in the two tubes stood at the same horizontal level, and at the same height as in the first trial.

That done, I fixed one of the tubes permanently in its vessel for continuous experiment. I marked on the glass the height of the quicksilver, and leaving that tube where it stood, I requested Father Chastin, one of the brothers of the house, a man as pious as he is capable, and one who reasons very well upon these matters, to be so good as to observe from time to time during the day any changes that might occur. With the other tube and a portion of the same quicksilver, I then proceeded with all these gentlemen to the top of the Puy-de-Dôme, some 500 fathoms above the Convent. There, after I had made the same experiments in the same way that I had made them at the Minims, we found that there remained in the tube a height of only twenty-three inches and two lines of quicksilver, whereas in the same tube, at the Minims, we had found a height of twenty-six inches and three and a half lines. Thus between the heights of the quicksilver in the two experiments there was found to be a difference of three inches one line and a half. We were so carried away with wonder and delight, and our surprise was so great that we wished, for our own satisfaction, to repeat the experiment. So I carried it out with the greatest care five times more at different points on the summit of the mountain, once in the shelter of a little chapel that stands there, once in the open, and once in the rain and fog which visited us occasionally. Each time I most carefully rid the tube of air; and in all these experiments we invariably found the same height of quicksilver. This was twenty-three inches and two lines, which gives the same difference of three inches, one line and a half with the twenty-six inches three lines and a half that had been found at the Minims. This satisfied us fully.[16]

On the way down, at a place call La Font de l'Arbre, Périer repeated the experiment twice "with the same tube, the same quicksilver, and the same vessel." Canon Mosnier then asked to perform the experiment himself and Périer let him do it. The mercury stood at 25 inches, which was 1 inch, three and a half lines less than at the Convent of the Minims, but 1 inch 10 lines more than at the top. "It increased our satisfaction not a little to observe in this way that the height of the quicksilver diminished with the altitude of the site," writes Périer. Upon returning to the Minims, they found that the quicksilver left behind in a vessel was at the same height as it had been in the morning. According to Fr. Chastin, it had stood at this height throughout the day, "although the weather had been very unsettled, now clear and still, now rainy, now very foggy, and now windy." We may wonder how often Fr. Chastin, who was a busy friar, actually cast an eye on the tube during their absence! Périer then repeated the experiment with the tube he had carried up the mountain, but, this time, in the vessel of mercury left at the Minims. The mercury fell as before to 26 inches three and a half lines. He closed the day's proceedings with a final experiment in which he used the tube, the quicksilver and the vessel that had been carried up the mountain. The quicksilver dropped once more to 26 inches 3 and a half lines. As Périer writes, "this was a final confirmation of the success of our experiment."[17]

On the next day, 20 September 1647, another theologian (an Oratorian this time), Father de la Mare, who had been in the garden of the Minims on the previous morning but had not been able to go up the Puy-de-Dôme, suggested making the experiment at the bottom and the top of the highest tower of the cathedral of Notre-Dame of Clermont. Périer first repeated the experiment in a private residence that stood on the highest ground of the

---

[16]*Œuvres de Pascal*, II, pp. 682–684, translated in I.H.B. and A.G.H. Spiers (eds.), *The Physical Treatises of Pascal*. New York: Columbia University Press, 1937, pp. 103–105.
[17]*Ibid.*, p. 685; Spiers trans., p. 106.

city, some six or seven fathoms above the gardens of the Minims, and he recorded that the quicksilver stood at about 26 inches 3 lines. He then proceeded to the top of the cathedral Notre-Dame, which he estimated at 26 or 27 fathoms (roughly 53 meters) above the garden of the Minims, and found that the mercury stood at about 26 inches 1 line.[18] These experiments enabled Périer to correlate the elevation with the height of the column of mercury, and his results can be tabulated as follows:

Location	Height of Mercury	Drop from the Height of the Mercury in the Garden
Garden of the Minims	26 inches, 3.5 lines	
Some 7 fathoms higher	26 inches, 3 lines	0.5 lines
Some 27 fathoms higher	26 inches, 1 line	2.5 lines
Some 150 fathoms higher	25 inches	15.5 lines
Some 500 fathoms higher (Puy-de-Dôme)	23 inches, 2 lines	37.5 lines (namely 3 inches, 1 line and a half)

Périer provides a good account of his experiment but he fails to tell us something important: did they walk or drive up the hill? This is relevant to the time actually left to make careful measurements. Périer says that if he had been given more time he would have made experiments at intervals of 100 vertical toises (roughly 200 meters). How would he have determined the altitude? Probably by measuring the length of the slope and the angle of elevation in order to know the perpendicular height (altitude), which could easily be computed from an elementary formula in trigonometry, $\sin \theta$ = perpendicular / hypothenuse.

A Cloud of Witnesses

The availability of prominent persons and the fact that Périer had "rectified" the mercury (i.e., purified or refined, usually by repeated distillation)

---

[18]*Ibid.*, p. 68; Spiers trans., pp. 106–107. On p. 18 of the first edition of the *Account* (1648) we find a line measuring 163 mm with the information, "Length of the half-foot used to make all the measurements recorded in Monsieur Périer's Account" (*Œuvres de Pascal*, II, p. 687, n. 1). Hence an old French inch was 2.717 cm and a foot 32.604 cm (compare with the modern English inch = 2.54 cm). A fathom (*toise*), which is 6 feet, is equal to 195.6 cm.

during three days before the experiment make it clear that the date of 19 September had been agreed upon, providing the weather was favorable. It is interesting that Périer should have considered it necessary to surround himself with influential clergymen and important laymen. Among the clerics, we find the former Superior of the Minims (the religious Order to which Mersenne belonged) and a Canon of the Cathedral, namely representatives from both the regular and the diocesan clergy, to which was added an Oratorian on the next day. The laymen included two lawyers, who were government officials, and a physician. This was not a French idiosyncrasy but a feature of the new science that had to be seen (and seen authoritatively) in order to be believed. A few years later, with the added technology of his marvelous pump, Robert Boyle repeated the vacuum-in-the-vacuum experiment. When the stop-cock was turned to let in a little air, the mercury that had fallen rose, and Boyle rightly viewed this as a confirmation of atmospheric pressure, but it was essential to produce reliable witnesses. This is why he repeated the experiment a few days later, "in the presence of those excellent and deservedly famous Mathematic Professors, Dr. *Wallis*, Dr. *Ward*, and Mr. *Wren*, who were pleased to honour it with their presence; and whom I name, both as justly counting it an honour to be known to them, and as being glad of such judicious and illustrious witnesses of our experiment."[19] Boyle had chosen them with care: John Wallis (1616–1703) was the Savilian Professor of Geometry at Oxford, Seth Ward (1617–1689) had been Savilian Professor of Astronomy at the same University, and Christopher Wren (1632–1723) had succeeded him in this post. Although Boyle's motivation was similar to Périer's, he appealed to a new professional class, not to witnesses from a wide variety of walks of life.

Boyle was concerned with showing that his experiments, unlike those of several alchemists, had actually been seen by expert witnesses. Périer, acting under Pascal's instructions, was simply concerned that they should be recorded as having been actually performed, unlike those of Noël, which were merely imagined. But both Boyle and Pascal insisted that witnessing was a collective act. In natural philosophy, as in criminal law, the reliability of a testimony depended upon its multiplicity, as Boyle went to great length to stress:

---

[19] Robert Boyle, *New Experiments Physico-Mechanical* in *Works*, vol. I, p. 34. On this experiment see Steven Shapin and Simon Schaffer, *Leviathan and the Air-Pump*. Princeton: Princeton University Press, 1985, pp. 40–47.

For, though the testimony of a single witness shall not suffice to prove the accused party guilty of murder; yet the testimony of two witnesses, though but of equal credit . . . shall ordinarily suffice to prove a man guilty; because it is thought reasonable to suppose, that, though each testimony single be but probable, yet a concurrence of such probabilities (which ought in reason to be attributed to the truth of what they jointly tend to prove) may well amount to a moral certainty, i.e., such a certainty, as may warrant the judge to proceed to the sentence of death against the indicted party.[20]

The number of witnesses lent weight to the testimony. As Steven Shapin and Simon Schaffer point out: "It was not merely that one was multiplying authority by multiplying witnesses (although this was part of the tactic); it was that *right action* could be taken, and seen to be taken, on the basis of these collective testimonies. The action concerned the voluntary giving of assent to matters of fact."[21] This is also why, like Pascal and Périer, Boyle elected to publish many of his experiments in the form of letters to other practicing or potential experimentalists. His *New Experiments* of 1660 originated as a letter to his nephew, Lord Dungarvan, and the various tracts of *Certain Physiological Essays* of 1661 were written to another nephew, Richard Jones.

Pascal Triumphs

Périer wasted no time in informing Pascal of the happy outcome of the experiment and, on 22 September, sent him the account from which we have quoted. Pascal was elated when he learned that the height of the column of mercury had dropped by half a line after Périer had ascended 7 fathoms (roughly 14 meters). This meant that he could perform the experiment in Paris where there was no shortage of tall buildings and high steeples. He promptly ascended to the top of the belfry of the church of Saint-Jacques-de-la-Boucherie, some 25 fathoms (about 50 meters) high, and found a difference of a little over 2 lines in the height of the mercury in the two positions. He then climbed 90 steps in a private house and the mercury

---

[20] Robert Boyle, "Some Considerations about the Reconcileableness of Reason and Religion" (1675) in *Works*, IV, p. 182, quoted in *Leviathan and the Air-Pump*, p. 56, n. 64.
[21] Steven Shapin and Simon Schaffer, *Leviathan and the Air-Pump*, pp. 56–57.

rose by half a line. This agreed perfectly with Périer's report and confirmed that the pressure of the air is greater at lower places. Pascal immediately saw that he could use this to know if two distant places are at the same altitude, "that is to say equally distant from the centre of the earth."[22] He also realized that thermometers left open at the top are unreliable because the level of the liquid is altered by changes in the pressure of the air.

It was important to publish these spectacular results, and within one month the *Account of the Great Experiment on the Equilibrium of Fluids Devised by Monsieur B. Pascal* was in print. It opens with a few introductory remarks by Pascal, and is followed by his letter to Périer of 15 November 1647, then comes Périer's account of 22 November 1648, and a brief outline of similar experiments made by Pascal in Paris. The work ends with an elegantly written section headed *To the Reader* in which Pascal discusses the steps that led to the overthrow of the ancient belief in Nature's horror of a void:

> Laymen and experts alike generally agree that nature would suffer her own destruction rather than the smallest void space. A few sharper minds took a less radical position and, although they still believed that nature is horrified by the void, they maintained that this repugnance is limited and can be overcome by force. But no one, as yet, has proposed this third position and said that nature has no repugnance for a void, makes no effort to avoid it, and even allows it without trouble or resistance. The experiments set forth in my *Abridgement* [i.e., his *New Experiments*] invalidated, in my judgement, the first of these positions, and I do not see how the second can withstand the evidence that I now present. Therefore, I now find no difficulty in accepting the third, namely, that nature has no repugnance to a vacuum, makes no effort to avoid it, and that all the effects ascribed to such a horror are due to the weight and the pressure of the air, which is their only real cause. Because this was not known an imaginary horror of the vacuum was invented in order to account for these effects. This is not the only instance when the inability to discern the real causes has led to the substitution of imaginary ones to which were attached specious

---

[22]*Œuvres de Pascal*, II, p. 687. It will later occur to Pascal to use the sea level as the starting point for measuring altitude, as he mentions in his *Treatise on the Weight of the Mass of Air* (*Œuvres de Pascal*, II, p. 1089). The fact of the matter is that pressure is not only less at higher elevations, but also, because air is compressible, its density decreases with elevation. For this reason, we cannot use the height of mercury in the tube for calculating the height of the atmosphere. Pascal was not aware of this difficulty.

names that fill the ears, but not the mind. Thus we hear that sympathy and antipathy in natural bodies are the real efficient causes in many cases, as if inanimate bodies were capable of sympathy and antipathy. It is the same with *antiperistasis* and several other chimerical causes that only bring a vain relief to our eagerness for concealed truths. Far from uncovering them, they only serve to cloak the ignorance of those who invent them, and feed the ignorance of their followers.

Nevertheless, it is not without regret that I abandon opinions so generally received. I only yield to the compulsion of truth. I resisted these new ideas as long as I had any reason for clinging to the old. The arguments that I used in my *Abridgement* sufficiently testify to this. In the end, however, experimental evidence compelled me to lay aside the views that I held out of respect for antiquity. I left them little by little, and I withdrew by degrees. For from the first of these three principles—that nature has an unconquerable horror of a vacuum, I passed on to the second—that she does feel that horror but not insuperably, and at last, have come to believe the third—that nature has no such horror at all.[23]

Seldom has a scientist illustrated so clearly the passage from an uncritical reliance on organic analogies to the recognition that experiments are crucial. The obstacles to scientific progress are the ascription of properties such as sympathy and antipathy to "inanimate objects" and the invention of theories like *antiperistasis*, which Pascal brands as chimerical but does not explicitly refute.

Fanciful Explanations

Appeals to sympathy and antipathy were still common in Pascal's day. In the *Compendium Musicae*, Descartes wrote: "The voice of a friend is more agreeable than that of an enemy because of the sympathy that we experience. For the same reason, a drum covered with the skin of a lamb ceases to vibrate and becomes silent when, as we are told, it sets up a res-

---

[23]*Œuvres de Pascal*, II, pp. 688–689; Spiers trans., pp. 110–111. Torricelli had written to Ricci in his letter of 11 June 1644: "Many have said that a vacuum cannot be created, and some that it can but only with difficulty and against the resistance of nature. I do not know of anyone who said that it can be created without difficulty and without any resistance from nature."

onance in another drum covered with the skin of a wolf."[24] His authority was probably Giovanni Battista della Porta, who gives two versions of this belief in his *Natural Magic*. The first runs as follows: "The wolf is hurtful and odious to sheep after he is dead: for if you cover a drum with a wolf's skin, the sound of it will scare the sheep, and if you hang several skins side by side, the wolf's skin will eat up the lamb's skin." The second is as follows: "There is Antipathy between sheep and wolves, as I said often, and it remains in all their parts, so that an instrument strung with sheep strings mingled with strings made of a wolf's guts, will make no music, but will jar, and make all discords."[25] There are other variants on this theme. For instance, Robert Burton applies it to Jan Zizka, the fifteenth-century national hero of Bohemia: "The great captain Zisca would have a drum made of his skin when he was dead because he thought the very noise of it would put his enemies to flight."[26]

If sympathy and antipathy were anthropomorphic categories, *antiperistasis* offered a strictly mechanical explanation. Noël had appealed to such a concept in his *Fullness of the Void*, and the ancient source is Plato's *Timaeus*, in which a circular thrust is invoked to explain respiration. The idea is that when breath is expelled from the mouth it dislodges the air just outside without leaving the space inside empty. Here is how Plato puts it: "Since there is no void into which any moving body could enter, and our breath moves outward, the consequence is plain to anyone: the breadth does not go out into the void but pushes the neighbouring body out of its place, and the body thus displaced in turn drives out the next, all this takes place *simultaneously like a revolving wheel because there is no void*."[27] Plato used this theory to explain not only respiration but also the action of a medical cupping glass and the motion of projectiles. It follows that a void is impossible since air or some other fluid is required for

---

[24]René Descartes, *Compendium Musicae* in *Œuvres de Descartes*, edited by Charles Adam and Paul Tannery, revised edition. Paris: Vrin, 1964–1974, vol. X, p. 90.

[25]John Baptista della Porta, *Natural Magick*. Anon. trans. London, 1658. Facsimile New York: Basic Books, 1957, book I, chapter 14, pp. 19–20, and book XX, chapter 7, p. 403. Della Porta ascribes this view to Pythagoras. The earliest source that I have come across is Fracastoro's book on *Sympathy and Antipathy* of 1550 in which he writes, "striking a drum made of the skin of a wolf will, they say, shatter drums made of the skin of lambs" (Girolamo Fracastoro, *De Sympathia et Antipathia Rerum*. Lyons, 1550, p. 22).

[26]Robert Burton, The *Anatomy of Melancholy*, first published in 1621. Everyman's Library reprints the 6th edition of 1651, 3 vols. London: Dent, 1932, vol. I, p. 38.

[27]Plato, *Timaeus* 79 B, see also 58 A.

projectile motion to occur. Aristotle took the theory over from Plato to explain what makes a javelin continue to move after it leaves the thrower's hand. Since the work is no longer being done by the thrower, it must, he thought, be done by the air that receives from the thrower the power of moving the projectile.[28] A sixth-century commentator, John Philoponus, rejected this theory, and had enormous fun in subjecting it to ridicule. Why, if this were the mechanism, he exclaims, all artillerymen would have to do to project stones would be to line them up on a wall and blow on them with bellows![29]

Aristotle's theory of *antiperistasis* was revived in the Middle Ages by John Buridan, who maintained that when an arrow swiftly leaves the place it occupies, nature, which does not permit a vacuum, just as swiftly sends in air behind it to impel the projectile further along.[30] It was only in the sixteenth century that it was gradually realized that in order to achieve this the air must perform three distinct motions: (1) it must be pushed forward by the arrow, (2) move back, and finally (3) turn around and go forward once more. But how can the air, which is pushed by an arrow, turn around and retrace its course? And if it turns around, how can it impinge precisely on the notched end of the arrow and push the arrow forward?

But if the air does not move the projectile, what causes its continued motion? Some revived Philoponus' explanation that an incorporeal kinetic force is impressed in the stone or the arrow. This force keeps the body in motion but is gradually spent out by the resistance set up by the weight of the body. However attractive, this view left open the nature of the relation of the force to speed and acceleration.

The End of Horror

We may doubt whether Pascal abandoned generally received opinions with as much regret and reluctance as he says, but it is true that he gradually passed from the view that nature has an invincible horror of a vac-

---

[28]Aristotle, *Physics*, book 4, chapter 8, 215a 14–19 and book 8, chapter 10, 266b27–267a20.
[29]John Philoponus, *In Aristotelis Physicorum Libros Commentaria*, quoted in Clagett pp. 508–509. There is an extended English translation of this passage in I. E. Drabkin in Cohen and Drabkin, *A Source Book in Greek Science*. New York: McGraw-Hill, 1948, pp. 221–222.
[30]*Quaestiones super octo phisicorum libros Aristotelis*. Paris, 1509, quoted in Marshall Clagett, *The Science of Mechanics in the Middle Ages*. Madison: University of Wisconsin Press, 1961, p. 508.

uum to the view that the horror is not invincible and, finally, that there is no such horror. Let us review the three different types of objections that Pascal raised against the notion of a *horror of the void*. The first is that since nature is not animated, it cannot feel horror. The second is that horror of the void is no explanation at all: it merely draws attention to certain phenomena that have to be explained. The third objection is that the horror is invalidated by experimental evidence.

As for the first objection, we have, in reality, no way of knowing whether or not nature has emotions. This is a philosophical point that is debated by holistic philosophers and deep ecologists, not by physicists. As we have seen, Aristotle and the Ancients preferred mechanical explanations whenever possible even if, as in the example of *antiperistasis*, their attempts seem crude to us. The proposition that nature has a horror was taken figuratively rather than literally by many natural philosophers, but even then they took it as expressing a real, measurable resistance to attempts to create a vacuum. Taken in this sense, the proposition merely gave a description of cases already observed and, as new cases arose, it was considered normal to modify the description to make it fit. For instance, after the vacuum-in-the-vacuum experiment, it was said (without doing away with the horror entirely) that the force needed to create a void was not as great when the void is surrounded by rarefied air as when it is surrounded by air of ordinary density. After the Puy-de-Dôme experiment, some still wanted to say that nature has a greater horror of the void at the bottom than at the top of a mountain. Pascal realized that the phrase *horror of a void* was no longer a physical theory but a mere metaphor.

The proposition that nature has a horror of the void could be taken in another sense, as a scientific generalization that transcends mere description of a limited number of actually observed cases and states that there is a causal connection between the vacuum and the force expended in raising the mercury. Put this way, the proposition can be considered a general law that describes the way nature works without making any assumptions about nature's emotions or feelings. It is in this physical sense that Pascal will consider the proposition a few years later in his treatise *On the Equilibrium of Liquids*, where we shall see him develop the principles of hydrostatics that are relevant to the pressure and equilibrium of liquids. In his second treatise, *On the Weight of the Mass of Air*, he will show how phenomena ordinarily attributed to the horror of the void can be explained by hydrostatic principles so that the horror of a void becomes redundant. In other words, the way nature reacts in the presence of a vacuum can be fitted into a system of more general mechanical laws of equilibrium. Since a unified science was as much a goal of Pascal's opponents as of himself, this was a powerful argument to dismiss the horror of the void as superfluous.

If we ask what was Pascal's "crucial argument," we might want to consider the experiment of a vacuum-in-a-vacuum as well as the one on the Puy-de-Dôme. In order to make the concept of the horror of the void fit the vacuum-in-the-vacuum experiment, traditional philosophers had to shift their ground and say that nature has a horror of very rarefied air as well as of a vacuum, and that as the air becomes denser her horror decreases. Could they use the same ploy for the Puy-de-Dôme experiment? If so, they would have to say that as we ascend a mountain the atmospheric air becomes more rarefied so that the horror is increasingly directed towards this outer air rather than against what is inside the tube and, as a result, the empty space above the column of mercury increases. Such a limited horror of the void might account for the facts, but the alternative explanation by the weight of the atmosphere is so obviously more compelling that it gradually won the day.

The Impact of the Great Experiment

The first known reaction to the publication of the *Account of the Great Experiment* is a letter of the French astronomer Ismael Boulliau to Johann Hevelius in Dantzig in which he dismisses Pascal's explanation as erroneous, but without offering an alternative.[31] Gassendi, who had left Paris before the experiment on the Puy-de-Dôme, was told about it by his Lyonnese publisher, François Barancy, who had been informed by Mosnier.[32] In a letter dated 13 December 1648, Gassendi summarizes the experiment and speaks in flattering terms of Pascal as an "incomparable young man."[33]

---

[31]Quoted in *Œuvres de Pascal*, II, p. 700.

[32]This is probably Claude Mosnier, a canon of the cathedral of Clermont, not Pierre Mosnier, the physician who published the works of his teacher, Honoré Fabri, in Lyons, but the question remains open since Pierre Mosnier was born in Clermont and almost certainly met Barancy in Lyons (see Sylvain Matton, "La *grande expérience* du Puy-de-Dôme revisitée," *Chrysopoeia* 2 (1988), pp. 319–326). As we know from Périer's account, Canon Mosnier was the person who performed one of the experiments on the way down from the Puy-de-Dôme.

[33]Quoted in *Œuvres de Pascal*, II, pp. 726–727. On Gassendi's contribution to the debate, see Simone Mazauric, *Gassendi, Pascal et la querelle du vide*. Paris: Presses Universitaires de France, 1998.

The data for the readings taken at three stations on the way up the Puy-de-Dôme were provided by Mosnier and are oddly at variance with those published by Périer and Pascal in their *Account of the Great Experiment*, as we can see by comparing them:

Place	Values given by Périer	Values given by Gassendi
Clermont	26 inches 3.5 lines	25 inches 5.5 lines
La Font de l'Arbre (half way up the mountain)	25 inches	24 inches 4 lines
Puy-de-Dôme	23 inches 2 lines	22 inches 5 lines

The discrepancy between the readings recorded by Périer and those provided by Gassendi could be explained by assuming that Périer used the "foot of Macôn," 33.6 cm long, when he made his measurements on the Puy-de-Dôme and that he then converted them into the "Parisian foot," 32.6 cm long. The two kinds of feet were divided into twelve inches but the Macôn inch was subdivided into six lines, whereas the Paris inch appears to have been subdivided into six or twelve lines. Assuming that Périer used a division into six lines for both kinds of inches, the discrepancy between the two sets of values is slight, as we can see:

Place	Values given by Périer and their conversion into cm (1 foot de Macôn = 33.6 cm)	Values given by Gassendi and their conversion into cm (1 Parisian foot = 32.6 cm)
Clermont	72.2 cm	72.5 cm
La Font de l'Arbre	67.9 cm	69 cm
Puy-de-Dôme	63.3 cm	63.9 cm

Some three years later, in February 1652, Gassendi repeated the experiment on the Mont Faron near Toulon and found, to his surprise he says, that when the sky was clear and a north wind was blowing, the mercury stood higher than when the sky was overcast and the wind came from the south. Gassendi conjectured that this was because, in the first case, less heavy vapors rose from the earth, and he forestalled criticism by claiming that thicker or coarser air is not necessarily heavier.[34]

---

[34] Letter of Gassendi to Bernier, 6 August 1652, published in Section I, book II, chapter 5 of the second part of Gassendi, *Syntagma* in his *Opera Omnia*, Lyons, 1658, vol. I, pp. 215–216.

Carrying several kilos of mercury up a mountain was a somewhat strenuous way of confirming a theory, and Pascal was anxious to find an easier method to convince his less athletic colleagues. When he returned to Clermont for a visit in the spring of 1649, he hit upon a simple experiment: fill a balloon with a small amount of air, just enough to make it flabby, tie it to a string, and begin the ascent of the Puy-de-Dôme. As the balloon rose, it swelled until it was quite round and hard. On the way down, it shrunk by the same degrees until at the foot of the mountain it was as flabby as before. This greatly impressed Gassendi, who saw it as a clever extension of Roberval's experiment with the bladder of a carp.[35]

Descartes' Reaction to the Puy-de-Dôme Experiment

Constantin Huygens, a famous Dutchman and the father of an even more famous son, Christiaan, corresponded regularly with French scholars. When he received a copy of Pascal's *New Experiments*, he passed it on to Descartes, who was then living in the Netherlands. Descartes assumed that it had been sent at the author's request, and he was annoyed for several reasons that he immediately communicated to Mersenne on 13 December 1647. The first was that Pascal said that his opponents were biased and that they placed in the empty space above the mercury something "that existed only in their imagination." Descartes read this as an attack on his subtle matter and requested that Pascal "put down his best arguments on the subject and not be surprised if, at the appropriate place and time, I defend myself accordingly." The next reason for his displeasure was the ignorance in which he claimed he had been kept by Mersenne: "You ask me to write about the experiments with mercury but you neglect to tell me about them, as if I could guess what they are. If I find the correct explanation, I don't want people to assume that I made the experiment, and if I fail to do so I don't want them to think less well of me. But if you will candidly let me know what you have observed, I will be much obliged, and if I use the information, I will not neglect to acknowledge my source."

Descartes was also irritated with what he considered Pascal's failure to comment on his suggestion that he should experiment to see whether quicksilver rises as high at the summit as at the base of a mountain.[36] The

---

[35]*Ibid.*, p. 215. This passage of the letter is also to be found in *Œuvres de Pascal*, II, p. 933.
[36]Letter of Descartes to Mersenne, 13 December 1647, quoted in *Œuvres de Pascal*, II, p. 549.

suggestion could only have made when Descartes called on Pascal in Paris in September 1647, but another person, Adrien Auzoult, later claimed to have given Pascal the same advice at about the same time. Pascal, however, was to remain adamant that the idea was his own. Although he never said that he had been the first to think of the experiment of a vacuum-in-a-vacuum (although he was credited for this invention by Father Noël), he always insisted that the Puy-de-Dôme experiment was his own.[37] We can only conclude that several persons, including Pascal, Descartes, Auzoult and Mersenne, thought of checking whether mercury rose as high at the summit as at the base of a mountain, and that each assumed the idea was original to himself. Let us recall that in his letter to Ricci of 28 June 1644, Torricelli had already conjectured that if the air weighing on the mercury in the bowl were rarefied, the column of mercury in the tube would stand lower, and that if all the air were removed, the column would drop altogether. Both the experiment of the vacuum-in-the-vacuum and that of the Puy-de-Dôme would be readily suggested by this passage. With hindsight, the experiment appears almost obvious.

Descartes may have been annoyed but he was not idle. He enclosed in his letter to Mersenne a strip of paper two and a half feet long with regular subdivisions between two feet two inches and two feet four inches. This was identical with the one he had been using to determine the height of the mercury on different days. He told Mersenne that on Monday, 9 December, the mercury had stood at two feet two inches, and on Thursday, 12 December, at two feet four inches. In return for this information, he asked Mersenne to observe how high the mercury rises when the weather is warm or cold, or when the northern or the southern wind blows. We might get the impression from this letter that Descartes had been making observations for some time, but he had just begun to do so.

Descartes was always willing to put other people to work, and he asked Mersenne to kindle a fire in an evacuated tube in order to see whether the smoke went up or down. He also wanted him to suspend a piece of sulphur from a string in a vacuum and light it with a magnifying glass. "I cannot do this here," he declared, "because the sun is not warm enough and I have been unable to find a tube fitted with a bottle." Whether the winter sun is much warmer in Paris than in Holland can be doubted, and Descartes had

---

[37]"I invented this experiment and therefore I can claim that the new knowledge that comes from it is entirely due to me" (letter of Pascal to Antoine de Ribeyre, 12 July 1651, *Œuvres de Pascal*, II, p. 813).

more means at his disposal to buy equipment than Mersenne, who was a poor friar. Descartes' high opinion of himself is manifested in the further reprimand with which he closes his letter to Mersenne: "I am surprised that this experiment should be four years old, as Pascal says, and that you should have failed to inform me before this summer, for as soon as you mentioned it I realized its importance and its usefulness in confirming what I have written about physics."[38]

Mersenne feared that the tube would break if he were to light a fire inside, but Descartes assured him that there was no danger as long as he did not use gunpowder. He warned Mersenne, however, that in a hermetically sealed room attempts to evacuate the tube would fail because all the space available would already be occupied by subtle matter, so that the mercury would have nowhere to drop! When Mersenne complied with Descartes' request and started making regular measurements of the height of the mercury, Descartes wrote again:

> I am happy that you should have observed the height of quicksilver in a tube that is fixed. It sometimes stays ten or twelve days without rising or falling but, on three different occasions, I have seen it rise within two or three days by more than one inch. I am convinced that it would have risen much more had the winter been colder, but this year has been the mildest that I have seen in this country. All this time, the quicksilver only reached two feet three inches, or a little more, and I am surprised that it was noticeably higher in Paris.[39]

On 7 February 1648, Descartes wrote again to say that he had been recording the height of mercury in a fixed tube for two months,[40] which proves that he had just begun to do so when he wrote the letter of 13 December to Mersenne that we quoted above. In his next letter, dated 4 April 1648, Descartes declares that he can explain Mersenne's experiments,

> at least those that are genuine for, concerning those that come from others, they are partly distorted by their fancy . . . But in order that you may know that it is not the air enclosed in the tube nor the heat

---

[38]Letter of Descartes to Mersenne, 13 December 1647, quoted in *Œuvres de Pascal*, II, pp. 549–550.
[39]Letter of Descartes to Mersenne, 31 January 1648, *Œuvres de Descartes* (edited by C. Adam and P. Tannery), vol. V, pp. 115–116.
[40]Letter of Descartes to Mersenne, 7 February 1648, *ibid.*, vol. V, p. 119.

that makes the mercury rise or fall according to the weather, let me tell you that, for more than six weeks, I have been using two tubes. The first, which contained no air whatsoever, was kept in a cold room high above the ground, and the second, which contained some air, was kept in a heated room where a fire was on at all times. The quicksilver rose or fell by the same amount in these two tubes when the weather changed. On 12 March, and again at the end of March, the quicksilver rose above two feet four inches and, in between, fell below two feet three inches. But this does not exclude that when the tube that contains air is heated vigorously, the mercury falls, as is the case with a thermometer.[41]

Descartes informed Mersenne that he would soon be in Paris and he arrived around the middle of May 1648 but failed to inform Mersenne, who was still waiting for him early in June. It was at this time that Roberval gave a number of public lectures in which he attacked the Cartesian notion of a plenum. The differences in outlook and temperament between Descartes and Roberval were such that a clash was unavoidable, especially since Roberval enjoyed hounding his adversary. On one occasion, at the house of a prominent person, Roberval was more domineering than usual, speaking as a master might to his scholars, and Descartes declared that he wished to carry on such disputations in writing only. After this, Roberval felt justified in going around saying that he had taught him a lesson.[42]

Mersenne died on 1 September 1648, a victim of abusive blood-letting. In the spring of 1649, Pierre de Carcavy wrote to Descartes to offer himself as his correspondent. Descartes replied outlining the services that had been rendered by Mersenne, such as informing him about current debates, new experiments, and new books. Descartes added, rather disingenuously, that he "never inquired about anything,"[43] but proceeded forthwith to break this rule by asking about the Puy-de-Dôme experiment:

---

[41]Letter of Descartes to Mersenne, 4 April 1648, *ibid.*, vol. V, pp. 141–142.

[42]The source is Adrien Baillet's *Vie de Monsieur Descartes*. 2 vols Paris, 1691. Facsimile reprint Geneva: Slatkine, 1970, vol. II, pp. 344–346.

[43]"Although I never made the slightest enquiry, I was always fully informed of what went on among the learned such that, if he sometimes asked me some questions, he paid very generously for the answers by informing me of all the experiments that he or others performed, all the inventions that had been sought or found, all the new books that were talked about and, finally, all the discussions among scientists" (letter of Descartes to Carcavy, 11 June 1649, *Œuvres de Pascal*, II, p. 717).

I am confident that you will not take it amiss if I ask you to inform me about the outcome of an experiment that I was told M. Pascal made or had made on mountains in the Auvergne in order to determine if quicksilver rises to a greater height at the base of a mountain than at the summit, and by how much. I could have expected Pascal himself, rather than you, to send me this information since it was I who advised him, two years ago, to make this experiment, and told him of the outcome, although I had not made it myself. But because he is the friend of Mr. R(oberval), who declares not to be mine, and also because I have already seen two or three-pages of printed text in which he attacks my subtle matter, I have reason to believe that he shares the sentiments of his friend.[44]

Carcavy provided, in July 1649, a clear summary of the experiment made by Périer on the Puy-de-Dôme, but made no reference to the *Account of the Great Experiment*, published in October 1648, from which he probably got the accurate figures he gave for the height of the columns of mercury. On 17 August 1649, Descartes acknowledged receipt and claimed once again that he had suggested the experiment two years earlier, and that he knew of the outcome because it agreed entirely with his theory. Because Pascal did not accept his philosophical principles, Descartes was convinced that he never could have thought of it alone![45]

Descartes also confused the position of Roberval, who talked so much, with that of Pascal, who had been rather quiet when they met in September 1647. As late as January 1648, Roberval and his friend Le Tenneur maintained that a column of mercury would stand at the same height at the top as at the bottom of a mountain.[46] Their friend, Mersenne, thought that making the experiment on a mountain, or in cities of different altitudes, *might* prove that atmospheric pressure was the cause of the suspension of mercury, but he was sceptical.

Rival Theories

The debate over the role of atmospheric pressure takes on its full significance when seen against the background of the two great rival theories

---

[44]*Ibid.*, p. 717.
[45]Letter of Descartes to Carcavy, 17 August 1649, *Œuvres de Pascal*, II, p. 719.
[45]Letter of Le Tenneur to Mersenne, 16 January 1648, *Œuvres de Pascal*, II, p. 492

that dominated the first half of the seventeenth century: the Aristotelian and the Archimedean, which may also be called the organic and the mechanical. The first explained the world with the aid of the language normally used to describe growth and decay, with the analogy of an acorn growing into an oak always ready at hand. What struck people who looked at the world in this way was not so much the regularity and uniformity of nature as its constant change. Yet within the process of change there was a consistency that had to be accounted for. Acorns do not grow into poplars. This led to the view that there was a potentiality or purpose built into all natural phenomena, a so-called "final cause," which dominated development. Natural growth was movement directed towards an end. Aristotelians saw this process repeated throughout nature, not merely in living things but in the movement of inanimate objects, and in chemical reactions. Bodies had a "natural" motion but they could be compelled to move in an "unnatural" way. A stone that fell straight down behaved "naturally," whereas a projectile hurled upwards was being moved "unnaturally." By contrast, the mechanical or Archimedean tradition rested on a view of nature in which the dominant model was the machine understood as an instrument to transmit force or move weights. The prime instances of such machines were the lever, the balance, and the pulley, and problems of equilibrium and displacements of forces were what mattered in this context.

The experiment carried out at the base and the summit of the Puy-de-Dôme was designed to test the mechanical hypothesis that a column of mercury in an evacuated tube is in equilibrium with the air weighing on the surface of the bowl of mercury into which the tube is inverted. The outcome was clear, and where weight would do, horror of the void became melodramatic.

CHAPTER 6

Why Pumps Work

The Treatise on the Equilibrium of Liquids

Pascal never completed the *Treatise on the Vacuum* that he intended to write, but he drafted two short works that were published posthumously by his brother-in-law, Florin Périer, in 1663 as the treatises *On the Equilibrium of Liquids and On the Weight of the Mass of Air. Containing the Explanation of the Causes of Various Effects of Nature Which Had Not Been Known Hitherto, and in Particular of Those Which Had Been Ascribed to the Horror of a Vacuum.*[1] Pascal's most valuable contributions to physics are found in these two works. In the first, he provides a complete outline of a system of hydraulics, the first in the history of science. In the second, he shows how the phenomena formerly attributed to nature's horror of a void are to be explained in terms of this system. Périer writes that the two treatises "were ready to be printed more than twelve years ago," namely before 1651, but that Pascal laid them aside because he came to see, after his religious conversion, "the vanity and emptiness of all such knowledge."[2] This is clearly an exaggeration, although it is true that Pascal spent most of his time on religious topics after 1651.

---

[1]The two treatises (pp. 1–140) are followed by two fragments from Pascal's intended *Treatise on the Vacuum* (pp. 141–163), his *Account of the Great Experiment on the Puy-de-Dôme* (pp. 164–194), a summary of Périer's observations (pp. 195–209), and a discussion of Boyle's recent work in England (pp. 210–232). The work is translated as *The Physical Treatises of Pascal* by I.H.B. and A.G.H. Spiers. New York: Columbia University Press, 1937.

[2]*Œuvres de Pascal*, I, p. 680; Spiers trans., p. IX.

Périer emphasizes Pascal's experimental caution and his careful usage of the French word *vide* (*empty* or *void*) in the experimental sense of a *vacuum*, namely a space empty of all matter with which we are familiar and that we can detect.[3] This contrasts with Descartes for whom le *vide* was the ontological *void* and, strictly speaking, *nothing*. Like Descartes' identification of matter with three-dimensional space, it belonged to the realm of metaphysics, not science.

The first treatise, *On the Equilibrium of Liquids*, which we will examine in this chapter, provides an account of the behavior of liquids in terms of forces in equilibrium. Pascal avoids the danger of barren observation by seeking the mechanical law behind the multiplication of forces, and by showing how the theory of machines makes sense of the experimental data. Simple machines (the lever, the wheel, the pulley, the inclined plane and the screw) had been known for centuries and were studied by Archimedes, who formulated the fundamental laws of statics, the branch of mechanics that deals with forces at rest or in equilibrium. The most important machine is the lever (of which the balance is a special case) and it became Pascal's main explanatory tool as he tells us in such statements as, "A body in water is supported in the same way as if it were in one of the trays of a pair of scales, while the other tray is weighted with an equal volume of water."[4]

How Liquids Weigh According to Their Height

Chapter I of *On the Equilibrium of Liquids* opens with six simple experiments, which are illustrated in Figure 1. In the top half we find: (I) an upright cylindrical tube, (II) a slanting cylindrical tube, (III) a large vessel with a very broad rim, (IV) a narrow vessel with a very small rim, and (V) a slender tube that ends in a short, broad base. In the lower half, three rectangular containers with two soldered pipes of different size on top are numbered VI–VIII. These eight receptacles were used to study the pressure exerted by a liquid on the containing wall.

---

[3]"Whenever the word *void* is found, it must not be supposed that M. Pascal intended to prove that there could be an absolute void: by this word he always means a space empty of all materials that are perceptible to the senses, as he states in many places" (*Œuvres de Pascal*, II, pp. 1041–1041; Spiers trans., p. XXII).
[4]*Œuvres de Pascal*, II, p. 1053; Spiers trans., p. 17.

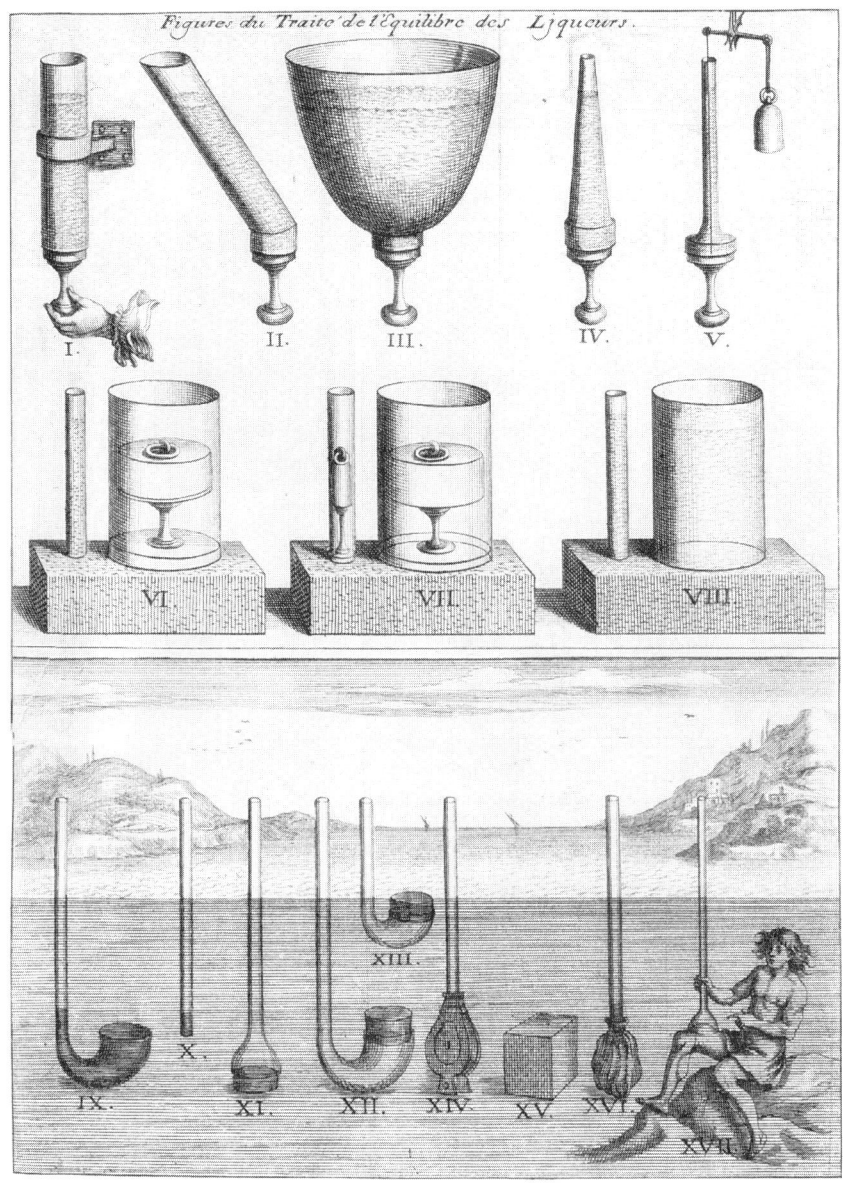

FIGURE 1 *The Experiments described by Pascal in his treatise* On the Equilibrium of Liquids.

The five first receptacles are different in shape and size, but at their base they have similar openings in which are inserted plugs. When they are filled to the same height with water, an equal force must be exerted under each of these plugs to prevent them from falling out. In order to measure this force, Pascal stopped the opening with a round piece of wood, surrounded by oakum like the piston of a pump, so that it could move easily in the opening but still prevent the water from running out. In this way he found that the force required is always equal to the weight of the water contained in the first receptacle that has the same diameter throughout (see Figure 1, illustration I). This may seem paradoxical for, as Pascal writes, "if the water weighs a hundred pounds it will take a force of one hundred pounds to hold up each of the plugs, even the one in the fifth tube, although the water in it weighs no more than one ounce."[5] To prove this, Pascal replaced the plug in the fifth receptacle with a piston fastened to a thread that passed through the tube and was tied to one arm of a pair of scales. When a hundred-pound weight was suspended from the other arm, it balanced the ounce of water in the small receptacle, and when this weight was diminished, the weight of the water brought the piston down.

After describing these experiments, but before proceeding to the second set of receptacles, numbered VI, VII and VIII, Pascal mentions, in the conditional mode, two other experiments that are legitimate extensions of the experiment made in the fifth container but may not have been actually performed. The first consists in making holes on the side or the top of any of the first four receptacles to show that the pressure of the water is uniform throughout the liquid. The second, which calls for special climatic conditions, I leave Pascal to describe in his own words: "If the water happened to freeze but did not stick to the tube (which it seldom does) a one-ounce weight would balance the ice, but if we applied heat to the tube and the ice melted, it would take a one-hundred pound weight to balance the weight of the melted water although it only weighed one ounce."[6] The freezing and thawing of the water renders more dramatic the result of the fifth experiment where a *column of water* was balanced by a *physical weight* of 100 pounds, although the water alone weighed only one ounce, namely 1600 times less.

---

[5]*Ibid.*, p. 1043; Spiers trans., p. 3.
[6]*Ibid.*, p. 1043; Spiers trans., p. 4.

The next three experiments were performed with the apparatus shown in illustrations VI, VII and VIII in which two cylinders communicate by way of a box-like base. If a piston is placed in the larger cylinder and water is poured in the smaller, a considerable weight must be placed on the piston to prevent the weight of the water in the smaller cylinder from driving it up. Pascal points out that this corresponds to the case considered above when a force of one hundred pounds was necessary to prevent the weight of the water in the receptacles from driving the stoppers down when the opening was in the base. When the height of the water in the smaller cylinder is doubled, the weight on the piston in the large cylinder must also be doubled. The is also the case when the width of the opening of the smaller cylinder is made twice as large.

Teasing out the Cause

We can summarize Pascal's procedure by noting that the first five receptacles have the *same height* and the *same size* of opening at the base. Other factors, such as width, shape and inclination are made as different as possible. Pascal's initial step is to show that the *force* that is necessary to keep the plugs from falling out is the same in all the receptacles, although the weight of the water varies between one hundred pounds and one ounce. The *force* required is the *weight* of the water in the first upright cylinder. This indicates that the height is the determining factor. A slight terminological ambiguity is created, however, by the use of "weight" where what is intended is "pressure downward." This difficulty is avoided when we write, in modern notation, $P = W/A$.

In other words, Pascal's discovery consisted in the realization that water acts like a lever, and that the force necessary to prevent water from running out at an opening is proportional to the height of the water and not to its volume. He clearly saw that this could be extended to every kind of liquid, but because he had used tubes that were more than half a centimeter in width, he was led to make a rash generalization: "If the tube that was filled with water were a hundred times smaller, then, so long as the level of the water in it remained the same, the same weight would always be required to balance it."[7] Périer pointed out that the statement is cor-

---

[7] *Ibid.*, p. 1044; Spiers trans., p. 5.

rect provided the tubes are at least "two or three lines in diameter" (5 to 7 mm). If the tube is narrower, water will rise to a higher level as a result of capillary action.[8]

Background Voices

Pascal did not invent all of this from scratch, and we can get the flavor of contemporary discussions by quoting from Mersenne's *Cogitata Physico-Mathematica*, published a few years earlier:

> VIII. The water that lies on a flat and even bottom will weigh as much as a column of water whose base covers the same area and whose height is the vertical distance from the bottom to the surface of the water.
>
> IX. The last statement will sound strange when we notice the following consequence: one pound of water can weigh on the bottom of a jar as much as a thousand pounds or even the whole ocean! Suppose that the ocean and the pound of water are placed in two jars whose bottoms are equal in size, and in such a way that the jar that contains the pound of water is fitted with a tube that is so narrow that the water rises as high as the ocean does in the other jar. This pound of water will press the bottom of the second jar as much as the ocean presses that of the first . . .

---

[8]*Ibid.*, p. 1041; Spiers trans., p. XXI. Périer owed this information to Jacques Rohault, whose work on capillarity was subsequently published in his *Traité de physique* in 1671 (part I, chapter XXII, sections 81–85). When Rouhault submerged a glass tube whose bore was so narrow that he could hardly pass a horsehair through it, the water rose to a height of one foot. He assumed that this happened because the air above the capillary tube weighs less there than elsewhere. The correct explanation had to await the Newtonian view of matter, according to which the cohesive power of water is less than its adhesive power to glass. Hence, water (1) clings to glass and wets it, (2) exhibits a concave meniscus, that it, its surface curves upwards near the glass, and (3) creeps up a fine tube. On the contrary, the cohesive power of mercury is greater than its adhesive power for glass. This is why mercury (1) does not cling to glass or wet it, (2) exhibits a convex meniscus, and (3) sinks below the gross equilibrium level in a fine glass tube. The height of a convex meniscus is easier to read, and hence mercury is preferable for thermometers because it expands regularly over a wide range and has a high boiling point. It is also better for barometers because it is very dense and has a very low vapor pressure or, as it is also called, saturation pressure.

XI. Suppose that we wish to plunge a stick into the jar that contains the ocean, and over which a cover has been placed to stop the water from rising. We will only be able to achieve this if the force at our disposal is equal to the product of the weight of the stick by the surface area of the ocean divided by the surface area of the end of the stick [namely F = W (of stick) × a (of sea) : a (of stick)].

XII. If the stick could be pushed in through a hole in the cover, the water under it would exert an upward pressure on the cover that would be equal to the pressure that it would experience from the top to the bottom if it had to sustain a wooden cylinder of the same height as the stick, and having a base of the same dimension as the cover placed on the jar that contains the ocean. Furthermore, the stick and the wooden cylinder would exert exactly the same pressure on the sides of the jar, and the same force would be required to stop a hole in the sides or the bottom of the jar or in the cover.

XIII. If the water were to freeze, the ratio of the motion and the speed, which we considered above, would no longer obtain.[9]

The last two references to holes in the sides and the bottom of a jar, and to water freezing, are particularly striking. Did Pascal come across these passages in Mersenne's *Cogitata*? We have no indication that he did, but it was not necessary for him to have read the book to have an idea of its contents. He belonged to the inner circle of Mersenne's friends, and could have heard these ideas mentioned in informal conversation. But whether Pascal was inspired by Mersenne or not, their scientific style is profoundly different. Mersenne is fond of startling descriptions and apparent paradoxes. Whereas Pascal contrasts the effective weight of one pound and a hundred pounds, Mersenne would have us compare the weight of one pound of water with the weight of the whole ocean. But as we shall see later, Pascal could also seek dramatic effect when he wanted to drive a point home.

---

[9] From Mersenne's *Cogitata Physico-Mathematica*, part entitled *Ars Navigandi, Hydrostaticae Liber Primus*. Paris, 1644, pp. 227–229, quoted in Pierre Duhem, "Le principe de Pascal," *Revue des Sciences Pures et Appliquées* 16 (1905), pp. 601–602, where the passage is translated into French.

Why Liquids Weigh in Proportion of Their Heights

The experiments in the opening chapter of the treatise *On the Equilibrium of Liquids* showed that a fine thread of water could balance a heavy weight. "It remains to demonstrate the cause of such multiplication of force," writes Pascal, and in the next chapter he proceeds to do so with the aid of a receptacle with which we are now familiar (see Figure 1, illustration VII). This time, the opening of the larger cylinder is a hundred times as large as the other. Tightly fitting pistons are adjusted to each of the two cylinders so that, whatever the relative areas of these openings, when the forces exerted on the pistons are in the same ratio as the areas, they balance one another. The analysis is livened by replacing mere weights with human exertion. "One man pressing on the smaller piston," Pascal writes, "will exert a force equal to that of one hundred men pressing on the larger, and will exceed that of ninety-nine men doing the same." Pascal is speaking rhetorically, of course. He did not have a group of a hundred assistants to hold the piston down! He now concludes: "Hence a vessel full of water is a new principle of mechanics and a new machine for multiplying forces to any degree we might wish." This new machine (our hydraulic press) exhibits the same law as the old machines such as the lever, the wheel, and the pulley, namely "the distance traversed increases in the same proportion as the force."[10] In other words, the principle of simple machines applies to hydraulics, and Pascal proceeds to show this with the aid of three demonstrations that are based on three different principles, namely (1) *the principle of virtual displacement* (known to Archimedes), (2) *the principle of hydrostatics* (Pascal's original contribution), and (3) *Torricelli's principle concerning the center of gravity*. We shall consider each in turn.

Pascal's First Demonstration

In his first demonstration that the distances traveled are inversely proportional to the forces, Pascal begins by noting that it is the same thing to lift a given weight through a given distance as to lift ten times that weight through one-tenth the distance. Therefore, when two weights are so ad-

---

[10]*On the Equilibrium of Liquids*, *Œuvres de Pascal*, II, p. 1045; Spiers trans., p. 6.

justed that neither can descend any distance without causing the other to ascend a certain distance, the second distance being to the first as the first weight is to the second, there is no reason why one weight should descend rather than the other. Therefore, the two weights are in equilibrium and no motion takes place. This is what was later called *the principle of virtual displacement*. Now consider, says Pascal, a man pressing down the piston a distance of one inch in the smaller cylinder (see Figure 1, illustration VII). The displaced water presses on the other piston over an area that is one hundred times greater and, hence, will only raise it by one-hundredth of an inch. "Thus," concludes Pascal, "the distance is to the distance as the force to the force, and such may be taken as the true cause of the effect. It is evident that it amounts to the same thing whether we make one hundred pounds of water move through one inch, or one pound of water move through one hundred inches."[11] Pascal's argument is cogent, but it would have been even easier to follow if the concept of *work*, namely the product of the force that acts on a body and the distance through which the body moves, had been available to him, and had been expressed in symbols such as $W = Fs$. Pascal saw that when a body is lifted vertically with a uniform speed through a height h, the force exerted upward is equal to the weight W of the body, and the Work done is equal to Wh, the product of the weight by the height through which it is raised. But Pascal uses the same word, *force*, for both Wh (*work*) and W/A (*pressure*). Wh is meant when he writes, "one pound has as much force to move one hundred pounds through one inch as one hundred pounds has to move one pound through one hundred inches"; W/A is intended when he discusses the *force* required to stop the plunger from falling out.

Virtual Displacements and Virtual Velocities: Galileo and Descartes

Before examining Pascal's second demonstration, it may be interesting to clarify his contribution by contrasting it with the somewhat different ways that Galileo and Descartes went about the same kind of task. Galileo, throughout his treatment of machines, uses the principle of *virtual ve-*

---

[11]*Ibid.*, pp. 1045–1046; Spiers trans., pp. 6–7.

locities, that is, he considers the relative speed with which displacements are made.[12]

Here is how Galileo applied the principle of virtual velocities to an apparent hydrostatic paradox. Water is poured into a large vessel EIOF connected to a small tube ICAB (see Figure 2) such that the level is the same in both tubes. The water is stationary, but because the water in the large tube is much heavier, let us *imagine* that it is depressed to QO. What would happen to the water in the small tube in such a case? It would have to rise to the height of AB and, as Galileo points out, "the rise LB would be greater than the fall GQ by as much as the breadth of the vessel CD exceeds the width of the tube LC. But since the moment of the speed of motion in one body compensates that of the weight in the other, what wonder is it, if the swift rising of the small quantity of water CL resists the very slow descent of the large quantity GO."[13]

What Galileo compared is the *moment of speed* with the *moment of weight*, and we can summarize his explanation as follows: if we consider the *virtually* simultaneous rise and fall of the water in the tube and in the vessel, we can argue that equilibrium (and hence rest) is obtained when the ratio of the speeds of rise and fall is inversely proportional to the amount (and hence the weight) of water in the tube and in the vessel. Galileo did not suggest how to determine whether the speeds are effectively as he states. Indeed this is not possible since the speeds are virtual. What can be measured is the weight of the water, the surface area of the tube and vessel and, significantly, the *"rise LB"* and the *"fall GQ."* Galileo's warrant

---

[12] The expression *virtual velocities* may have originated with Johann Bernoulli, "Discours sur les lois de la communication du mouvement," in *Opera Omnia*, edited by J. E. Hofmann, 4 vols., Geneva, 1742, facsimile reprint Hildesheim: Georg Olms, 1968, vol. III, p. 23. See Ernst Mach, *The Science of Mechanics*, trans. by Thomas J. McCormack. La Salle, Ill.: Open Court, 1960, pp. 63–78.

[13] Galileo, *Discourse on Bodies that Stay Atop Water or Move in It* in *Opere*. Florence: Barbera, 1899–1909, vol. IV, p. 78. English trans. in Stillman Drake, *Cause, Experiment and Science*, 1981, p. 50. See William R. Shea, *Galileo's Intellectual Revolution*. London: Macmillan, 1972, pp. 14–48. The term *moment* originated among mathematicians investigating centers of gravity as a term denoting in effect the product of a weight and the length of a lever arm. This is still called *moment* (*moment of force*, *static moment*, or *torque*) and is defined as the product of the force and the perpendicular distance from the axis to the line of action of the force. Galileo added to the purely static concept the idea of a very small (or "virtual") moment, foreshadowing the concept of momentum as the product of a mass (for Galileo, a weight) and a velocity (for Galileo, a speed). See Richard S. Westfall, "The Problem of Force in Galileo's Physics" in Carlo L. Golino (ed.), *Galileo Reappraised*. Berkeley: University of California Press, 1966, pp. 67–95.

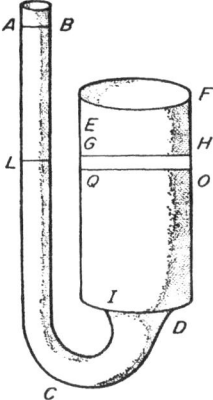

FIGURE 2   *Galileo's illustration of an apparent hydrostatis paradox.*

is the explanation of equilibrium in the *Questions of Mechanics*, a treatise that was generally ascribed to Aristotle. When we move a lever in which the fulcrum is not at the center, the longer arm describes a greater circle than the shorter arm. Hence, since both ends of the lever move *in the same time*, the extremity of the longer arm moves *faster*. If the weights at the ends of the lever balance, there is no motion because the product of the weight and the speed at one end is equal to the weight and the speed at the other. If we write m for weight and v for speed, we have mv, the modern formula for what we know as *momentum*. We have slipped from *moment* (or *torque*) to *momentum*. Since both ends of the lever move in identical time without acceleration, the same result is obtained whether one uses the *virtual displacements* of the two weights or their *virtual velocities*. But there is a crucial difference between the two concepts. Pascal, following Simon Stevin, strongly objected to reasoning from speed when no body actually moves.

Descartes on Mechanical Principles

We can also compare Pascal's approach with the one that we find in the short treatise on simple machines that Descartes wrote in 1637, shortly after completing his *Discourse on Method*. Here is how Descartes states his main principle: "These machines are all based on one principle alone, which is that the same *force* which can raise a weight, for example, of a hundred pounds to a height of two feet can also raise one of 200 pounds to the height of one foot or one of 400 to the height of a half foot, and so with others, provided it is applied to the weight." In a version sent to Mersenne, he expressed the principle in more general terms:

> The demonstration depends on only one principle that is the general foundation of statics, namely that the force required to raise a heavy body to a certain height is neither greater nor smaller than the one required to raise a lesser weight to a height that is greater by as much as the weight is smaller, or to raise a heavier weight to a height that is proportionally less.[14]

When considering simple machines, Galileo had used velocity and displacement interchangeably, but Descartes rejected the identification of virtual displacements with virtual velocities, and insisted that displacements alone could serve. Reference to velocity in this context he saw as an error that is more dangerous because it is more difficult to recognize, "for it is not the difference of velocity that determines that one of these weights must be double the other, *but the difference of displacement.*" Descartes realized that we do not need a force that is exactly double to raise a given weight twice as swiftly, but that if we want to raise it, with the same speed, twice as high, then a force that is exactly double is required.[15]

Descartes distinguished between a *two-dimensional force* used to raise a weight to a certain height and a *one-dimensional force* that supports a weight (for instance, a nail in a wall). Consideration of velocity would have required the addition of a third dimension whose exclusion he justified on the grounds that before we can discuss velocity intelligently, we must understand weight, and to understand weight we must understand the whole system of nature. Descartes' problem with velocity can be seen in his reply to Mersenne, who contended that if a given force can raise a weight a certain distance in one interval of time, then twice that force will

---

[14]Descartes, *Œuvres*, edited by C. Adam and P. Tannery. Paris: Vrin, 1964–1974, vol. I, pp. 435–436; vol. II, p. 228. The treatise, which was only published posthumously, was sent to Constantin Huygens, the father of the great Christiaan, on 6 October 1637 (*ibid.*, I, pp. 435–447), and to Marin Mersenne on 13 July 1638 (*ibid.*, II, pp. 222–245). Additional comments were added in a letter, also to Mersenne, on 12 September 1638 (ibid., II, pp. 352–362).
[15]Letter to Mersenne, 12 September 1638, in Descartes, *Œuvres*, vol. II, p. 354.

raise the weight twice as high in the same interval. Descartes thought that an experiment could show that this was not the case. "Take a balance that is in equilibrium," he writes,

> and place in one tray the least weight that will make it turn. It will turn very slowly, whereas if you put twice as much weight in it, it will turn more than twice as fast. By contrast, take a fan in your hand without employing any force beyond what is necessary to support it, and you will be able to raise and lower it at the same speed with which it would fall by itself in the air if let go. But in order to raise and lower it twice as fast, you will have to employ a force that is more than double the first since that one was nil.[16]

The product of size times velocity, derived from the law of the lever, bedeviled mechanical discussions. It was valid for virtual, but not for real velocities. Speed and distance can be considered as equivalent only in the case of the lever where a mechanical connection insures that the bodies at both ends move for the same time and in which, because of equilibrium, the motion involved is virtual, not accelerated.

Pascal's Second Demonstration and His Hydrostatic Principle

In the light of our discussion of the work of Galileo and Descartes, we are now in a better position to appreciate the originality of Pascal's second demonstration that the principle of simple machines applies to hydraulics. Pascal's original insight was that "the water in these tubes behaves exactly as would pistons of the same weight." Whereas a mechanical connection in the form of a rigid bar or an analogous device is found in simple machines, in the case of the hydraulic press it is the water that transmits the pressure equally throughout the entire liquid. As Pascal puts it: "It is impossible to move one weight without moving the other because of the continuity of the water between the piston." This is what leads him to develop his second and most original demonstration of the law that governs the functioning of the hydraulic press. It differs from the other two inas-

---

[16]Letter to Mersenne, 2 February 1643, *ibid.*, vol. III, p. 614. On the notion of force in the seventeenth century, see Richard S. Westfall, *Force in Newton's Physics*. New York: Elsevier, 1971.

much as it does not appeal to any of the more general laws of statics but explains the transmission of pressure in a liquid enclosed in a vessel as a result of the continuity and fluidity of water. For instance, if a piston carrying a weight of one pound is placed on a container full of water that has an opening, one inch in size, the weight will press against all parts of the vessel generally, because every part, one square inch in area, is subjected to the same pressure as if it were pushed by the weight of one pound.[17]

It follows immediately that if a second opening is made in the vessel, the force that is to be applied at this opening is to the force applied to the first opening as the area of the second opening is to the first. It does not matter whether the *pistons* are solid or liquid. What is essential is that the matter that extends from one opening to the other be a liquid, "for this is the true cause of the multiplication of the force." Stevin had considered the pressure due to the weight of the liquid, but it had not occurred to him to ask whether a confined liquid could be subjected to an additional pressure by the application of an external force. If he had, he might have found the principle of the hydraulic press, and been the first to state that *the pressure applied to an enclosed fluid is transmitted equally throughout the fluid*. But in order to arrive at that result, Stevin would have had to think, as Pascal did, of connecting two cylinders, one of small cross-sectional area, a, and the other of large cross-sectional area, A, each fitted with a piston (see Figure 3). By exerting a force, F, on the small piston, an additional pressure ($P = F/a$) is produced. This pressure is transmitted throughout the liquid and hence acts on the larger piston of area A and weight W. Hence, $W = P/A = F A/a$, and $W/F = A/a$. The weight that can be lifted by a hydraulic press is determined by multiplying the force F by the ratio of the areas of the two pistons. Pascal saw this clearly, but he lacked convenient symbols for force, pressure, weight, and area, and thus could not write the simple physical equations that contribute so much to ease of comprehension.

Pascal's Third Demonstration: The Center of Gravity

Pascal's third explanation of the hydraulic press is more technical. It rests on the principle that *a body never moves by its own weight without low-*

---

[17]*Treatise On the Equilibrium of Liquids*, chapter 2, *Œuvres de Pascal*, II, p. 1046; Spiers trans., pp. 7–8.

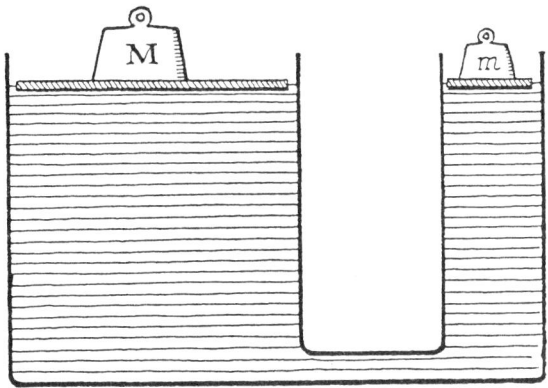

FIGURE 3 *Illustration of the transmission of pressure throughout a liquid.*

ering its center of gravity, something that Torricelli had already used to demonstrate the law of the lever.[18] The extension to the hydraulic press was Pascal's own idea, and his proof that the two pistons are in equilibrium runs as follows: the common center of gravity of the pistons is clearly located at the point that divides the line joining their individual centers of gravity at a distance inversely proportional to their weights. Suppose now that the pistons were to move. Then the distances they would cover would be inversely as their weights. But if we take their common center of gravity in this second situation, we will find it precisely in the same point as before, for it will always be located at the point that divides the line that joins their individual centers of gravity in the proportion of their weights. Now comes a *reductio ad absurdum*: "Therefore the two pistons, considered as a single body, will have moved without lowering their common center of gravity, and that is contrary to the principle. Therefore they cannot move, and, therefore, they will remain at rest, that is to say in equilibrium, which is what was to be proved."[19]

---

[18]*Ibid.*, II, p. 1046; Spiers trans., p. 8. See Pierre Duhem, *Les origines de la statique*, 2 vols. Paris: Hermann, 1906, vol. II, p. 2. The reference is to Torricelli's *De motu gravium naturaliter descendentium et projectorum* published in his *Opera Geometrica*, Florence, 1644, book I, p. 99. Pascal praised Torricelli for his "work in geometry that goes beyond anything in Antiquity" (letter to Antoine de Ribeyre, 12 July 1651, *Œuvres de Pascal*, II, p. 809).

[19]*Ibid.* p. 1047; Spiers trans., p. 9.

Pascal did not wish to convince only those who could follow a geometrical demonstration, and he immediately reverted to a more intuitive approach to drive home the notion that "liquids weigh according to their height." For instance, the rectangular container full of water in Figure 1, illustration VI, is connected to a small and a large cylinder. A piston is placed inside the large cylinder, and water is poured into the smaller one. The water and the piston are in equilibrium if their weights are to each other as the areas of the openings. What is essential is that there be a *continuous fluid connection* between one opening and the other. To make this clear, Pascal has us return to the apparatus described in Figure 1, illustration V. As we already know, even if the water weighs only one ounce, it will balance a hundred-pound weight, namely the weight of the water contained in the first cylinder, which has the same height and is cylindrical. Now let us imagine that the water in the slender tube in the upper part of the apparatus freezes, while the water in the bottom part remains liquid. It will still take one hundred pounds to counterbalance the weight of the ice and water. But if the water at the bottom freezes (whether the rest freezes or not), one ounce will now suffice to balance it. "From this," concludes Pascal, "it is clearly evident that it is the fluidity of the substance by which one opening communicates with the other that is the cause of this multiplication of force. The fundamental reason is, as we have said, that a vessel full of water is a mechanical machine for multiplying force."[20] The pressure between two columns of liquid, free to move vertically upward, is communicated equally in all directions. This basic principle of equilibrium for liquids depends on a fact of everyday observation. If two columns of the same liquid are in direct connection and both open to the outside air, they are at rest only when the height of each is the same. For practical purposes, height in this context usually means vertical distance above the ground, but if we consider the water of the oceans, it is evident that vertical height implies distance from the center of the earth.

Pascal's demonstration was not completely unheralded. In a letter to Giovanni Paolo Capra, included in his *Diversarum Speculationum Liber* (1585), the Italian physicist, Giovanni Battista Benedetti (1530–1590), considers two connected cylinders of different cross-sectional areas. He replaces the column of water by a piston of the same weight, first in the narrow cylinder, then in the wider one. If he had made this substitution in both cylinders at the same time, he would have been the unequivocal in-

---

[20]*Ibid.* p. 1048; Spiers trans., p. 10.

ventor of the hydraulic press. The French historian Pierre Duhem, who always looked for direct influences even where independent results were more likely, makes three interesting comparisons. First, Pascal, like Benedetti, saw the need of a tightly fitting piston. Second, Benedetti explained that unequal quantities of water in connected cylinders of different cross-sectional area are in equilibrium because the weight is divided in the ratio of the surface at the bottom of the vase or, as Pascal expressed it, "the water is equally pressed upon under the two pistons, for though one of these is one hundred times as heavy as the other, it is, on the other hand, in contact with an area a hundred times greater." Third, Benedetti explained how a pump works by applying the law that governs the equilibrium of water in communicating cylinders. Pascal went about it the other way round: he started from the principle of the hydraulic press and proceeded to deduce the law of hydrostatical equilibrium in two communicating cylinders or vases.[21]

Examples of the Equilibrium of Liquids

In Chapter III of the treatise we are considering, Pascal illustrates the equilibrium of liquids with a receptacle (Figure 1, illustration VIII) similar to the one in illustration VII but with the pistons removed. Water is poured into the rectangular vessel and rises in the two cylinders. The two columns of water are in equilibrium because they stand at the same height and behave like two pistons whose weights are proportional to the openings. This explains why water can mount as high as its source.

For the next experiment, Pascal would have us go to a river and lower a tube, curved into a U shape and full of mercury, into the water is such a way that its upper end remains above the surface (see Figure 1, illustration IX). When the plug is removed from the bottom, the mercury runs out until it reaches a height that is equal to one fourteenth of the height of the

---

[21]See Pierre Duhem, "Le principe de Pascal," *Revue Générale des Sciences Pures et Appliquée* 16 (1905), pp. 609–610. Pascal may have been indirectly influenced by Simon Stevin's *Practice of Hydrostatics*, a work in which he applies in a number of ways the principles established in his *Elements of Hydrostatics*, first published in Flemish in Leyden in 1586 and included in his *Mathematical Memoirs* of 1608. This collection was translated into Latin by Willebrord Snell, and into French by Albert Girard as *Les Oeuves Mathématiques de Simon Stevin de Bruges*. Leyden, 1634. The *Practice of Hydrostatics* appears as Book V of *De la Statique* in Girard's edition.

FIGURE 4 *Forces Acting on Vessels of Different Shape and Size Filled to the Same Height.*

water above the curved end. If the tube is sunk in deeper, the mercury will rise because the weight of the water increases; if the tube is raised, the mercury will fall because its weight is now greater than that of water. If a straight, fourteen-feet long tube (see Figure 1, illustration X) is used, so long as the upper end is out of the water, the mercury will sink to a level of one foot above the lower end, and will be held there by the weight of the water. "It is obvious," comments Pascal, "that, if there is no quicksilver in the tube, the water will enter it and rise to a height of fourteen feet. Therefore, since one foot of quicksilver weighs as much as the fourteen feet of water that it replaces, it will be in equilibrium with the water that balances fourteen feet of water in the tube."[22] Such experiments could have been carried out in a tank of moderate depth with much shorter tubes. When Pascal gives the distance below the surface of the water to the curved end as 14 feet, he is using a rhetorical device to render more striking results arrived at by simpler and less spectacular means.

At this point, it might be useful to take a step beyond Pascal and offer a non-technical but more modern analysis of the forces acting on vessels of different shape and size that are filled with water to the same height (see Figure 4). *First Case*: in a regular cylinder, the weight of the cylindrical column of water is the only force acting downward on the bottom. It exerts a force on the walls of the cylinder, which, in turn, exert equal forces in the opposite direction on the water. But since the walls are vertical, the forces are horizontal and have no components in the vertical direction. *Second Case*: if the walls of the vessel are made to flare inwards, they exert forces that have vertical components directed downwards. Although the weight of the water in this vessel is less than in the regular cylinder, it can

---

[22]*Ibid.*, p. 1051; Spiers trans., pp. 13–14. The specific gravity of mercury is 13.6.

be shown that the vertical component of the force exerted by the walls is just equal to the difference between these weights. Hence the force on the bottom of this vessel is equal to that on the bottom of the cylindrical vessel. *Third Case*: when the sides of the vessel flare outward, they contribute a force that has a vertical component upward to support the weight of water that is in excess of that of the cylinder in the first case. Hence, the force on the bottom of this vessel is also equal to that of the cylindrical column of water of the same height.[23]

Liquids and Solids in Equilibrium

Chapter IV of *On the Equilibrium of Liquids* illustrates how water exerts pressure in all directions. To demonstrate upward pressure, Pascal recommends a twenty-foot-long tube with a funnel-like end into which a cylinder of copper is fitted so that it can slip out without allowing any water to pass (see Figure 1, illustration XI). The tube is lowered into a river, once again, so that the small end of the tube remains above the level of the water. The cylinder is held in place by the pressure of the water from below when it is sunk to a depth that is nine times the height of the cylinder, because copper is nine times as heavy as water. If the tube is lowered to a greater depth, it becomes more and more difficult to wrench the cylinder from the funnel-like bottom because of the greater weight of the water. If water is allowed to enter the tube, the cylinder falls by its own weight. If the lower end of the tube is curved so that the mouth points upward and is fitted with a wooden cylinder (see Figure 1, illustration XII), the cylinder does not rise but is forced down into the mouth. Finally, when the tube is turned sideways and immersed, the cylinder remains pressed into the mouth. This shows that an immersed solid is subjected to the same pressure from all directions.

Needless to say, none of these experiments could have succeeded unless the surface of the water was perfectly calm, which makes it most un-

---

[23]We are reminded of the way Simon Stevin proceeds in his *Elements of Hydrostatics* (partly translated in an appendix to *The Physical Treatises of Pascal*, pp. 135–158. The source is *Les Œuvres mathématiques de Simon Stevin de Bruges augmenté par Albert Girard*, Leiden, 1634). Stevin demonstrated that the bottom of a rectangular container full of water supports a weight that is equal to the weight of the vertical column of water just above. In the corollaries he deforms the shape of the container to show that the water presses from below as well as from above.

likely that they were carried out in a river. Here again, Pascal had worked in laboratory conditions, and he enlarged the scale of his actual experiments to render them more vivid. To understand why it was so important to stress that pressure is exerted in all directions, we must recall that Archimedes had considered that the weight of a fluid was exerted *vertically only*, and that this was believed well into the seventeenth-century. Giovanni Battista della Porta, for instance, thought he could show this with the following experiment. Take a vase that has a small opening on the side, and divide it into two equal parts with a vertical partition. Fill the vase with water and measure the distance the water shoots out when the stopper is removed from the opening on the side. Fill the vase once more, but this time remove the partition. You will find that the water goes the same distance, a decisive indication, according to della Porta, that pressure is only exerted in the vertical direction.[24] We can see from this how novel was Pascal's idea that pressure is exerted in all directions, and why he devised striking though impractical illustrations.

Bodies Wholly Immersed in Water

In Chapter V Pascal explains, with the aid of Archimedes' principle that a submerged body is buoyed up by a force equal to the weight of the fluid displaced, why a body can be completely submerged without sinking to the bottom (see Figure 1, illustration XV):

> Since the water covers all the sides to the same height, it will press upon them equally. The immersed body, consequently, will receive no particular impulsion towards any side any more than would a weather vane between two equal winds. But as the height of the water is greater at the bottom than at the top, it will obviously press the body more upward than downward; and since the difference between these heights of water is the height of the body itself, it will be readily understood that the water presses it upward and not downward, and with a force equal to the weight of a volume of water equal to that of the body.[25]

---

[24]Archimedes, *On Floating Bodies*, book I, postulate I in *The Works of Archimedes*, p. 253. Giambattista della Porta stresses this point in *I tre Libri de' Spiritali* (cap. 10, pp. 25–26, quoted in de Waard, *L'expérience barométrique*, p. 47, n. 1.

[25]*Treatise on the Equilibrium of Liquids*, *Œuvres de Pascal*, II, p. 1053; Spiers trans., p. 16.

This is as close as Pascal comes to the Principle that bears his name: *Pressure applied to an enclosed fluid is transmitted equally throughout the fluid*. He usually explains buoyancy, along Archimedean lines, as a case of equilibrium. A solid body is supported in water in the same way as if it were in one tray of a pair of scales, while the other tray was weighted with an equal volume of water. A body that is heavier than water, such as copper, sinks because its weight overbears that of the counterpoise; a lighter body, like wood, rises with the force by which the weight of the water exceeds its own; and a body of equal weight, such as wax, neither sinks nor rises.

Let us see how Pascal applies this to the explanation of why it is easier to lift a pail full of water when it is still submerged in a well, before it is taken out of the water. This is not because a given volume of water weighs less when surrounded by water than by air, says Pascal, but because the water acts as a counterweight when the pail is still under the water. In other words, Pascal does not account for the apparent loss of weight of the immersed body by saying, as in our modern textbooks of physics, that the weight of the body is reduced by the upward thrust of the water, but by comparing weights balanced at the ends of a lever. In like manner, Pascal explains that a shallow dish of metal floats in water because its shape displaces a volume of water that weighs more than the dish itself.[26]

Immersed Compressible Bodies

The last two chapters of the treatise *On the Equilibrium of Liquids* deal with the effect of liquid pressure on compressible bodies. Pascal takes special pains to explain, first, that bodies immersed in a fluid are compressed towards their center because a fluid transmits pressure in all directions, and second, that the effect of pressure exerted at all points on the surface of the body is different from the effect of the same pressure exerted on only

---

[26]*Ibid.*, pp. 1053–1055; Spiers trans., pp. 17–19. Pascal could have given the submarine as an illustration. Between 1620 and 1624, Cornelius Brebbel, a Flemish inventor, successfully manoeuvred a waterproof craft at depths of four to five meters beneath the surface of the Thames. This performance was much discussed by Mersenne and his circle. The modern submarine is equipped with tanks, which can be filled with water or emptied by means of a store of compressed air. When the thanks are quite full the total density of the submarine is slightly greater than water. The expulsion of a small quantity of water is then sufficient to make the submarine rise to the surface.

part of the surface. He has us return to the river for four additional experiments to illustrate these two points. In the first experiment, a bellows with a tube twenty feet long and with no other opening is immersed in water with its end protruding into the air (see Figure 1, illustration XIV). The bellows are difficult to open because of the pressure of the water on the wings, and the deeper the bellows, the harder it gets. If this experiment occurred in the open air, it might conceivably be attributed to the horror of the void, but under water no void is created, and the force at work is clearly the weight of the water.

In the second experiment, a balloon is attached to the end of a 20-foot tube, which is filled by way of the tube with mercury (see Figure 1, illustration XVI). Tube and balloon are then immersed so that the end of the tube sticks out above the level of the water in the air. The mercury in the tube rises to a certain height in the tube because it is pressed by the water at all points and pushed towards the mouth of the tube where there is no pressure. The mercury ceases to rise when the pressure that it exerts is balanced by the pressure exerted by the water without. Pascal does not give the actual height of the mercury and it is hard to see how he could actually have experimented with a 20-foot tube. The next trial is even more clearly a thought experiment. In Figure 1 above, we see in illustration XVII a man sitting at the bottom of a river and pressing against his thigh the end of a 20-foot tube whose upper end emerges above the surface of the water. His flesh rises "as if suction were applied to the spot," writes Pascal, "but it is evident that this swelling is not due to horror of a void, for the tube is open and there would be no swelling if there was only a little water in the tank." The cause is the weight of the water that presses on every part of the body, save at the bottom end of the tube. Perforate the tube and the swelling ceases.

A simpler experiment consists in blowing up a balloon very hard and immersing it in a tank full of water. The balloon does not change in apparent size. Yet if a finger is applied to one spot, the balloon is compressed at that place. This is because the rest of the surface of the balloon is not under this new pressure and so can accommodate what is displaced by the finger. This is confirmed by taking the balloon out of the water, and squeezing it between both hands. Try as we may to cover every part, there will always be some that slip out between our fingers, and at these points swelling will appear. Likewise, an inflated body (a football or a fish) that is plunged under water is not deformed because the pressure is equal on all sides. This is why when we swim under water we feel no pressure.

Aquatic Creatures

A small balloon full of air, a second balloon half full, and a fly are dropped in a tube full of water. When a piston is placed on the opening of the tube and pressed down hard, the flaccid balloon is visibly compressed, but the full balloon and the fly are not affected. The fly, which apparently suffers no damage, flies off as soon as it is released from its prison. Should an even more conclusive demonstration be desired, adds Pascal, we can remove the piston and fill the tube with a quantity of water that weighs as much. Even if the water above the fly freezes, so long as there remains just enough liquid to surround it on every side, the fly will not feel the weight of the ice any more than it felt the weight of the water. Indeed, the water in a pond could freeze to within one foot from the bottom without the fish feeling the weight of the ice any more than that of the water into which the ice eventually melts. So the determining factor is not whether the weight is solid or liquid, "the reason is solely that the animal is surrounded with water." "Let us then," concludes Pascal, "no longer give as the reason for this that water in water has no weight: for it weighs everywhere alike; or that it weighs otherwise than do solids, for all weights are alike." This is a polemical remark aimed at those who, like Galileo and Mersenne, maintained that air does not weigh in air or water in its own milieu.[27]

Thought-Experiments and Their Critic

Pascal does not clearly state that he actually performed all the experiments that he describes. In general, he merely says that if one does so and so, the result will be this or that. Robert Boyle, who was one of the first to attempt to replicate the experiments, objected that some of them could not be performed. In his *Hydrostatical Paradoxes* (published in 1666 but earlier presented to the Royal Society in May 1664), Boyle criticizes the recently published *On the Equilibrium of Liquids* on three counts:

> First, because though the experiments he mentions be delivered in such a manner, as is usual in mentioning matters of fact, yet I re-

---

[27]*Ibid.*, p. 1058; Spiers trans., p. 23. Galileo, in his early *De Motu*, argued against Aristotle's position that air has weight in its own place: "In their own places the elements are neither heavy nor light. For if a part of the water were heavy in water, it would sink. But it does not do this. And if it were heavy, how, when we swim under water, would we not feel the weight of so vast an amount of water?" (*Opere di Galileo*, I, p. 288).

member not, that he expressly says, that he actually tried them, and therefore he might possibility have set them down, as things, that *must* happen, upon a just confidence, that he was not mistaken in his ratiocinations. And of the reasonableness of this doubt of mine, I shall ere long have occasion to give an instance.

Secondly, whether or no Monsieur *Pascal* ever made these experiments himself, he does not seem to have been very desirous, that others should make them after him. For he supposes the phenomena he builds upon to be produced fifteen or twenty feet under water. And one of them requires, that a man should sit there with the end of a tube leaning upon his thigh; but he neither teaches us, how a man shall be enabled to continue under water, nor how, in a great cistern full of water, twenty foot deep, the experimenter shall be able to discern the alterations, that happen to mercury, and other bodies at the bottom.

And thirdly, these experiments require not only tubes twenty feet long, and a great vessel of, at least, as many feet in depth, which will not in this country be easily procured; but they require brass cylinders, or plugs, made with an exactness, that, though easily supposed by a mathematician, will scarce be found obtainable from a tradesman.[28]

In brief, Boyle complained that Pascal did not expressly state that he performed the experiments; that he failed to explain how a man could sit under 20 feet of water with a tube resting on his thigh; and that he underestimated the difficulty of finding a 20-foot-long tube, a sufficiently deep tank, and water-tight plugs.

Boyle also had problems with two other experiments. It appeared to him (but here he was indulging in his own thought-experiment!) that if the mercury in Figure 1, illustration XVII were to drop to one fourteenth of the height of the water above the curved end, it would fall so fast that it would flow out of the pipe. He did not attempt to carry out this experiment, but he repeated the one with a fly under water, only to find that it presently drowned. Boyle's strictures should be taken seriously, but he failed to recognize Pascal's aptness for dealing with practical problems in mechanics that is attested by his success in performing other experiments of considerable difficulty, and by his invention and production of the Calculating Machine, which required unusual ingenuity, skill, and patience.

---

[28]Robert Boyle, *Hydrostatical Paradoxes Made Out by New Experiments* (1666) in *Works*, ed. by Thomas Birch, 6 vols., London, 1772. Reprinted Hildesheim: Georg Olms 1965, vol. II, pp. 745–746.

William R. Shea

The Mechanics of Fluids: Pascal's Achievement

The system of hydrostatics developed in *On the Equilibrium of Liquids* led to a second treatise, *On the Weight of the Mass of Air*, where Pascal extended his results to the atmosphere and liquids in general. The two treatises were almost certainly intended as two sections of a single work. But before we examine this second treatise, let us summarize what has been achieved in the first. The most notable contribution is the deduction of the law governing the hydraulic press from three principles: the principle of virtual displacements, the principle that the pressure applied to an enclosed fluid is transmitted equally throughout the fluid (which is Pascal's own), and the principle that a body never moves by its own weight without lowering its center of gravity, something that Torricelli had grasped.

The connection between Pascal's own principle and the general rule that concerns the hydraulic press can briefly be expressed as follows. In quantitative terms, Pascal's principle states that if a unit force is exerted over a unit area of the surface of a liquid enclosed in a vessel that force will be transmitted throughout the liquid so that each unit area of the enclosing walls of the vessel will be subjected to unit force. It follows from this (and now we come to the hydraulic press) that if two forces are exerted over two different areas of the surface of an enclosed liquid, the forces being proportional to the openings, then the forces will be in equilibrium. Pascal showed that the shape of the vessel is a matter of indifference since, as he put it, "liquids weigh in proportion to their depth, not in proportion to their volume." Finally he deduced the laws of floating bodies that had been known since Archimedes and are based on the experience that when we raise ourselves while taking a bath we notice that our body seems to be lighter when immersed than it is outside the water. The downward pulling power on a body immersed in a liquid is therefore less than it would be in air. If a weight is immersed in water, a smaller weight hanging freely in air will balance it. The difference is equivalent to the weight of the water displaced, and a body therefore loses weight in a liquid by an amount equivalent to that of the same volume of liquid.

The reader who is familiar with the Newtonian terminology that became common a few years after Pascal may find it useful to express what he achieved with the aid of Figure 5, which shows two columns of a fluid contained in two tubes connected by a cross-piece so that the height, h, of the fluid in each is the same. The bore of each tube is uniform. The areas of cross-section (A and a) are different. The height of fluid above any hor-

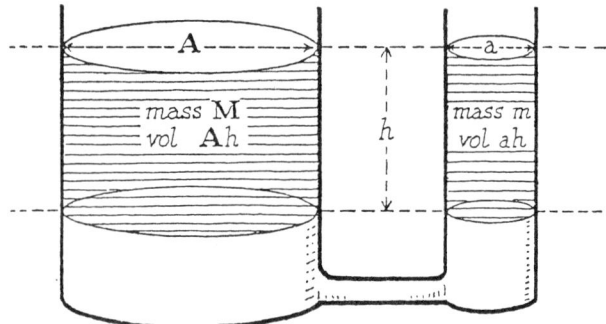

FIGURE 5   *Two columns of a fluid contained in two tubes connected by a cross-piece.*

izontal level in either column is the volume per unit area of cross section ($h = v \div a$), and the volume is proportional to the mass if the density in both columns is the same (since $m = ahd$). So if the height is the same in both columns, the mass per unit area ($m \div a$) is also the same in each column at the same level. Therefore, the weight (after Newton we would say the force of gravity) acting on unit area ($mg \div a$) is also the same in each column at the same level. Force per unit area is pressure. So two columns are balanced when the pressure at one and the same horizontal level is the same in both. Pressure is communicated equally in all directions, and equilibrium between two connected columns of the same fluid, if both are free to move vertically upward, is attained when both are at the same vertical height. This implies that the force on unit area is the same at any level in each; hence a column of large sectional area will support a proportionately larger weight than a column of smaller area. Thus, a small force acting on a column of small sectional area will cause a connected column of large sectional area to exert a proportionately larger force in the opposite direction. Notice that a large downward displacement of *m* would be necessary to produce a small upward displacement of M. Here the notion of *work* as the product of the *weight* and the *distance* through which it moves would have helped Pascal clarify the relationship with the lever, the pulley, and other machines. But his new understanding of hydraulics was sufficient to enable him to re-examine the old problem of the weight of the air in his treatise *On the Weight of the Air*, to which we turn in the next chapter.

CHAPTER 7

Submerged in a Sea of Air

The Treatise on the Weight of the Mass of Air

The weight of air had been discussed since Antiquity, notably by Aristotle, who found that an inflated bladder weighs more than an empty one. In the sixth century (nearly a thousand years later), the Greek commentator Simplicius repeated the experiment and claimed to have found that an inflated bladder weighs the same (not more!) than a deflated one. He believed that Aristotle must have got it wrong because some dampness was introduced into the bladder when air was breathed into it, and he concluded that air has no weight. This was generally accepted in the Middle Ages, although there was enough disagreement on the point to keep discussion alive.

Aristotle had also mentioned that a pint of water can cool ten pints of air, and some took this to mean that one measure of water is equivalent to ten measures of air and, hence, that the density of air compared to water is as 1:10.[1] During the Renaissance, Girolamo Cardano suggested that the ratio of the density of air to water could be determined by firing a small

---

[1] Aristotle, *On the Heavens*, book IV, chapter 4, 311b9–11; *On Generation and Corruption*, book 2, chapter 6, 333b22–25. One to ten is the value given by Robert Cornier in his letter to Mersenne of 15 November 1627, where he speaks of *density* (*Correspondance du P. Marin Mersenne*, I, p. 592). In the *Two New Sciences*, First Day, Galileo writes that water is not 10 times heavier than water, as "Aristotle seems to have believed," but about 400 times heavier (*Opere di Galileo*, VIII, p. 124).

cannon ball into the two different media. He claimed to have witnessed such an experiment and that the cannon ball traveled 65 feet in the air, but only a foot and a half in water during the same time interval. From this he concluded that the correct ratio was 1:50.

The problem of the weight of air was closely linked to the problem of weight in general, and to the effective weight of falling bodies in particular. Mersenne realized that heavy bodies fall at the same speed regardless of their weight, but he concluded, erroneously, that their motion is not accelerated. He offered as evidence that a cannon ball that was dropped from a fifty-foot building covered the first twenty-five feet as fast as the last twenty-five. When a correspondent, Jean Rey, queried his assertion on the grounds that acceleration is hard to measure, Mersenne replied, somewhat petulantly, that his experiment had been witnessed by "persons of quality," and that he "knew for sure that if you drop a piece of lead and a piece of charcoal from a high window, both fall as fast."[2]

Mersenne also experimented with air. He weighed an empty flask, and after heating it found that it had become less heavy. He plunged the flask still warm into a bucket of water, where it sucked in as much water as air had gone out. He then weighed the water and estimated that it was 255 times as heavy as air. We know today that he would have obtained better results had he sealed the opening of the flask when it was hot, weighed the flask again after it had cooled, and then broken the seal when the flask was under water, but this procedure did not occur to him.

The Weight of Air and Galileo

As we have seen in Chapter 2, Galileo estimated the weight of air at about 400 times less than water (the correct figure is 1:773), but he never grasped the connection between the weight of the air and the possibility of a vacuum. His friend Baliani recognized that water and air can be treated in a similar way and, hence, that he could reason from the effects of weight in one case to the effects of weight in the other. Baliani "imagined himself" at the bottom of a lake with ten feet of water above him, and

---

[2]Letter of Mersenne to Rey, 1 April 1632, *Correspondance du P. Marin Mersenne*, vol. III, p. 275. See Pierre Duhem, "Le P. Marin Mersenne et le poids spécifique de l'air," *Revue Générale des Sciences Pures et Appliquées* 17 (1906), pp. 769–782, 809–817.

he asked why he did not feel its weight. It is this approach that Torricelli and Pascal were to follow.

In 1639 Mersenne published a paraphrase of Galileo's *Two New Sciences*, in which he states that his own experiments in Paris indicated that lead covers 12 feet in water in the time that it falls 48 feet in the air. Taking Galileo's value of 400:1 for the ratio of the density of water to air, and the value of 12:1 for the relative density of lead compared to water, Mersenne concluded that lead was $400 \times 12 = 4800$ times lighter than air. Since lead had fallen 48 feet in the air in a given period of time, Mersenne inferred that its speed in water during the same interval should be reduced by $4800 \div 400$, namely by a factor of twelve. Now one twelfth of 48 is 4, so lead should have covered in water a distance of $48 - 4 = 44$ feet, not 12 feet as he had found. Rather than query his results and make the experiment over again, Mersenne preferred to claim that it was Galileo who had gotten his data wrong.[3]

In the Netherlands, Isaac Beeckman recognized like Baliani that we live in a sea of air. He showed Pierre Gassendi, who visited him in 1624, "that air has weight, that we are equally pressed by it on all sides and hence do not feel it, and that it is the cause of what is called the flight from the void."[4] We could not have a more pithy or correct formulation. Descartes was familiar with Beeckman's position, but he took an entirely different tack. Replying to a correspondent who had asked why mercury does not flow out of an inverted tube, he wrote: "Imagine air to be like wool and the ether in its pores to be like whirlwinds moving about in the wool." Descartes used the sketch in Figure 1 to show that the air at the bottom is pressed down by the layers of air on top, and is therefore much heavier although this goes unnoticed: "If we push the air at E towards F," he wrote, "the air at F will move in a circle in the direction GHI and return to E, so that its weight is not felt, just as the weight of a rotating wheel is not felt if it is perfectly balanced on its axle."[5]

---

[3] *Les Nouvelles Pensées de Galilée*, Paris, 1639. Facsimile Paris: Vrin, 1973 (with notes by P. Costabel and M.-P. Lerner), pp. 67–69. When he published his *Cogitata Physico-Mathematica* in 1644, Mersenne provided a new ratio for the weight of water and air, namely 1356:1 (quoted in P. Duhem, "Le P. Marin Mersenne et la pesanteur de l'air," p. 782).

[4] Quoted in *Œuvres de Descartes*, vol. X, p. 37, note b. Cornélis de Waard published several pages from Beeckman's *Journal* that show his clear understanding that we live in a sea of air (*L'expérience Barométrique*. Thouars: Imprimerie Nouvelle, 1936, pp. 149–162).

[5] Letter of Descartes to a correspondent, probably Henri Reneri, 2 June 1631, *Œuvres de Descartes*, vol. I, p. 206.

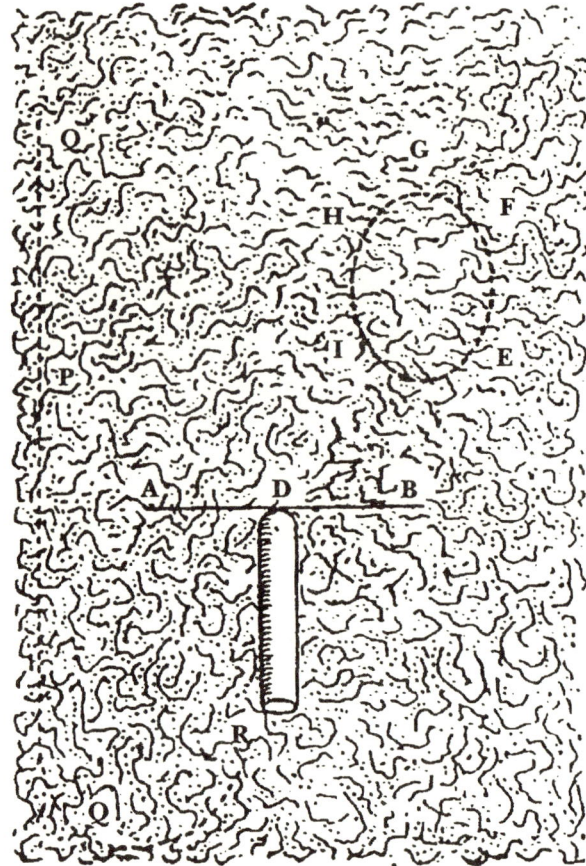

FIGURE 1 *Descartes' sketch to show how air is pressed down by the layers on top.*

Descartes thought that a football becomes hard when air is pumped into it because the particles of air are compressed and behave like little *springs* that try to regain their original shape. Although Descartes missed the key notion that air exerts *pressure* in all directions, he called for more experiments.[6] In a rare political aside, he lamented that Cardinal Riche-

---

[6]Letter of Descartes to Mersenne, 19 January 1642, in Descartes, *Œuvres*, III, pp. 483–484. Descartes made one important discovery in the process: he realized that water condenses as it cools and then expands after reaching a certain point. The density of water is indeed at a maximum at 4°C. But this was an age when experimentation was in its infancy, and Descartes

lieu, who had just died, had not left part of his fabulous fortune "to investigate the properties of bodies, something that would be much more useful than going to war."[7]

Pascal's Treatise

By 1651, when Pascal drafted his treatise *On the Weight of the Mass of Air*, it was generally recognized that air has weight. "A football is heavier when inflated than when empty, which is proof enough," he declares. Now Aristotelians, still numerous in educational establishments, did not deny the evidence. They queried its relevance. *Pure* air, they said, is *igneous* air, a substance that is naturally light and goes straight up. If the air that is compressed into a football makes it heavier, this merely shows that it contains impurities. To this ploy Pascal replied that he wanted to deal with empirical reality, not theoretical constructs: "I am not acquainted with *pure* air, and I doubt whether it could be found. In this treatise, I speak only of the air that we breathe, regardless of whether it has component elements or not. Whether it be a simple substance or a mixture, this is what I call air, and I declare it to have weight. This is not controversial and nothing more is required for what follows."[8]

If we were to ascend a mountain with a balloon that is half inflated, it would grow in size because the atmospheric pressure decreases. "Now since experience is far more convincing than argument," adds Pascal, "I have no doubt that everyone will want to see this confirmed by experiment." But what if the half-inflated balloon did not expand as it was carried up the mountain? The question is merely rhetorical, of course. Pascal

---

offered elsewhere, as an equally valid observation, something that is perfectly spurious: "Experience also shows that water that has been kept on a boil for a long time freezes more rapidly, and the reason is that those of its parts that can least cease to bend evaporate while it is being heated" (*Les Météores*, Descartes, *Œuvres*, VI, p. 238). The "reason" expressed in the second part of the sentence provides the rationale for the claim in the first, allegedly experimental, part.

[7]Letter of Descartes to Mersenne, 4 January 1643, *Œuvres de Descartes*, III, p. 610.

[8]*Treatise on the Weight of the Mass of Air*, *Œuvres de Pascal*, II, p. 1062; Spiers trans., pp. 26–27. A more scathing variant of this declaration is found in Pascal's *Letter to Le Pailleur*: "Fr. Noël bases his invisible matter on experiments that are wrong in order to explain experiments that he has not understood. Hence it was right that he should have used a material that can neither be seen nor understood to account for experiments that he has neither seen nor understood" (*Œuvres de Pascal*, II, p. 574).

knew the outcome, and he had himself carried a flabby balloon to the top of the Puy-de-Dôme. It had swelled of its own accord as he ascended, and when he brought it down it gradually shrunk by the same degrees. At the foot of the mountain, it had resumed its former condition.[9]

The Weight of the Air Expels the Horror of the Void

Pascal next proceeds to examine the phenomena that had been allegedly explained by the horror of a void. The first is the cohesion of polished bodies in close contact, for instance two marble slabs. In the first century B.C., Lucretius had already argued that when the slabs are pulled apart, air cannot instantly rush in to fill the empty space and a vacuum must occur for a brief time. Galileo offered the same example in the First Day of his *Two New Sciences*.[10] Some replied that the two slabs could be separated gradually by lifting the end of one of them, thereby creating a slight angle between the surfaces. According to them, this allowed air to slip in and prevent the formation of a vacuum. Others argued that perfectly smooth and parallel surfaces do not exist, and that some air is always trapped between two slabs of marble or two sheets of glass. When the surfaces are pulled apart, the thin film of air expands before any vacuum can be formed. They saw no dire effects following from the formation of a momentary void, whereas for Galileo there was no such danger. Nature sought to fill a vacuum as soon as it was produced, but its occurrence was not calamitous. In this way, nature's horror of a void became simply aversion, well on its way to becoming mere indifference.

Pascal analyzes the phenomenon in the same way as Lucretius and Galileo, but he goes beyond them and sees the cohesion of polished bodies as "a particular case of the general rule that applies to the pressure exerted by fluids on a body when they touch on one side and not on the other." When two smooth surfaces are in close contact and the uppermost is raised, the lower remains suspended to it because the air below pushes

---

[9]*Ibid.*, pp. 1064; Spiers trans., pp. 30–31. Pascal does not tell us that he ascended the Puy-de-Dôme himself but we know that he did from a letter that Pierre Gassendi wrote to François Bernier on 6 August 1652 (*Œuvres de Pascal*, II, p. 933). Mesnard gives May 1649 as the date of the ascent (*ibid.*, II, p. 647). Pascal does not say whether the balloon was held by someone who walked up the mountain or by someone who was sitting in a carriage.

[10]Lucretius, *De rerum natura*, book I, lines 385–397; Galileo, *Two New Sciences*, *Opere di Galileo*, vol. VIII, p. 59; Drake trans., p. 20.

it up.[11] Since atmospheric pressure decreases with altitude, Pascal declares that polished plates are easier to separate on a steeple than at street level. This is an experiment that is difficult to make because great care is required in placing the plates together, and there is no observable force until they are in very good contact. We may wonder whether Pascal actually made it.

Across the Channel, Robert Boyle "conjectured" that if the plates of marble were suspended in a receiver from which air was pumped out, the lower one would fall as the pressure decreased.[12] He set out to test this with two small marble squares of the same length and width, one half an inch thick and the other one-quarter. Immediately, difficulties arose: they could not be ground smoothly enough to adhere in the air above a minute or two, a much shorter time than it took to exhaust the receiver. Boyle moistened the interior surfaces of the squares with alcohol and got them to stick. He then attached a weight of four ounces to the narrower plate and lowered the pair by means of a string into the receiver, and started pumping. All to no purpose: the marble squares did not separate. Boyle suggested two reasons for the failure: first, the alcohol might have acted as a glue rather than merely as a means of excluding intervening air and, second, the pump might have leaked and air allowed in by the porousness of the sealing material or the looseness of fit between sucker and cylinder. This enabled Boyle, as Shapin and Schaffer point out, "to recuperate the apparent *failure* as an actual success."[13]

Boyle mentioned another experiment in which the water in a barometer inside the evacuated receiver could not be made to fall below one foot on evacuation. Might this not show, he asked, that air could exercise enough pressure to keep two smooth marbles cohering? But Boyle was not satisfied, as we can see from his wistful comment that students of the Jesuit Nicolò Zucchi (whom he calls an *expert*) were unable to lift a brass plate that stuck to a marble table. That Boyle, the English champion of experimental science, should appeal to a Jesuit to illustrate a result that he could have replicated himself is surprising. The reason is clearly that the production of the phenomenon was not easy and commonly failed.

---

[11]*Treatise on the Weight of the Mass of Air, Œuvres de Pascal*, II, p. 1071; Spiers trans., pp. 38–39.
[12]Robert Boyle, *New Experiments*, p. 69.
[13]Shapin and Schaffer, *Leviathan*, p. 191.

Boyle warned would-be experimenters that they should arm themselves with patience: "We have never yet found," he writes, "any sort of experiments, wherein such slight variations of circumstances could so much defeat our endeavour; which we therefore mention, that in case such experiments be tried again, it may be thought the less strange if others be not able to do as much at the first or second, or perhaps the tenth or twentieth trial, as we did after much practice had made us expert in this nice experiment."[14] The major problem was securing slabs of marble or sheets of glass of sufficient smoothness: irregularities in their surface would permit the entry between them of a small amount of air, and they would quickly fall apart.

Stubborn Pistons

The second phenomenon that Pascal investigated is the raising of water by suction. Let us imagine, he says, a syringe with a ten-foot-long piston, hollow throughout and fitted at its base with a valve opening downwards but not upwards, so that it cannot suck water above the level of the liquid because air enters freely through the hollow of the piston. The mouth of the syringe is plunged into a vessel of mercury, and the whole apparatus is placed in a tank full of water so that the top of the piston just emerges. When the piston is drawn, the mercury rises behind it, as though adhering. This does not happen out of fear of a vacuum, but because water presses the mercury from every side except at the mouth of the syringe. This can be confirmed, says Pascal, by replacing the water by sand, and bearing down on it with our hands. The mercury will be driven up the syringe until it is high enough to balance the extra pressure: "Thus it is clear that the rising of water in syringes is but a particular case of the general rule that any liquid, pressed on every one of its parts save one by the weight of another liquid, is thereby driven towards that part against which no pressure is exerted."[15]

Pascal stresses the role of the respective weights of the liquids involved. Let us consider, he says, a twenty-foot tube soldered to a siphon

---

[14]Robert Boyle, "History of Fluidity and Firmness," published in the collection *Certain Physiological Essays* (1661), in *Works*, vol. I, pp. 407–408.

[15]*Treatise on the Weight of the Mass of Air*, Œuvres de Pascal, II, p. 1073; Spiers trans., p. 40.

and filled with mercury. The legs, which are respectively 12 and 13 inches high, dip into two identical vessels filled with mercury but with one raised an inch above the other. The whole apparatus is placed in a tank filled with water to a depth of sixteen feet, leaving the upper end of the open tube out of the water.

The water exerts pressure on the mercury in each vessel but not on the mercury inside the legs that dip into it. Because sixteen feet of water is more than what is required to push up 12 or 13 inches of mercury, the mercury will be driven up each leg, but the outcome is surprising. The pressures in the two legs oppose each other at the top, and the stronger must prevail. Since the water above the lower vessel is higher by one inch, it will drive up the mercury in the longer leg with more force than in the other leg by the margin of power derived from one inch of height. We might have expected the mercury to rise from the longer leg into the shorter, but the opposite occurs. As Pascal explains:

> We must take into consideration that the weight of quicksilver in each leg opposes the effort of water to push it up, and that these two resistances are not equal. Since the quicksilver in the longer leg is deeper by one inch, it offers a resistance that is greater by the force derived from one inch of height. Although the quicksilver in the longer leg is driven up by a force of water that is greater by a height of one inch, it is borne down by its own weight, that is, by the excess weight of one inch of quicksilver. Now one inch of quicksilver weighs more than one inch of water. Therefore, the quicksilver in the shorter leg is driven up with greater force and consequently must continue to rise so long as there is quicksilver in the vessel into which it dips.[16]

The proof is completed by varying the conditions. When mercury, which is heavier than water, is replaced by oil, which is lighter, just the contrary happens. Because one inch of oil is lighter than one inch of water, it is the oil in the longer leg that is driven up more forcibly than in the other, so that the oil flows from the lower vessel to the higher. When the siphon is filled with a liquid that weighs as much as water, there will be no exchange. If we calculate the forces involved, we shall find that they cancel one another, and that equilibrium prevails.

Pascal does not say that he actually performed these experiments, which are intended to show how the behavior of water, oil, or mercury can

---

[16]*Ibid.*, pp. 1076–1077; Spiers trans., p. 44.

be explained by the same principle. We are dealing with "thought-experiments." Indeed, the experiment with oil in the siphon cannot be performed as it is described. If Pascal had made the attempt, he would have found that as soon as the siphon is immersed in the tank, the oil rises above the water. If the vessels are sealed with a membrane, the pressure of the water pushes up the oil in the vertical tube that is open at the top, and the siphon is rendered useless. The only way the experiment can be made is by doing away with the vertical tube and the vessels, and plunging a siphon full of oil in the tank of water. The oil then flows from the longer to the shorter leg, and rises to the surface.

How Sucking Works

Another phenomenon that was commonly explained by the horror of the void is sucking. "When someone places his mouth on water and sucks," writes Pascal,

> the water rises to his mouth for we know that the weight of the air presses the water on all sides except where his mouth is since that is the only place that the air does not touch. Hence when the breathing muscles expand the chest and increase the capacity within, the air inside has more room to fill than before, and less force with which to oppose the entrance of water into the mouth than the force which the outside air, pressing on that water from all sides save one, has to drive it in. That is the reason for this attraction, which is exactly the same as for syringes.[17]

Fluids are drawn into the mouth by creating a vacuum pressure in the oral cavity, and infants rely on this method of food ingestion until they are capable of eating more solid substances. Pascal was not aware, however, of the physiological mechanism whereby a partial vacuum is created in the oral cavity by retracting the tongue to the back of the mouth. The rear portion of the tongue seals against the roof of the mouth, allowing liquids to be drawn into the front region, and when the oral cavity is full, the tongue relaxes, the fluids flow back to the throat, where they are swallowed.

Pascal also tried to explain the practice, common in his day, of removing pus from infected tissue by placing a candle on the sore and cov-

---

[17]*Ibid.*, p. 1078; Spiers trans., pp. 46–47.

ering it with a cupping glass so that the flesh rose when the flame died. He believed that the air in the cupping glass was rarefied by the flame and then condensed so that the flesh was sucked up. From our modern point of view, the air is not rarefied and then condensed; rather the pus rises because the process of combustion draws oxygen from the air, thereby reducing its volume and making room for the pus to rise.

Breathing and the in-drawing of air are also explained by the weight of the air. "When the lungs open," says Pascal,

> the air near the nose and mouth, driven by the weight of its whole mass, enters and drops down by the natural and necessary action of its weight. This is so intelligible, easy and simple, that it is strange that recourse should have been had to the horror of a void, occult qualities, and other far-fetched and chimerical causes to account for them. It is just as natural for air to enter and drop into the lungs when they open, as for wine to drop into a bottle when it is poured in.[18]

Pascal does not seem to have been aware that the lungs are rhythmically inflated and deflated by the up-and-down movement of the muscular diaphragm or midriff that separates the chest cavity or thorax from the abdomen. During inspiration, the midriff is depressed so that fresh air rushes through the windpipe into the lungs. When it is elevated during expiration, some of the foul air is expelled. The internal capacity of the chest cavity is also increased during inspiration by the expansion resulting from the contraction of muscles that lie between the ribs and force the latter to bulge outwards.

The Vacuum-in-a-Vacuum Revisited

The pressure of the air decreases as it gets thinner and would vanish entirely if all the air could be removed. As we have seen, Pascal had found an ingenious way of showing this by producing a vacuum-in-a-vacuum. Roberval and Auzoult made improvements to his apparatus but it remained difficult to replicate. In *On the Weight of the Mass of Air*, Pascal describes a device that is a considerable improvement over the three models that we discussed in Chapter 5. It is illustrated in Figure 2.

---

[18]*Ibid.*, p. 1080; Spiers trans., pp. 47–48.

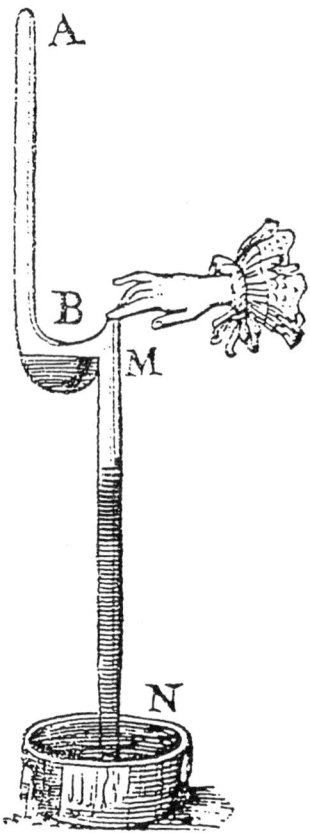

FIGURE 2 *Pascal's improved version of the experiment of a vacuum-in-a-vacuum.*

A tube, which is closed at A, is open at B and N.[19] After the opening at B is stopped with a fingertip, the tube is filled with mercury and inverted in a bowl of the same liquid. The mercury in the upper part of the tube falls into the curve and overflows into the lower part of the tube. No mercury stays in the upper part, but the mercury in the lower part remains sus-

---

[19]*Ibid.*, p. 1087; Spiers trans., p. 57. The figure is not correctly drawn. The bent part at B is not curved enough and should be in contact with the residual mercury. Otherwise when the finger is removed, the air allowed in would not drive the quicksilver up the tube AB but would freely enter the tube through the passageway. This was pointed out by Florin Périer in his 1663 edition of Pascal's treatise *On the Weight of the Mass of Air*, (*Œuvres de Pascal*, II, p. 1042; Spiers trans., p. xxii).

pended at the usual height of about 760 mm. The reason why the mercury falls in the upper part of the tube is that there is no air pressing upon the mercury at B. When the finger is removed from B, the air rushes in and the mercury is pushed up to the usual height. Pascal adds an interesting observation:

> But because nothing is ever lost in nature, although the quicksilver in the curved part does not feel the weight of the air, excluded as it is by the finger, the finger itself suffers sharply, for it bears the whole weight of the air pressing it from above, while it is unsupported from below. Consequently it feels as if it were pressed against the glass and drawn or sucked into the tube, and a swelling rises as if the place had been cupped. The weight of the air presses the finger, the hand, and the whole body everywhere except at this opening to which it has no access; hence that spot swells and aches.[20]

So what is experienced as an *attraction* coming from *inside* the tube is really *pressure* exerted from the *outside*! The weight of the air presses on every part of the human body except for one point alone at the opening at B, and this is why the finger aches. Pascal mentions this in such detail because he had described the finger as "drawn or sucked inward" in his *New Experiments*.[21] The point is not trivial: we can easily be deceived by our senses into believing that our finger is being pulled from below when it is really being pressed down from above.

Shortly after Pascal had completed his two *Treatises* but before their publication, Robert Boyle had an air pump or "pneumatical machine" constructed by the instrument-maker Greatorex and Robert Hooke in 1658. "The principal fruit" that Boyle promised himself from his air pump was the realization of Pascal's vacuum-in-a-vacuum. By placing the tube of mercury and its dish in an evacuated receiver, Boyle was effectively taking it out of the atmosphere. His equipment consisted of a three-foot-long glass tube in which the mercury subsided to a height of about 29 inches when inverted in a dish full of mercury. Boyle noted that after a quarter-

---

[20]*Ibid.*, pp. 1087–1088; Spiers trans., p. 57.

[21]When the plunger is pulled back, "the finger feels a strong and painful attraction," (*Œuvres de Pascal*, II, p. 502). Giovanni Battista Baliani got it right much earlier when he wrote to Mersenne that the finger "feels an inclination to move downwards, but what you call *a strong attraction of the vacuum*, I consider a strong push for the air above" (letter to Mersenne, 23 February 1648, *Correspondance de Mersenne*, vol. XVI, p. 116).

hour's pumping (how many sucks is not recorded), the mercury would fall no further and remained about an inch above the level of the liquid in the dish. This argued for a small but significant leakage.[22] The vacuum-in-a-vacuum was a tricky business.

A Modern Critique of Pascal's Experiment

Pascal's experiment has recently been submitted to a critical appraisal by the Japanese historian Kimoyo Koyanagi, who believes that it was merely a *thought-experiment* on the strength of the following considerations.[23] First, the three earlier experimental set-ups that we outlined in Chapter 5 are described by witnesses, this one is not; second, the curved part of the tube in the figure is unsatisfactory, as Périer pointed out in his edition of Pascal's text and, third, Koyanagi's own attempts to reproduce the experiment met with failure. Her criticism of Pascal deserves close scrutiny. The apparatus (see Figure 3) that she used was designed by Akira Kurobe in 1974, and the experiments were presumably carried out in that year. The outside diameter of the tube was 10 mm; the inside diameter 6 mm. AB and MN were both 800 mm long.[24]

In the process of repeating the experiment, Koyanagi made five interesting observations. First, once filled the apparatus could not be inverted without breakage unless it was fixed to a wooden board. Second, when she tried to pour mercury into the inverted tube, it would not go in because the air that was trapped in the tube could not escape, and in order to get rid of this air an opening had to be left at the top, at A, where a stopcock was installed. Third, it was impossible to keep a fingertip on the opening at M while the apparatus was being turned upside down, and it had to be

---

[22]Robert Boyle, *New Experiments*, pp. 33–34.

[23]Kimiyo Koyanagi, "Pascal et l'expérience du vide dans le vide," *Japanese Studies in the History of Sciences*, no. 17 (1978), pp. 105–127; "Les expériences du vide dans le vide," *Courrier du Centre International Blaise Pascal*, II (1989), pp. 3–23. In this second article, Kimiyo Koyanagi provides an account of her replication with Masao Uchida of the vacuum-in-a-vacuum experiments of Pascal, Roberval, Auzoult, and Rohault. See also Masao Uchida, "De la faisabilité des *Expériences du vide dans le vide* rapportées par Pascal et ses contemporains," *Bulletin de la Faculté des Lettres et Sciences Humaines de l'Université Wako* 17 (1982).

[24]According to the figure on p. 123 of the article, although on p. 124 she states that the total length of the apparatus was approximately 1800 mm, when we would have expected 1600 mm.

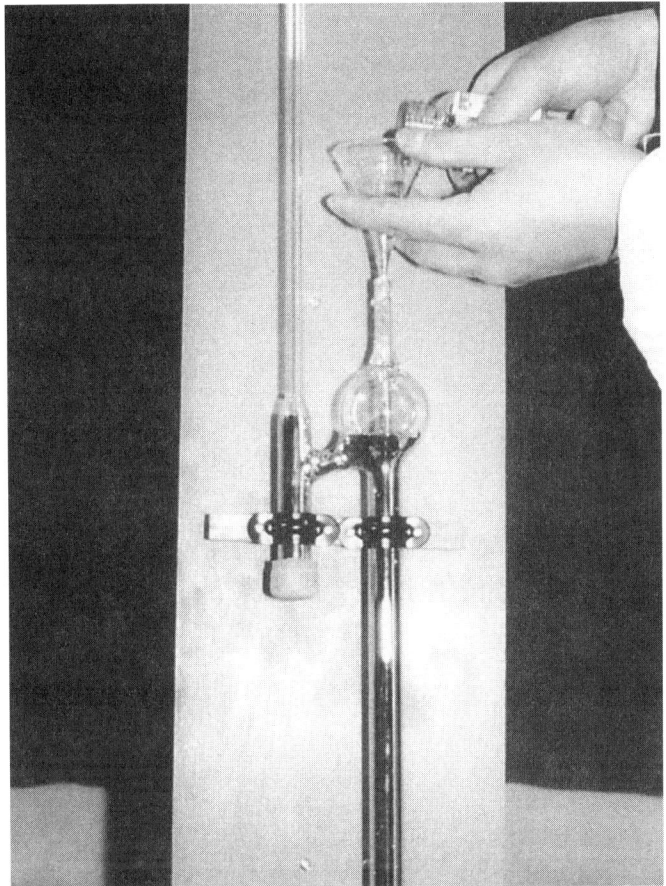

FIGURE 3   *Photograph of the modern equipment used to replicate Pascal's experiment.*

closed with a membrane, or a stopcock had to be installed. Fourth, the curved part at B had to contain enough mercury to allow it to rise to a height of 760 mm, but in Pascal's diagram this curved part is much too narrow. Koyanagi enlarged it almost six times, giving it an interior diameter of 35.2 mm, and then narrowed the part near M so that the opening left was small enough to be stopped with a finger. If this had not been done, Koyanagi claims that the bent part at B would have to be curved downward some 360 mm to contain enough mercury to rise to 760 mm in AB in the second part of the experiment. Such a curved part would also be extremely difficult to fill with mercury. Fifth, the air had to be let in very gently, otherwise the mercury shot up and broke the end of the tube at A. Pascal had mentioned that the mercury overshot the mark, but his tube was probably longer than Koyanagi's.

Jacques Rohault's Version

Koyanagi's strictures are serious, but we must recall that Pascal's experiment was repeated successfully in the seventeenth-century, for instance, by the versatile and popular experimentalist Jacques Rohault in Paris. This may indicate that a considerable amount of time is required to acquire the necessary skill in handling the apparatus. Rohault must have met with initial difficulties since he devised an improved version of the experiment of a vacuum-in-a-vacuum illustrated in Figure 4.[25]

The crucial difference with Pascal's apparatus is the narrow tube DE that Rohault placed in the upper tube. The opening at G was closed with a hog's bladder and the apparatus inverted so that it could be filled through the opening of the narrow tube at E. When the level of mercury reached F, it began to pour through a hole at this place into the larger tube that was gradually filled. The opening at E was then sealed with a piece of bladder, and the rest of the apparatus filled through the opening at C, which was afterwards sealed in turn. The apparatus was inverted again so that the end C dipped in a vessel of mercury. The remainder of the experiment followed the procedure outlined by Pascal.

It is interesting to note that Rohault did not believe that a vacuum had been produced when the bladder was removed from C and the mercury fell to its usual height of 760 mm. He was convinced that the space left vacant by the mercury was not really void of matter because an object placed behind it remained visible. It seemed to him that a genuine vacuum would not be able to transmit light.

---

[25] *A System of Natural Philosophy* by Jacques Rohault, volume I. A facsimile of the edition and translation by John and Samuel Clarke, published in 1723 in London. Johnson Reprint Corporation, New York and London, 1969, pp. 73–75. The text is quoted by Koyanagi, *art. cit.*, pp. 116–117. It should also be noted that Pascal's experiment was replicated by Roland Jouanisson in 1981 (see Dominique Descotes, "L'expérience du vide dans le vide à Clermont-Ferrand," *Courrier du Centre International Blaise Pascal*, II (1989), p. 24.

FIGURE 4 *Diagram of the apparatus used by Rohault in the seventeenth century*

How High Can Water Be Raised?

We return to the treatise *On the Weight of the Mass of Air* and Pascal's realization that we can estimate the altitude of a place by how high water can be raised there. In Paris, for instance, a pump can never raise water above thirty-two feet, but a siphon ten feet tall will work on any mountain in the world because none is high enough to prevent it from doing so. The level to which water is raised indicates the altitude so that we can determine, from the difference of level in two places, by how much higher one is than another. All that is needed is a tube four or five feet long filled with mercury. Pascal took an important step towards the definition of standard atmospheric pressure by starting his measurements at sea level, "which is everywhere the same, that is to say equally distant from the center of the earth in every place." But he neglected the importance of temperature, and assumed that the level to which water rises in a barometer is the same for all places by the sea, a height he estimated to be 31 feet 2 inches.[26]

Pascal was aware, however, that the degrees of heat marked on the thermometer in use in his day were unreliable. This was the air-thermometer, the original type of which was invented by Galileo around 1600. According to his student Benedetto Castelli, who described it around 1638, this consisted of a bulb of glass about the size of a hen's egg, sealed to the top of a tube the size of a straw and about 55 cm long that was dipped vertically into a vessel of colored water. The air in the bulb, which was warmed by the hand, contracted as it cooled, and thus sucked colored water into the stem where it fell and rose as the air in the bulb was warmed or cooled by changes in the environmental temperature.[27] The result was the opposite of the one we observe in modern thermometers, since the level of the water rose when the weather became cold and fell when it became warmer. Since the vessel of colored water was open to the air, it was subject to atmospheric pressure and was not entirely dependable. The first hermetically sealed thermometers, which contained alcohol, were made by the Florentine Academy, and were first described by them in the *Saggi di Naturali Esperienze* published in 1667, five years after Pascal's death.

Weighing the Atmosphere

Pascal extended his quantitative analysis to the determination of the weight of the atmosphere from the fact that air at sea level weighs as much as a column of water 31 feet high. Since a pump can raise water by as much as 1 foot 8 inches higher in some weather conditions than in oth-

---

[26]*Treatise on the Weight of the Mass of Air*, *Œuvres de Pascal*, II, pp. 1089–1090; Spiers trans., p. 59. The location where water was pumped to 31 feet 2 inches is Dieppe. From the information provided in *The Account of the Great Experiment* (II, p. 687), we know that 31 feet 2 inches = 10.16 meters, which translates, using 13.6 for the specific gravity of mercury, into 745 millimetres of mercury. This is rather low, since the modern standard atmospheric pressure is defined as the pressure equivalent to that produced by a column of mercury 760 mm high at 0°C.

[27]See Benedetto Castelli's letter of 20 September 1638 to Ferdinando Cesarini (*Opere di Galileo*, XVII, p. 377).

ers, Pascal claimed that this indicated that the densest vapors weigh as much as a column of water 1 foot 8 inches. It follows that if all the vapor held by the air over a given area turned into rain, it would not cause a rise of more than 1 foot and 8 inches of water (this would have interesting implications for the biblical account of the flood!). Furthermore, if the whole sphere of the air were compressed into the smallest possible space by condensing into water, the water level on earth would rise by 31 feet. "Consequently," concludes Pascal,

> the whole mass of the air, free as it is, may be treated as though it had formerly been a mass of water thirty-one feet high all round the earth, and had then been rarefied and very greatly expanded until it changed into what we call air, occupying a greater space, but conserving exactly the same weight as water thirty-one feet high.[28]

This is more interesting than meets the eye because it assumes the principle of the conservation of mass, which was not generally accepted at the time.

Pascal's next step was to compute the actual weight of the atmosphere, something he plainly enjoyed. "Nothing is easier," he writes, "why a child who knows how to add and subtract could do it!" And he does it himself, just "for the fun of it." The volume of a sphere is πd^2, where π is the ratio of the circumference of a circle to its diameter, and d is the diameter of the Earth, which was estimated by contemporary geographers at 2,291 leagues (11,203 km). This makes 16,495,200 square leagues for the total spherical surface of the Earth, and since 1 league = 2.25×10^9 square feet, this gave Pascal a figure of 3.71142×10^{16} square feet. A cubic foot of water weighs 72 pounds, so that a prism with a base of one square foot and a height of thirty-one feet will weigh $72 \times 31 = 2.232$ pounds. If the Earth were covered with 31 feet of water there would be as many prisms of water as there are square feet on its whole surface, and the whole mass of water would weigh $2.232 \times 3.71142 \times 10^{16} = 8.28388944 \times 10^{16}$ pounds. If we assume that Pascal was using the Parisian pound, which was worth 490 grams, this works out to roughly 4.06×10^{16} kilograms.[29] The reader may

---

[28]*Treatise on the Weight of the Mass of Air*, *Œuvres de Pascal*, II, p. 1093; Spiers trans., pp. 64–65.

[29]*Ibid.*, p. 1094; Spiers trans., p. 66. Pascal anticipated a possible objection against this simplifying assumption: "I am well aware that these would not be prisms, but sectors of a sphere, but I purposely neglect this refinement or precision."

wish to know that the modern value for the weight of the atmosphere is 5×10^{18} kilograms.

The Demise of Horror and Anthropocentrism

In his *New Experiments Concerning the Vacuum*, Pascal had described nature as an active agent. For instance, he gave as one of his maxims: "All bodies have a *repugnance* to separate themselves from one another, and *to allow* a vacuum in the interval between them; that is to say that *nature abhors a vacuum*."[30] Although he had changed his mind after hearing about Torricelli's hypothesis in the summer of 1647, he did not "consider it permissible to discard the maxims handed down to us by the Ancients unless obliged to do so by proofs that are compelling and cannot be doubted."[31] In the General Conclusion to his *Two Treatises*, Pascal reverts to this topic and offers his finest and most devastating critique of the abuse of anthropomorphic language. "It is not difficult," he declares,

> to demonstrate that nature does not abhor a vacuum and that this manner of speaking is improper since created nature, which is the nature under consideration, is not animated and can have no passions. Such language is metaphorical, and means nothing more than that nature makes the same efforts to avoid a vacuum as if she abhorred it. Those who use this phrase mean that it is the same thing to say that nature abhors a vacuum as to say that nature makes great efforts to prevent a vacuum. Now, since I have shown that nature does nothing at all to avoid a vacuum, if follows that it does not abhor it. To stick with the same analogy: just as we say of a man that something is indifferent to him when his actions never betray a desire or a dislike for it, so we should say of nature that it is supremely indifferent to a vacuum, since it never does anything either to seek or to avoid it.

Pascal as Historian of Science

Pascal is willing, however, to allow that there was a time when evidence favored the notion that nature abhors a vacuum and he shows, in a re-

---

[30]Pascal, *New Experiments*, II, p. 507; Popkin trans., p. 40.
[31]Letter of Pascal to Florin Périer, 15 November 1647, *Œuvres de Pascal*, II, p. 679; Spiers trans., p. 99; Popkin trans., p. 43.

markable passage that could stand as a model for historians of science, how this idea was gradually abandoned:

> It is perfectly true (and this is what misled the Ancients) that water rises in a pump when air has no access to it, that a vacuum would result if the water did not follow the piston, and that water ceases to rise as soon as any crack develops by which the air can get in to fill the pump. Thus it looks as though the water rose merely for the purpose of preventing a vacuum, since it only rises when otherwise a vacuum would be created.[32]

Scientific knowledge, at any given period, depends on a network of assumptions that allow certain questions but preclude others, which are dismissed as wrong-headed or irrelevant. Progress hinges on the development of new theories (what Pascal calls *principles*) and hence the willingness to entertain new questions. Because air had initially been considered as essentially light, weight had not been taken into account, and when air was recognized as being heavy, the belief that "the elements do not weigh in their own milieu" lingered on. It is only when air was recognized as being heavy in whatever milieu it is placed that it became possible to think of pressure as a force that acts in all directions and not only downwards.

What delayed the acceptance of atmospheric pressure, explains Pascal, was a widespread error concerning the use of pumps. It was assumed that water could be raised as high as one pleased, without limit, and no one asked whether air, whose weight is clearly limited, could be the cause of such an impressive effect. The error was eventually toppled over by fountain cleaners in Florence from whom Galileo learned more than from books of natural philosophers. But the road had not been completely cleared: Galileo did not go beyond concluding that water could not be raised above 18 *braccia* because this was the "natural" height to which it could ascend. He did not consider that altitude might play a crucial role, and it took a Pascal not only to suspect that it might, but to show that it did.

Three obstacles stood in the way of a correct understanding of the nature of atmospheric pressure according to Pascal. The first was the belief that air has no weight and this was destroyed, not by philosophical dis-

---

[32]*Conclusion of the Treatises*, Œuvres de Pascal, II, p. 1095; Spiers trans., p. 68.

putations, but by experimental evidence. The blindness Pascal imputes to the followers of Aristotle is no ordinary lack of insight but failure to imagine experimental situations to test current assumptions. The second obstacle was the tenet that elements do not weigh in their own milieu on the grounds that we do not feel the weight of a pail of water as long as the pail is immersed in water. The argument is ridiculed by Pascal: "Why should water in a bucket have weight when taken out of the water but no weight when poured back in, or why should it lose its weight when added to water, and recover it when lifted above the surface. Strange are the means people employ to cover their own ignorance!"[33]

The third obstacle was the notion that the horror of a vacuum was a universal force. This is what led Hero of Alexandria to claim that water from a river can be raised and made to pass over a mountain by means of a siphon, however high the mountain.[34] He had seen ordinary suction pumps that were roughly 4 meters high work beautifully, and he never imagined that there was a limit beyond which things were otherwise. So positively was this believed that it became the foundation of all treatises on the vacuum. "It is taught every day in classrooms all over the world," writes Pascal "and ever since books have been written, everyone has firmly believed it." The error was finally exposed by simple workmen, a fact that "could open the eyes of those who dare not doubt an opinion that has always been entertained."[35] All that was required was to try to draw water above 10 meters.

The next step, for which Pascal takes credit, was to show that the weight of the air is greater at some places than at others and, more specif-

---

[33]*Conclusion of the Two Treatises*, *Œuvres de Pascal*, II, pp. 1097–1098; Spiers trans., pp. 69–70. The view that water does not weigh in water was argued by Jean Rey in his *Essays* of 1630 (facsimile reprint: London Edward Arnold, 1951, pp. 46–50).

[34]*Ibid.*, pp. 1098–1099; Spiers trans., p. 71. Although Hero places no limitations on the efficacy of the siphon he describes, such a statement has not been found in his *Treatise on Pneumatics*. According to Pascal, "all those who have written on the subject have said the same thing, and even now water engineers swear that they can make suction pumps that will raise water as much as sixty feet if required" (II, p. 1099; Spiers trans., p. 72). Agricola is more modest in his *De Re Metallica* (1555), and mentions twenty-four feet as the greatest height water was drawn by a suction pump (p. 186 of the English trans. of H.C. Hooker and L.H. Hooker [1912], reprint New York: Dover, 1950).

[35]*Ibid.*, p. 1099; Spiers trans., pp. 72–73. These artisans are the water engineers mentioned by Galileo. Although Pascal refers to Galileo's *Two New Sciences*, there is no evidence that he read the book itself, and his source is almost certainly the passage quoted in Marc-Antoine Dominicy's introduction to his 1647 edition of Pierre Petit's letter to Chanut (see II, p. 545).

ically, that it is greater at the foot than at the top of a mountain, Once this was done, the shoe was on the other foot. Pascal's opponents now had to offer a better explanation than the weight of the air for a large range of phenomena that included: (1) the difficulty in opening a sealed bellows, (2) the reason why a pump that works at sea level does not work at the top of a mountain, (3) the strength required to separate two polished bodies in close contact, and (4) the attraction produced by sucking.

Pascal concludes his analysis with a rhetorical flourish that extols the necessity of experimentation:

> Does nature abhor a vacuum more in the highlands than in the lowlands? In damp weather more than in fine? Is not its abhorrence the same on a steeple, in an attic, or in a yard? Let the disciples of Aristotle collect the best that their master and his commentators have written to account for these things by abhorrence of a vacuum if they can. Otherwise, let them acknowledge that experiments are the true masters that one must follow in physics, and that the one that was carried out on a mountain has overthrown the universal belief in nature's horror of a void, and given us the knowledge, never more to be lost, that nature has no horror of a void and does nothing to avoid it, and that the weight of the mass of the air is the true cause of all the effects hitherto ascribed to this imaginary cause.[36]

The *Two Treatises* thus end on a triumphant note. The procedure followed by Pascal may be said to illustrate the *principle of systematic variation*, which consists in varying the factors that are suspected of causing or influencing the outcome, and can take one of the following logical forms: (1) if one factor only is suppressed and the effect ceases, we can conclude that the effect was influenced by that factor; (2) if all the factors, except one, are changed and the effect still occurs, we can conclude that the unchanged factor was the cause of that effect; or (3) if a factor cannot be suppressed, it should at least be maintained constant while other factors are made to vary.

In the experiment of a vacuum-in-a-vacuum, one factor only, atmospheric pressure, was removed. When the mercury fell all the way down, Pascal concluded that it had been upheld by that pressure. In the experiment conducted on the Puy-de-Dôme, all the factors remained unchanged

---

[36]*Ibid.*, p. 1101; Spiers trans., p. 75.

except altitude, and since nature cannot be said to experience a greater abhorrence for a vacuum at the foot of a mountain than at the top, the cause of the drop of mercury must be the altitude, the only factor that varied. At the same altitude, Pascal varied the liquids used (for instance, water and mercury), compared the respective heights to which they fell, and showed that they were in the ratio of their specific gravity.

Systematic variation went hand in hand with *the rule of parsimony* that enjoins to keep things simple. Pascal used only a few instruments: the syringe, the siphon, and the barometric tube, and he did not attempt a rigorously quantitative analysis when he could make his point without giving precise numbers. For instance, when he went up the Puy-de-Dôme with a flabby football to see whether it progressively became rounder, he did not worry by exactly how much.[37]

Variations Due to Weather

Périer made known in an appendix to the posthumous treatises *On the Equilibrium of Liquids* and *The Weight of the Mass of Air*, two fragments that Pascal had intended to publish. The first deals with meteorological conditions, the second with attempts to determine the force required to overcome the creation of a vacuum. We shall examine them in turn.

Pascal's instrument was a glass tube sealed at the upper end, open and recurved at the lower one, which he set up in a room in a place where it could be conveniently seen without being disturbed. A strip of paper, ruled off in inches and lines, was glued along its length, as was done with thermometers. "This experiment," he informs his eventual reader, "is called the continuous experiment, because it can be observed continuously." The readings showed that the height of the mercury was "more or less in proportion as the air was more or less charged," where "charged" indicates "charged with moisture," i.e., humid, hence heavy.[38]

---

[37]In the *Pensées*, Pascal will use a similar method. The principles of justice, which should not be affected by changes of time, place, or circumstance are altered by changes in climate and geography. "Three degrees of latitude upset the whole of jurisprudence and one meridian determines what is true . . . It is a funny sort of justice whose limits are marked by a river; true on this side of the Pyrénées, false on the other" (fragment 60).

[38]*Œuvres de Pascal*, II, p. 788; Spiers trans., p. 80.

In this first fragment, Pascal relies on the recorded height of mercury in Dieppe. This is surprising since we would expect him to use readings taken in Paris or in Clermont where he spent time with his family between May 1649 and November 1650. The reason is almost certainly the greater variation in Dieppe, where his father's friend, Pierre Petit, went regularly as royal engineer in charge of fortifications. When the sky was completely overcast, the mercury stood at 28 inches 4 lines but it could drop by as much as four lines when the weather cleared. Pascal describes possible fluctuations, but he speaks hypothetically as if the general outline was certain but precise measurements were not available: "The next day," he adds, "it may have dropped by 10 lines; sometimes an hour later, it will have gone up 10 lines; some time thereafter it may go up or down according as the weather is overcast or clear." The recorded height of the mercury at Dieppe is high compared to the 26 inches 3 lines that Pascal gave as the height mercury reaches when the piston of a syringe is raised. The explanation probably lies in the way the paper was pasted to the recurved end of the tube. It would have been glued above the curved part, but the ruling started below the level of mercury in the open end, hence the greater height.

Pascal was eager to stress the practical implications of the "continuous experiment," which enabled him to infer "with certainty" when pumps would raise water higher and when bellows would be more difficult to open. Here is what will happen, he tells us, if we take a small bellows roughly three inches in diameter, seal it hermetically, and fasten one of the sides to a beam on the ceiling, and attach to the other side an iron chain whose links trail on the floor. The links, without counting those on the floor, should weigh approximately 120 pounds, which is enough to open the bellows. The side to which the dragging chain is attached will come down, the chain will drop, and the links that previously hung nearest to the floor will now rest on the floor so that their weight no longer pulls on the bellows. The wider the bellows opens, the fewer the links that remain hanging. "And the most amazing thing is this," pursues Pascal,

> when the weather clears, and when consequently a lesser weight will suffice to open the bellows, the hanging links, which weigh 113 pounds and were in equilibrium with the air when it was most heavily charged, will become too heavy because the air has cleared. They will therefore pull down the side of the bellows and open it wider until

the links that remain suspended balance the weight of the air above in its new condition; and the lighter the air becomes the more links will drop. But when the air charge increases, the bellows will, on the contrary, be seen to close up seemingly of its own accord, and doing so, will pull the chain up again, until the hanging links are once more in equilibrium with the weight of the air above in this new condition. Thus the chain will rise and fall, and the bellows will open or close more or less as the air charges or discharges itself; and at all times the hanging links will be in equilibrium with the outer air.[39]

Pascal's excitement shines through the dryness of the quantitative description. "If you want to get more enjoyment out of these observations," he adds, "you can carry out three or four such operations at once," for instance, with the following four pieces of apparatus: a tube filled with mercury, a sealed bellows, a suction pump 35 feet long, and a siphon with legs of 31 and 35 feet. When the weather is *most heavily charged*, we find that the mercury stands at 28 inches 4 lines, the hanging links weigh 113 lbs., the water level in the pump reaches 32 feet, and the siphon works because its short leg is under 32 feet.

We can also know, from inside a house, whether the pump in the courtyard will function or not just by glancing at the level of mercury in the room. "Whenever you see the mercury rising in the room," he writes, "you may be sure, without looking, that the siphon is working in the yard outside. And when you notice the mercury drop, you may be sure, again without looking out, that the siphon has stopped, because it works in immediate response to the weight of the air."[40]

Weather Forecasting

Although the phenomena may sometimes appear arbitrary because of the variations in the condition of the air, Pascal thinks we can be guided, on the whole, by three key observations:

1. Mercury usually falls in fine weather and rises when the weather becomes cold or overcast.

---

[39]*Œuvres de Pascal*, II, pp. 788–789; Spiers trans., pp. 81–82.
[40]*Ibid.*, II, p. 791; Spiers trans., p. 83.

2. Mercury is usually highest in winter and lowest in summer. Pascal specifies that the variation is less at the solstices than at the equinoxes, a surprising comment since he could only have collected data between 1647 and 1651, too brief a period to venture a generalization. Aware of the problem, he immediately qualifies his statement: "Not but that the mercury is sometimes high in summer, low in winter, irregular at the solstices, and steady at the equinoxes, for there is no regularity in the matter. But usually things go as we have said because usually, though not always, the air is most charged in winter and the least in summer. It is most variable in March and September, and most constant at the solstices."[41]
3. Mercury fluctuates between upper and lower limits. This is because there is normally a certain range in the *charge of the air*, but Pascal adds that "there may occur some accident in the air that charges it excessively, in which case the mercury rises exceptionally high; but such an occurrence is so rare that one cannot explain it by rule." The acknowledgement that rules apply only *in ordinary cases* may seem trivial, but it is an important change from the standard approach that we find in Galileo and other scientists for whom any physical situation was strictly determined by mathematical laws. The major figures of the Scientific Revolution lived in a realm of geometrical precision; Pascal alone inhabited a world of probabilities. For Galileo, God does not play dice; for Pascal, the very act of creation is like casting a die.

From these considerations, Pascal makes predictions in four cases: (1) when the sky becomes overcast and the mercury falls, we can expect fine weather; (2) when the weather is fine and the mercury is high, there are scattered vapors about, and though unseen, they will soon produce rain; (3) when the mercury is low and the weather is fair, the fine weather will last; and (4) when the air is charged and the mercury is high, the bad weather will continue.

A word of caution, however: a sudden wind may bring these forecasts to naught. Like all meteorologists, Pascal will not venture too far in time:

---

[41]*Ibid.*, p. 792; Spiers trans., p. 85. The last word in French is *équinoxes*, an obvious slip for *solstices*. This is not mentioned by Mesnard, who keeps équinoxes; the English translation amends the text.

"This knowledge can be very useful to farmers, travelers, and others for ascertaining the present state of the weather and its immediate outlook, but not for telling the weather three weeks hence."[42]

Measuring the Resistance to a Vacuum

The second fragment is more theoretical and consists of tables purporting to determine the values of the force necessary, at different altitudes and during different weather conditions, to overcome the resistance to the creation of a vacuum. Pascal estimated Clermont to be 400 fathoms above Paris, or roughly 780 meters (the correct value is 401 meters), and the Puy-de-Dôme at 500 fathoms above Clermont. This would give a height of 1750 meters for the mountain (which is actually 1465 meters). La Font de l'Arbre is assumed to be 250 fathoms above Clermont in the tables, although it was known to be closer to the base than the summit of the Puy-de-Dôme. The observations, which are not dated, are tabulated in three columns that indicate whether the air is *heavily*, *moderately* or *less charged*, qualitative appraisals that are by nature imprecise. We have no way of knowing whether the observations were made on the same day or at the same time. What interested Pascal was the determination of the force necessary to separate two bodies in contact but he does not specify the nature of these bodies or how well they were polished. We shall examine three of the tables that he compiled.

Table I: Force Required to Separate Two Plates

In the first table Pascal determines the force in pounds necessary to separate two spherical plates under different atmospheric conditions at four locations when the diameter of the plates is one foot (about 220 cm).

---

[42]*Ibid.*, p. 793; Spiers trans., p. 86. Pascal uses the term "mercure" in these fragments, whereas he had used *vif-argent* before. Pierre Petit in his letter to Hector-Pierre Chanut of 26 November 1646 used *mercure* (e.g., II, p. 350), a term that made Roberval's eyebrows rise (in a letter to Desnoyers he refers to "the wonderful properties of Mercury (for this is how they call quicksilver)," *Œuvres de Pascal*, II, p. 469)

Location	Heavily Charged Air	Moderately Charged Air	Less Charged Air	Difference
	Pounds	Pounds	Pounds	Pounds
Paris	1808	1761	1714	94
Clermont	1675	1628	1581	94
La Font	1579	1532	1485	94
Puy-de-Dôme	1483	1436	1389	94

Note that the force required is considerable even at the top of the Puy-de-Dôme and in fair weather, since 1389 pounds (roughly 680 kilograms) are necessary to pull the plates apart.

Table II: Weight of a Column of Mercury

The second table is intended to determine the height to which mercury remains suspended when the tube is evacuated in the standard way at the four different locations and under the three kinds of weather conditions.

	Heavily Charged Air			*Moderately Charged Air*			*Less Charged Air*			*Difference*	
Location	*ft.*	*in.*	*line*	*ft.*	*in.*	*line*	*ft.*	*in.*	*line*	*in.*	*line*
Paris	2	4	4	2	3	7	2	2	10	1	6
Clermont	2	2	3	2	1	6	2		9	1	6
La Font	2		9	2			1	11	3	1	6
Puy-de-Dôme	1	11	3	1	10	6	1	9	9	1	6

It is curious that the same difference in height (1 inch 6 lines) should be recorded between all four elevations. If we subtract the successive values in the "heavily charged" column, we obtain 1 inch 6 lines for the difference between Clermont, La Font, and the Puy-de-Dôme, but 2 inches 1 line for the difference between Paris and Clermont. The discrepancy is probably an oversight. It is also surprising that the highest value for Clermont should be 2 feet 2 inches 3 lines, whereas Périer recorded a higher value of 2 feet 2 inches 11 lines in the observations he made between 1649 and 1651. In any case, Pascal meant to revise his tables for publication but never got around to doing so. Clearly some of the values are computed and not the result of actual observations. This is confirmed by the next table that purports to give the height to which water can be suspended at different altitudes. We are told, for instance, that when the air is *heavily charged*, the column of water will stand at 26 feet 3 inches on top of the Puy-de-Dôme where Pascal had experimented with mercury but not with water, which would have made the experiment too costly and too difficult.

Table III: The Height of a Column of Water

Location	Heavily Charged Air		Moderately Charged Air		Less Charged Air		Difference	
	ft.	in.	ft.	in.	ft.	in.	ft.	in.
Paris	32		31	2	30	4	1	8
Clermont	29	8	28	10	28		1	8
La Font	28		27	2	26	4	1	8
Puy-de-Dôme	26	3	25	6	24	7	1	

Why did Pascal, who made the experiments with mercury, choose to express the differences in height in terms of water? The most likely answer is that the differences are greater, and hence more striking, when water rather than mercury is used.

We may summarize the results of Pascal's observations of the column of mercury during different kinds of weather and at different localities by saying that the height of the column of mercury is usually greater when the air is more humid or colder. Sometimes, however, the mercury is low when the weather is cloudy, and high when it is fair. This, he concluded, is due to a difference between the upper and lower regions of the air. On a cloudy day, it sometimes happens that the clouds do not extend upwards very far and have no great weight, in which case they usually soon dissipate and the weather becomes clear. So, when the mercury stands high on a clear day or low on a cloudy day, we may expect a change in the weather. Our predictions of the weather can, however, extend only a short way into the future.

Pascal believed that the weight of atmosphere increased when the air is "more charged with vapor." The matter is complicated, but strictly speaking, we would want to say the opposite of what Pascal affirms. Water vapor is less dense than air. So a mixture of water vapor and dry air is less dense than pure air. Since the height of the atmosphere may be taken to be nearly constant, changes in the atmospheric pressure at any particular place chiefly depend on moisture and temperature. A fall of pressure means that the air is less dense, and this generally indicates that it is more moist. Hence rain is to be expected. This rule is not highly reliable, because changes in pressure also result from unequal heating of the earth's surface. Warm air is less dense than cold air, and the atmospheric pressure, therefore, depends partly on the temperature. It is constantly changing because of the winds so produced. A further complication arises from the fact that more water vapor is required to saturate warm than to saturate cold air. In an island climate, the direction of the wind is especially important. In England, rain may be anticipated when cold winds are blowing over the sea from the north-east, even if the barometer is rising. In the weather forecasts now published, most importance is attached to the relative distribution of pressure based on satellite observations that make it possible to record how zones of high and low pressure are shifting from day to day. A chart constructed on the basis of simultaneous records of pressure at different stations frequently show closed areas of high pressure (anti-cyclones) from which dry winds are blowing spirally outwards, or closed areas of low pressure (cyclones) where wet winds are blowing spirally inwards. Their position changes from hour to hour, so that it is possible to foretell within short limits where they will next be. The approach of a cyclone betokens wet weather, that of an anti-cyclone dry weather.

CHAPTER 8

Pascal on the Advancement of Learning in Science and Religion

The treatises *On the Equilibrium of Liquids* and *On the Weight of the Mass of Air* were originally intended as parts of a *Treatise on the Vacuum* that Pascal announced as forthcoming in his *New Experiments* of 1647 but which was never completed. It was to have opened with a *Preface* that is extant in which Pascal criticizes reliance on authority in the natural sciences and vindicates the progressive character of human knowledge. We shall examine this *Preface* and discuss Pascal's conception of religious knowledge before turning, in the next chapter, to his essay *On the Geometrical Mind*, in which he develops his views on the nature of scientific demonstration.

The Preface of the Treatise on the Vacuum

Pascal distinguishes between two kinds of knowledge. The first, which we would call *scholarship*, rests on the availability and the reliability of the sources, which are generally written ones. For instance, the question, Who discovered America? is settled by examining the historical records. The prime example of this kind of discipline is theology that depends on sacred books for information about things that the human mind is too weak to reach unaided. In this case, innovation is precluded, for we can neither add to nor subtract from the authority of Scripture and the Church Fathers. Matters are entirely different in mathematics and the experimental sciences where knowledge is acquired through experience and reasoning. "Here authority is useless," and the human mind enjoys "complete freedom," says Pascal.[1] In these disciplines, real growth is possible and successive generations can build on the achievements of those who came before them.

The "bane of the age," as Pascal sees it, is that people appeal to authority in physics, where it is irrelevant, and spurn it in theology, where it is essential. "This is why," he writes, "we must pity the *blindness* of those who offer authority instead of reasoning or experiments in physics . . . and *confound the insolence and rashness* of those who produce novelties in theology."[2] Pascal was not merely laying down a general rule. He had real persons in mind. The first, *blind* adversary is Father Noël, who handled physical problems as though they were theological questions, and of whose science Pascal could have said what he later declared of Descartes': "*It is a novel about nature not unlike the story of Don Quichote.*"[3] The second, *insolent* adversary, is Jacques Forton, whom Pascal denounced to the bishop of Rouen for treating theology as a speculative science. Where reasoning and experiment are required, there is no role for authority. If the Ancients had not added to what they had received, we would be the poorer. It is up to us, urges Pascal, to use the contribution of our predecessors as the *means*, not the *end* of our studies, and thus "to surpass them by imitating them."

The Growth of Mankind

Pascal did not always speak of history in the same way, and he used different models of historical explanation. Indeed, we find three. The first is

---

[1] Pascal, *Preface to the Treatise on the Vacuum*, in Pascal, *Selections*, edited by Richard H. Popkin. New York: Scribner/Macmillan, 1989, p. 63. In the *Œuvres de Pascal*, edited by J. Mesnard, volume II, p. 779.

[2] *Preface to the Treatise on the Vacuum*, *Œuvres de Pascal*, II, pp. 777–779. Popkin trans., pp. 62–63. The *Preface* was first published by Bossut in 1779. It is found in the *Œuvres de Pascal*, II, pp. 777–785, and is translated by Richard H. Popkin in Pascal, *Selections*. London: Collier Macmillan, 1987, pp. 62–66.

[3] The source of this saying is the French Protestant physician Antoine Menjot, who met Pascal towards the end of his life (*Œuvres de Pascal*, I, p. 831).

borrowed from Montaigne, for whom history is a sequence of fortuitous events. In this vein, Pascal speaks of the shifting variety of customs and beliefs in different ages, and of the odd disproportion between cause and effect. For instance, had Cleopatra's nose been shorter, the whole face of the world would have been different! The second model is sacred history, which undergirds the apparent chaos of secular history. "How good it is," exclaims Pascal, "to see with the eyes of faith Darius and Cyrus, Alexander, the Romans, Pompey and Herod, working unwittingly for the glory of the Gospel." This history has a beginning and an end, and its center is Jesus Christ, "towards whom both Testaments are turned; the Old as its hope, the New as its model."[4]

The third model is exemplified in the development of science, where the secrets of nature are disclosed through experiments, "the only principles of physics."[5] If we can see further than our predecessors, that is because we are standing on a higher rung of the ladder of knowledge. This echoes what Lucan wrote in the first century, "Pigmies on the shoulders of giants see more than the giants themselves."[6] In Pascal's time, this famous metaphor of *the dwarf on the giant's shoulder* worked both ways: his position on the shoulders of the giant pointed to the broader scope of his knowledge, but the fact that the word *pigmy* was used argued the superiority of the Ancients. Godfrey Goodman, chaplain to Queen Elizabeth I and author of a gloomy study of decay entitled *The Fall of Man, or the Corruption of Nature Proved by the Light of our Natural Reason* believed the figure to be a recent conceit:

> But this great learned age hath found out a comparison, wherein we might seeme to magnifie the Ancients, but indeed very cunningly do

---

[4]*Pensées*, trans. by A. J. Krailsheimer. Harmondsworth: Penguin, 1966. Recent scholarship has made it possible to arrange the fragments of the *Pensées* into something resembling Pascal's actual intentions at the time of his death, and this English translation follows the new numbering given in the *Oeuvres Complètes*, edited by Louis Lafuma. Paris: Intégrale, Editions du Seuil, 1963. We shall quote the Pensées according to this method. The reference here is to fragment 317 and fragment 388.

[5]*Œuvres de Pascal*, II, p. 781, Popkin trans., p. 64. This is echoed in a lapidary sentence at the end of the treatises *On the Equilibrium of Liquids* and *The Weight of the Mass of Air*: "Experiment is the true master that one must follow in physics" (II, p. 1101; Popkin trans., p. 61; Spiers trans., p. 75).

[6]Lucan, *The Civil War*, book II, line 10. This is the source of Newton's remark in a letter of 5 February 1676 to Robert Hooke: "If I have seen further it is by standing upon the shoulders of Giants."

presse them down, making them our foot-stooles; preferring ourselves before them, extolling and exalting ourselves above measure; for thus it is said, that we are like dwarfs set upon the shoulders of Gyants, discerning little of ourselves, but supposing the learning and groundworke of the Ancients, we see much further than they, which, in effect, is as much as that we prefer our own judgments, before theirs.[7]

Pascal will have none of this pessimism. For him it is a characteristic of human nature that the achievement of one generation can be handed on to the next, which, in turn, passes on its own improvements. To ignore this distinguishing mark of reason is to reduce it to animal instinct that can only teach the same thing generation after generation. Beehives are as hexagonal now as they were a thousand years ago. Human beings alone grow in intellectual stature. "But it is not only that each person advances day by day in the sciences, today," writes Pascal,

> all mankind continually makes progress as the world grows older, because the same thing happens in the succession of human beings as in the different ages of an individual. The entire succession of human beings, during the course of so many centuries, ought to be considered as one and the same person who remains alive and never ceases to learn. From which we see how wrong we are to reverence philosophers for their antiquity. Since old age is the period most removed from infancy, who does not see that old age, in the case of this universal person [the human race], should be sought not in the period close to his birth but in those that are further removed? Those whom we call Ancients were inexperienced in all things and, properly speaking, constituted the infancy of mankind. Since we have added to their knowledge the experience of the centuries that have intervened, it is with us that the antiquity that we revere is to be found.[8]

This vindication of progress, and the idea that the sciences can be *augmented* had been stressed by Bacon in his *De Augmentis et Dignitate Scientiarum* (1623), which was translated into French by le sieur de Golefer, a member of an Auvergnat family acquainted with the Pascals. Bacon identified the exaggerated respect of Antiquity as a major obstacle to progress, and he used the analogy of the age of the world to claim that we

---

[7]Quoted in Richard Foster Jones, *Ancients and Moderns*. Berkeley and Los Angeles: University of California Press, 1965, p. 27.
[8]*Œuvres de Pascal*, pp. 782–783; Popkin trans., p. 65.

are older, hence wiser. In a famous aphorism in the *Novum Organum*, he writes:

> With regard to authority, it is the greatest weakness to attribute infinite credit to particular authors, and to refuse his own prerogative to time, the author of all authors, and, therefore, of all authority. For truth is rightly named the daughter of time, not of authority. It is not wonderful, therefore, if the bonds of antiquity, authority, and unanimity, have so enchained the power of man, that he is unable (as if bewitched) to become familiar with things themselves.[9]

Descartes echoed the same sentiment: "We do not honor the Ancients because of their antiquity for it is rather we who are ancients. The world is older now and we have more experience."[10] It is clear therefore that the idea was in the air when Pascal began to write, but the immediate source of his critical observation is probably Saint Augustine, who declares in the *City of God*: "The true education of the human race, as far as God's people is concerned, is like that of an individual. It advances step by step in time, as in the case of an individual when he reaches a new stage of life."[11] Pascal may have had this passage in mind when he wrote, "All mankind continually makes progress as the world grows older," but he transformed the Augustinian image by secularizing it and *introducing the idea of progress*. For Saint Augustine, the whole of history was a mere succession of events, and he even reintroduced the ancient Greek notion of cycles into the Judaeo-Christian concept of time, which is linear.

---

[9]Francis Bacon, *Novum Organum*, book I, aphorism LXXXIV. The source, not mentioned by Bacon, is the second-century writer, Aulus Gellius: "An old poet, whose name I forget, said that truth is the daughter of time" (*Noctes Atticae*, bk. XII, ch. 11). Since Gellius ascribes the saying to a poet of old, it can hardly be called an innovation. Another parallel can be found in Bacon's *Advancement of Learning*: "And to speak true, *Antiquitas saeculi, juventus mundi* (What we call antiquity is the youth of the world). These times are the ancient times, when the world is ancient, and not those which we account ancient *ordine retrogrado*, by a computation backward from ourselves" (Francis Bacon, *The Advancement of Learning*, book I, in Francis Bacon, *A Critical Edition of the Major Works*, edited by Brian Vickers. Oxford: Oxford University Press, 1996, p. 145).

[10]Quoted by Adrien Baillet, *La Vie de Monsieur Descartes*, vol. II, p. 531.

[11]Augustine, *City of God*, book X, chapter 14 (London: Heinemann, 1968, vol. 3, p. 313). Elsewhere, he writes, "The history of mankind from Adam to today is like the life of one person" (*De Vera Religione*, 27, n. 50, quoted in Philippe Sellier, *Pascal et Saint Augustin*. Paris: Albin Michel, 1995, p. 438).

Real Novelty

In his *Treatise on the Vacuum*, Pascal intended to show that progress implies continuous advance and that no step is ultimate. Just as Galileo and Torricelli would have been delighted to acknowledge that he went beyond them, so nothing would give him more pleasure than to see someone run ahead of him.[12] The surviving *Preface* offers as an instance of novelty the discovery of the nature of the Milky Way. The Ancients attributed its whiteness to a greater density of matter in that part of the sky. They described what they saw, but it would be unforgivable for us to hold the same view now that the telescope has shown that the Milky Way is composed of innumerable small stars. Cleo must have smiled at Pascal's oversimplified version of astronomical history, for several Ancients, including Aristotle, had already rejected the view that the Milky Way was a denser part of the heavens. In the Middle Ages, Averroes and Albert the Great had argued that it was a swath of densely packed stars, as Democritus had held before them.[13] Why did Pascal choose this example of novelty? The answer is that a *new* instrument, the telescope, put an end, "with ocular certainty," to empty disputation.[14] The telescope fascinated Pascal and he used it to query the size of things. Who can tell, he asks, whether the telescope offers an enlarged image or merely restores objects to their natural size? Consider what happens when we look at things through the wrong end of the telescope!

---

[12]Letter of Pascal to Antoine de Ribeyre, 16 July 1651, *Œuvres de Pascal*, II, p. 813.

[13]Aristotle maintained in his *Meteorology* that the Milky Way was formed, like comets, from hot vapors that rose from the uppermost part of the terrestrial sphere (345b31–346b15). Averroes and Albert the Great argued that the Milky Way was a collection of small stars because it showed no parallax, motion or change. Plutarch ascribes this view to Democritus (*De Placitis* III, I, 6), as does Macrobius in his *Commentarium in Somnium Scipionis*, XV, 6. At the end of the sixteenth century, the astronomer Tycho Brahe maintained that the Milky Way was made of the same matter as the stars but was more "diffused" in character. Christopher Clavius, an influential Jesuit who taught at the Roman College during this period, considered it more likely that the Milky Way was a denser part of the firmament that could absorb light from the sun, the view that the telescope was just about to discredit. On the history of the Milky Way, see Stanley L. Jaki, *The Milky Way, an Elusive Road to Science*, New York: Science History Publications, 1972, and the note of Isabelle Pantin in her edition and translation of Galileo's *Sidereus Nuncius*. Paris: Les Belles Lettres, 1992, pp. 82–83.

[14]Galileo, *Starry Messenger* (1610), translated in Stillman Drake, *Telescopes, Tides and Tactics*. Chicago, Chicago University Press, 1983, p. 18. In the *Opere di Galileo*, vol. X, p. 60.

Seventeenth-century writers often celebrate these new discoveries. For instance, Pascal's older contemporary, Tommaso Campanella, writes: "Those who think we should stay with Aristotle and other ancient philosophers, and not bother to inquire any further are either envious or have little insight and faith in God. This is especially the case after the light of the Gospel and the discovery of new stars that were unknown to the Ancients whose knowledge was elementary whereas ours is perfect. So that if we do not neglect the grace in us, we shall be better able to understand the world, which is the book of God."[15] Campanella comes perilously close to equating the new science with the Gospel. For him, *grace* was given in order to read the book of nature, which is the book of real wisdom. This religious slant is absent from the scientific writings of Pascal, who limits himself to drawing lessons in logic. For instance, he points out that whereas one disconfirming instance is enough to falsify a hypothesis, no limited number of cases, however large, can render a universal proposition absolutely certain. When we say that diamond is the hardest of all bodies, and gold the heaviest, we mean of all the bodies that are known, not all bodies necessarily. This is relevant to the vacuum:

> When the Ancients asserted that nature did not allow a vacuum, they meant that it did not allow it in all the cases they had seen, but they could not, without being rash, extend this statement to cases they did not know about. Had they known about them, doubtless they would have drawn the same conclusion we have, and would have sanctioned them, by their approval, with that antiquity that some now want to make the sole principle of science. Hence, without contradicting them, we can affirm the opposite of what they said. Whatever strength antiquity may possess, truth, even though just discovered, should always have more, because it is older than all the opinions about it. To imagine that it began to exist at the time that it began to be known is to ignore its very nature.[16]

Pascal's opponents thought he was brash and that he made exaggerated claims for the new science. They still clung to the myth of a Golden

---

[15]*Apologia pro Galileo*. Frankfurt, 1622, p. 20, facsimile reprint. Turin: UTET, 1968; English translation, *A Defense of Galileo*, by Richard J. Blackwell. Notre Dame: University of Notre Dame, 1994, p. 64.
[16]*Preface, Œuvres de Pascal*, II, pp. 784–785; Popkin trans., p. 66.

Age and the belief that nature was in its dotage. This was a deeply ingrained conviction that was not limited to conservatives in the academic world. The great poet, John Donne, weaved it in and out of his metaphorical conceits:

> So did the world from the first hour decay,
> That evening was the beginning of the day,
> And now the Springs and Sommers which we see,
> Like sonnes of women after fiftie bee.
> And new philosophy calls all in doubt,
> The Element of Fire is quite put out;
> The Sun is lost and th'earth, and no mans wit
> Can well direct him where to look for it . . .
> Prince, Subject, Father, Sonne, are things forgot,
> For every man alone thinks he hath got
> To be a Phoenix, and that they can bee
> None of that kinde, of which he is, but hee.[17]

Pascal did not share Donne's nostalgia for a lost paradise, but he considered the original sin of Adam and Eve an historical event that had greatly reduced the intellectual attainment and the moral excellence of mankind. He found in Scripture and the Church Fathers the moral truths that human beings require but are incapable of attaining by their natural powers. Although he stressed the difference between scientific inquiry and religious quest, his scientific outlook cannot be completely divorced from his religious ideas. The reason is intensely personal.

Pascal and the Religious Dimension of Human Experience

On the night of 23 November 1654, when Pascal was thirty-two years old, he had a religious experience that immediately and decisively determined the course of his few remaining years. A record of this experience survives in his own hand. It is a small piece of parchment that was found sewn in the lining of his doublet after his death, and which he carried with him at all times.

---

[17] John Donne, *An Anatomy of the World: The First Anniversary* (1611) in *Complete Verse and Selected Prose*, edited by John Hayward. London: Nonesuch Press, 1972, p. 202.

The year of grace 1654.
Monday, 23 November, feast of Saint Clement, Pope and Martyr,
    and of others in the Martyrology.
Eve of Saint Chrysogonus, Martyr and others.
From about half past ten in the evening until half past midnight.

Fire
"*God of Abraham, God of Isaac, God of Jacob*" [*Exodus* III. 6], not
    of philosophers and scholars.
Certainty, certainty, heartfelt joy, peace.
God of Jesus Christ.
God of Jesus Christ
"*My God and your God*" [*John* XX. 7].
"*Thy God shall be my God*" [*Ruth* I. 16].
The world forgotten, and everything except God.
He can only be found by the ways taught in the Gospels.
Greatness of the human soul.
"*O righteous Father, the world had not known thee, but I have
    known thee*" [*John* XVII. 25].
Joy, joy, joy, tears of joy.
I have cut myself off from him.
"*They have forsaken me, the fountain of living waters*" [*Jeremiah*
    II. 13].
"*My God wilt thou forsake me?*" [*Matthew* XXVII. 46].
Let me not be cut off from him for ever!
"*And this is life eternal, that they might know thee, the only true
    God, and Jesus Christ whom thou hast sent*" [*John* XVII. 3].
Jesus Christ.
Jesus Christ.

I have cut myself off from him, shunned him, denied him, crucified
him.
Let me never be cut off from him!
He can only be kept by the ways taught in the Gospel.
Sweet and total renunciation.
Total submission to Jesus Christ and my director.
Everlasting joy in return for one day's effort on earth.
"*I will not forget thy word*" [*Psalm* CXIX. 16]. Amen.[18]

---

[18]*Pensées*, fragment 913, pp. 309–310. The basic work on Pascal's relation to Holy Writ is J. Lhermet, *Pascal et la Bible*. Paris: Vrin, 1931. I have also found useful Hélène Bouchilloux, *Apologétique et raison dans les Pensées de Pascal*. Paris: Klincksieck, 1995.

As we can see from the passages in italic, which are all direct quotations from the Bible, Pascal's familiarity with Holy Scripture was such that it became both the source and the medium of his personal experience.

The Real Human Condition

Some time after his conversion, Pascal formed the idea of writing an *Apology for the Christian Religion* but he was not able to complete it. The fragments that were published as the *Pensées* are still bought by thousands of people each year and have assured Pascal's abiding fame. The audience that Pascal had in mind were the sophisticated freethinkers (*libertins*) who preferred polite agnosticism to militant atheism. Their ideal was the gentleman (*l'honnête homme*) whose code of behavior was etiquette rather than ethics. Pascal had admired their wit and liveliness, but he now saw their refined manners as a thin veneer for moral emptiness. What they, and himself before his conversion, took for cleverness was merely the avoidance of the implications of the human condition they claimed to analyze so sharply.

The gentleman's manual of right thinking was Montaigne's *Essais*. Pascal himself was not far from considering it a secular bible, and it was his contention that when read with a critical spirit it led straight to the truths of Christian revelation. Montaigne repeatedly stresses the frailty and futility of human life and argues that man's besetting sin is arrogance, based on exaggerated claims made for human reason, but easily confounded since the greatest thinkers throughout the ages were never able to agree among themselves. This shows that truth is not accessible by the light of reason alone and that man should give up all forms of intellectual pride. Pascal agreed with this analysis, but objected to the conclusion drawn by Montaigne, for whom the best policy is to stop worrying, follow our inclinations, and take things as they come. Montaigne gave up too soon. His realization of human weakness should have led him to look for an answer beyond mere humanity, not to concede defeat. For Pascal, weakness does not constitute the whole of the human condition, for it is inseparably, if paradoxically, linked with a greatness that cannot be denied or ignored. Hence the plan of the *Apology*: "First Part: Wretchedness of man without God. Second Part: Happiness of man with God."[19]

---

[19]*Pensées*, fragment 6. See also fragment 695.

As Alban Krailsheimer remarks, "Pascal's reaction to Montaigne's flight to a hammock of ignorance and indolence is to rock it so vigorously that even the bare ground of wretchedness below seems preferable. He uses every device of style to instil in his reader a feeling of desperate insecurity."[20] In a long fragment in a section entitled *Contradictions*, Pascal moves back and forth between the dogmatists and the sceptics to show that everyone is obliged to take sides because "anyone who imagines he can stay neutral is a sceptic *par excellence*."[21] We are a mystery unto ourselves. We know that we want truth and happiness, but we are unable to attain either. Our intellect is impaired and our emotions are crippled. The answer to this riddle is to be found in the biblical account of the fall of Adam and Eve. However far-fetched and revolting, the doctrine of original sin is the only one that can explain our predicament:

> Without doubt nothing is more shocking to our reason than to say that the sin of the first man has implicated in its guilt men so far from the original sin that they seem incapable of sharing it. This flow of guilt does not seem merely impossible to us, but indeed most unjust. What could be more contrary to the rules of our miserable justice than the eternal damnation of a child, incapable of will, for an act in which he seems to have so little part that it was actually committed 6,000 years before he existed? Certainly nothing jolts us more rudely than this doctrine, and yet, but for this mystery, the most incomprehensible of all, we remain incomprehensible to ourselves. The knot of our condition was twisted and turned in that abyss, so that it is harder to conceive of man without this mystery than for man to conceive of this mystery itself.[22]

The metaphysical and cognitive paradox of original sin is stressed in another fragment:

> Original sin is folly in the eyes of men, but it is put forward as such. You should therefore not reproach me for the unreasonable nature of this doctrine, because I put it forward as being unreasonable. But this

---

[20] Alban Krailsheimer, *Pascal*. Oxford: Oxford University Press, 1980, p. 52.
[21] *Pensées*, fragment 131.
[22] *Ibid.*, translation revised. Pascal collected quotations from the *Talmud* to show that the Jews believed in original sin (see *Pensées*, fr. 278, pp. 116–117). His source is the *Pugio Fidei*, written by Raymond Martini in the 13th century but only published in 1651. It was edited by Joseph de Voisin, a friend of Antoine Arnauld.

folly is wiser than all men's wisdom, *it is wiser than men*. For without it, what are we to say man is? His whole state depends on this imperceptible point. How could he have become aware of it through his reason seeing that it is something contrary to reason, and that his reason far from discovering it by its own methods draws away when faced with it?[23]

According to Pascal, to be truly rational is to recognize that there is an infinity of things beyond reason: "For instance, it is equally incomprehensible that God should exist or that he should not; that the soul should be joined to the body or that we should have no soul; that the world should be created or that it should not; that original sin should exist or that it should not."[24]

Reliable Witnesses

The origin of man's fruitless search for happiness lies in the past history of mankind, not in that of the individual. Pascal, like most of his contemporaries, was a fundamentalist. He believed that there was a first couple created by God in a state of innocence, and that their act of disobedience corrupted their nature and that of their descendants. God in his mercy promised a Redeemer, foretold in the Old Testament and proclaimed in the New. But the audience of freethinkers that Pascal had in mind did not believe that the Bible is inspired, and he could not get started without first establishing the reliability of Scripture as a source of knowledge. Here he faced a twofold problem: Is the Bible a trustworthy document and, if so, How do we interpret it? The stirrings of modern historical scholarship can be seen in a prior question: Is the Bible *one book* that has the same meaning throughout or is it a *collection of books* in which the words acquire different meanings or take on new colorations? A gifted and subtle writer, Pascal was conscious of the many lives of the same word: "Different arrangements of words make different meanings, and different arrangements of meanings produce different effects."[25] He compared his own dextrous placing of words to well-aimed shots at tennis: "Let no one say that

---

[23]*Pensées*, fragment 131 and fragment 695.
[24]*Pensées*, fragment 809.
[25]*Pensées*, fragment 784.

I have said nothing new; the arrangement of the material is new. In tennis both players use the same ball, but one places it better."[26]

As did most Christians of his time, Pascal found in the books of the Bible a fundamental unity of plan and purpose that indicated that they are by one hand. He could not state outright that God is their author since he was writing for agnostics, but he could claim that they were written by a *unique* people, which both *made* and *was made* by the Bible: "There is a great difference between a book made by one person and given to a people, and a book that makes a people."[27] The Jews are the "oldest known people" and "are descended entirely from one individual."[28] They are the only nation with an uninterrupted history and, hence, the ability to bear a continuous testimonial to past events.

> When the creation of the world began to recede into the past, God provided a single contemporary historian, and charged an entire people with the custody of this book, so that this should be the most authentic history in the world, and that all men could learn from it something that it was so necessary for them to know, and which could only be known from it.[29]

One of the reasons why we are not well acquainted with the history of our ancestors is that we hardly knew them. The longevity of the Patriarchs is an argument for the authenticity of biblical accounts, for when men lived so long (Methuselah was credited with 969 years), children had ample time to learn from their forefathers. For Pascal, Adam was known to Lamech, who was known to Shem who, in turn, had met Jacob who knew people who spoke to Moses. "Therefore," he concludes, "the Flood and the Creation are true." In another fragment, he argues that the ancient Hebrews were interested in what their ancestors had to say because they

---

[26]*Pensées*, fragment 696.

[27]*Pensées*, fragment 436. The crucial passage here is "a book that makes a people." I have amended Krailsheimer's translation in the light of the second of two contemporary copies of the *Pensées* (both made between 1662 and 1663), following Philippe Sellier in his edition of the *Pensées* (Paris: Mercure de France, 1976, p. 373). The warrant for this reading is textual but there is also a parallel passage in fr. 481, "Difference between a book received by a people and one that fashions a people."

[28]*Pensées*, fragment 451.

[29]*Pensées*, fragment 474.

had no other cultural or scientific interest. "Thus we see that people at that time took particular care to preserve their genealogy."[30]

To be reliable, a witness must be knowledgeable and sincere, and the Jews fulfil both conditions. The first because of the *contemporary* character of Jewish scriptures, which contrasts with narrations found among other nations:

> I am not surprised that the Greeks composed the *Iliad*, nor the Egyptians and Chinese their histories. You have only to see how this came about. These historians of fable were not contemporary with the things they wrote about. Homer composed a tale that was offered and accepted as such: for no one ever doubted that Troy and Agamemnon had ever really existed any more than the golden apple. He never meant to write history but to provide entertainment. He was the only writer of his times, and the beauty of the work made it endure. Everyone learns it and talks about it; it is something that has to be known and everyone knows it off by heart. Four hundred years later the witnesses of these things are no longer alive; no one knows any longer from his own knowledge whether the work is fable or history. It has simply been learned from earlier generations, and can pass for truth. Any history that is not contemporary is suspect: thus the Sibylline books and those of Trismegistus, and many others which have enjoyed credit in the world are false and have been found to be so in the course of time.[31]

The Jews also fulfilled the second condition, and their sincerity is demonstrated in the way they transmitted unflattering words about themselves:

> Lovingly and faithfully they hand on this book in which Moses declares that they have been ungrateful towards God throughout their lives, that he knows they will be still more so after his death, but that calls heaven and earth to witness against them that he told them so often enough.[32]

The Jewish people constitute virtually *one* personal witness to the genuineness of scriptural promises, and this continuity guarantees invariance and the absence of change.

---

[30] *Pensées*, fragment 296, and fragment 290.
[31] *Pensées*, fragment 436.
[32] *Pensées*, fragment 452.

Hermeneutic Rules

But granted that the Bible is a reliable historical document, how is it to be interpreted? First and foremost, says Pascal, the Bible must be taken as a single work with a coherent structure and a unity of purpose: "Every author has a meaning that reconciles all contradictory passages, or there is no meaning at all."[33] It follows that "to understand Scripture a meaning must be found that reconciles all contradictory passages; it is not enough to have one that fits a number of compatible passages, but one that reconciles even contradictory ones."[34] The only reconciliation possible is the prophetic interpretation of Scripture as pointing to Christ's life, death and resurrection. But not any figurative interpretation will do since this would open the door to the wildest opinions. "Some figures are clear and conclusive, but there are others that seem somewhat strained and only convince those who are already converted."[35] Pascal had in mind the authors of apocalyptic works, but the problem is much more general and had already been discussed by earlier writers. Saint Augustine had suggested that a non-literal interpretation of a biblical passage is warranted in three circumstances, namely when the literal sense is: (1) incomprehensible, (2) at variance with facts, or (3) in disagreement with what we know from the New Testament. Saint Augustine drew his criteria from the goal of Scripture as he understood it from a Christian standpoint, but Pascal is careful to avoid this committed stance. He tries to argue from what he finds in the text, and he agrees that the literal meaning must sometimes be denied "because the prophets themselves said so."[36] His point is that the Bible itself (and not only Christian apologists) invites the reader to look for a spiritual meaning whenever the literal meaning is unsatisfactory.

---

[33]*Pensées*, fragment 257.
[34]*Ibid.*
[35]*Pensées*, fragment 217.
[36]*Pensées*, fragment 272. The *order* of Scripture is one of Pascal's recurring concerns: "Against the objection that there is no order in Scripture. The heart has its order, the mind has its own, which uses principles and demonstrations. The heart has a different one. We do not prove that we ought to be loved by setting out in order the causes of love; that would be absurd. Jesus Christ and St Paul possess the order of charity, not of the mind, for they wished to humble, not to teach. The same with St Augustine. This order consists mainly in digressions upon each point which relates to the end, so that this shall be kept always in sight" (fragment 298).

In a passage in the *Pensées*, Pascal says that what we find in Montaigne is in us rather than in Montaigne.[37] The same is true of the Bible. After stating that the passage through the Red Sea, the conquest of Palestine, and the manna from heaven are all figures of redemption, Pascal adds a comment that reveals his underlying assumption:

> In these promises each person finds what is at the bottom of his own heart: either temporal or spiritual blessings, God or creatures. But with this difference, that those who are looking for creatures there find them indeed, but with many contradictions: they are forbidden to love them, bidden to worship and to love God alone (which comes to the same thing), and they find that the Messiah did not come for them. But those who are looking for God find him, without any contradictions, and find that they are bidden to love God alone and that a Messiah did come at the time foretold to bring them the blessings for which they ask.[38]

Knowledge can never be severed from experimentation in science or personal experience in religion. In physics, as we have seen, Pascal had devised a number of ways to investigate the space above the mercury in an evacuated tube. He showed that this space was empty of any material that could be lodged there (the candidates were air, water vapor, gases, or Descartes' subtle matter), but he was careful not to go beyond the evidence at hand. Although the facts pointed to a real vacuum, he did not affirm it without reservation. Let us recall how he summarized his position in the *New Experiments about the Vacuum*: "After having demonstrated that none of the materials which are perceived by our senses, and of which we have any knowledge, fills this apparently empty space, my view will be, until someone shows the existence of some material which might fill it, that it is truly empty and void of all matter."[39] Pascal saw physics as co-

---

[37] *Pensées*, fragment 689.

[38] *Pensées*, fragment 503. See also: "Two kinds of men in every religion. Among the heathen those who worship animals, and those who worship the one God of natural religion. Among the Jews those who were carnal and those who were spiritual, the Christians of the old law. Among the Christians the gross ones, who are the Jews of the new law. The carnal Jews awaited a carnal Messiah, and the gross Christians believe that the Messiah has dispensed them from loving God. True Jews and true Christians worship a Messiah who makes them love God."

[39] Pascal, *New Experiments about the Vacuum*, translated by Richard H. Popkin in *Pascal, Selections*. New York: Macmillan, 1989, p. 40.

extensive with actual experimentation and remained open to new and startling evidence.

If a scientist calls upon his eyes to witness the results of experiments, it is on his "heart" that an honest seeker after religious truth relies to discriminate between conflicting creeds. The pitfalls to be avoided in this case are excessive credulity and unreasonable skepticism. The first is exemplified in pagan religions, the second in the radical stance taken by people who are too clever by half. Pascal was familiar with this second attitude among some of his worldly friends. There was the Chevalier de Méré, who prided himself on his rigor in demanding a proof for every axiom in geometry but failed to realize that by doing so he precluded the very possibility of mathematics.[40] Next was a group of political scientists who undermined the customs that they claimed were essential to law and order. Finally, there were the freethinkers who failed to realize that to admit nothing but reason is to be most unreasonable.[41]

Pascal compares the Bible to a cipher that has an apparently plain meaning but is really meant to convey a secret message that has to be decoded:

> When we come upon an important letter, whose meaning is clear, but where we are told that the meaning is veiled and obscure, and that it is hidden so that seeing we shall not see and hearing we shall not hear, what else are we to think but that this is a cipher with a double meaning?[42]

Pascal was so convinced that Christians have the key to virtually every obscure passage of Scripture that he slipped unawares into the very excesses that he denounced when he claimed: "If the Jews had all been converted by Christ we would only have suspect witnesses left; if they had been wiped out we would have none at all."[43] Or again, in another fragment: "If they had taken these spiritual promises to their hearts and kept

---

[40]Antoine Gombaud, Chevalier de Méré, considered himself the epitome of the *honnête homme* and a paragon of common sense. He marveled at mathematicians who assumed that a straight line could be indefinitely divided, and asked for proof. Pascal tried to make him understand the role of axioms in geometry but gave up in sheer despair (see letter of 29 July 1654 to Pierre Fermat in *Œuvres de Pascal*, II, p. 1142).

[41]"Two excesses: to exclude reason, to admit nothing but reason" (*Pensées*, fragment 183).

[42]*Pensées*, fragment 260.

[43]*Pensées*, fragment 592.

them free from corruption until the coming of the Messiah, their testimony would have had no validity, because they would have been on his side."[44] This fanciful interpretation works against Pascal's own insistence on the reliability of the Jewish testimonial, for he can hardly claim that they are good witnesses because they were not converted. If this were the case, the testimony of those who acknowledged Christ, like the apostles, would be worthless. The argument is so unsatisfactory that it can be turned round with the same result. Indeed, Pascal does just this in another fragment, when he writes, "What could the Jews, his enemies, do? If they accepted him, they proved him by their acceptance, for it would mean that those entrusted with the Messianic hope were accepting him, and if they rejected him they proved him by their rejection."[45]

Pascal sometimes invokes the goal of Scripture as if he grasped God's design to the point of knowing what God *had* to do: "To inspire faith in the Messiah, previous prophecies were necessary and these had to be handed down by witnesses above suspicion."[46] Even if allowances are made for the rhetoric of the pulpit, this seems to imply that God could have only one plan of salvation, something Pascal denies elsewhere.

But what is the historical evidence for Jesus' mission? Pascal recognizes that it depends very largely upon the authors of the four Gospels and the apostle Paul. What if they conspired to deceive the first disciples? Pascal considers this totally unlikely:

> The hypothesis that the Apostles were rogues is quite absurd. Follow it out to the end and imagine these twelve men meeting after Jesus' death and conspiring to say that he had risen from the dead. This means attacking all the powers that be. The human heart is singularly susceptible to fickleness, to change, to promises, to bribery. One of them had only to deny his story under these inducements, or still more because of possible imprisonment, tortures and death, and they would all have been lost. Follow that out.[47]

This is a traditional argument for the authenticity of the miraculous and prophetic revelation of Jesus, but Pascal's clinching argument is a moral one: the proof that the biblical doctrine is genuine lies in the up-

---

[44]*Pensées*, fragment 502.
[45]*Pensées*, fragment 262.
[46]*Pensées*, fragment 572.
[47]*Pensées*, fragment 310.

lifting experience that accompanies its acceptance. It is a life-giving miracle, and it speaks to the heart before addressing the mind.

Science and Religion

Galileo was condemned in 1633 when Pascal was ten years old and this sent shock-waves throughout the European scientific community, but French Catholics generally regarded the Roman intervention as an administrative rather than a doctrinal matter. When Pascal mentions the Copernican theory for the first time, it is not to brand it as suspect but to point out that it explains the motions of the planets as well as the rival geocentric systems of Ptolemy or Tycho Brahe. But this does not give it greater cogency, because agreement with facts is not enough for a theory to be true. Pascal illustrates the underlying logic with the aid of a homely analogy: if we find a hot stone in the street, we might suppose that it has just been taken by carriage from a house on fire but we cannot affirm this until we examine other possibilities such as the heat of the midday sun or the friction caused by the rubbing of carriage wheels. Any bright person can come up with a clever hypothesis; the snag is to make it stick.[48]

Humility as well as prudence is required when examining biblical statements about natural events that are not easy to ascertain. In some instances, the last word of science is found to echo old truths from the Bible. Here is Pascal's illustration: "How many things unknown to earlier philosophers have telescopes revealed to us! We boldly took Scripture to task over the great number of stars, saying: 'There are only 1,022 of them; we know.' "[49] Pascal had in mind the passage in *Genesis* 22,7 where God blessed Abraham and promised: "I will multiply your seed as the stars of the heavens, and as the sand that is on the seashore." The source for the number of stars is the *Almagest*, where Ptolemy lists 1,022 stars: 706 in the Northern Hemisphere (*Almagest*, bk. VII, ch. 5) and 316 in the southern (bk. VII, ch. 1). In Ptolemy's day, science seemed to go against the biblical estimate of the number of stars, but recent astronomical progress shows that Ptolemy was wrong and the Bible right. Note how Pascal explodes (and ridicules) the pretensions of science by adding two telling

---

[48]Letter to Etienne Noël, 29 October 1647, in Pascal, *Oeuvres de Pascal*, II, p. 524.
[49]*Pensées*, fragment 782.

words to a matter-of-fact report: "There are only 1,022 of them; *we know*." As is frequently the case with Pascal, the infinitely large brings to mind the infinitely small, and he continues in the same blend of irony and awe: "There is grass on earth; we can see it. From the Moon it could not be seen. And on this grass there are hairs, and in these hairs little creatures, and beyond that nothing? Presumptuous man!"[50]

Science without Fetters

Between 23 January 1656 and 24 March 1657, Pascal wrote 18 letters in defense of his Jansenist friends who had been accused of heresy by the Jesuits. The last of these letters, which are known as the *Provincial Letters*, contains a vigorous defense of the autonomy of scientific research.

> How then do we learn what the facts are? From our eyes, Father, which are the rightful judges of facts, as reason is of natural and intelligible things, and Faith of things supernatural and revealed. But since you compel me to do so, Father, let me tell you that in the opinion of two of the greatest doctors of the Church, Saint Augustine and Saint Thomas, these three principles of knowledge, the senses, reason and Faith each have separate objects, and are certain within their range.[51]

> Faith is so far from destroying the certainty of the senses that it would, on the contrary, destroy Faith to cast doubt on the faithful evidence of the senses. No one, not even the Pope himself, could make decrees that run counter to ascertained facts. Hence the futility of the condemnation of Galileo, for the motion of the earth is a scientific theory that might one day be confirmed:

> > It was in vain too that you obtained from Rome the decree against Galileo, which condemned his opinion regarding the movement of the Earth. It will take more than that to prove that it keeps still, and if there were consistent observations proving that it is the earth that

---

[50]*Pensées*, fragment 782.

[51]Pascal, *Lettres écrites à un Provincial*. Paris: Garnier Flammarion, 1967, p. 265. Galileo made a similar point in his *Letter to Christina of Lorraine*, which was written in 1615 but only published by the Elzeviers in 1636.

goes round, all the men in the world put together could not stop it turning, or themselves turning with it.[52]

When a question arises it is therefore essential to know by what principle it is to be judged:

> If it is something supernatural, we will judge it neither by our senses nor by reason but by Scripture and the rulings of the Church. If it is something that is not revealed and is adapted to natural reason, then natural reason will be its first judge. And if it is a matter of fact, we will believe the senses by which we naturally know such things.[53]

Facts are God's infallible words, says Pascal, and when a literal interpretation of Scripture is at variance with facts, a new interpretation is called for. Scripture can be interpreted in different ways, "but the testimony of the senses is unequivocal, and we must, in these cases, take as the true interpretation of Scripture the one that agrees with the faithful testimony of the senses."[54] In the first chapter of *Genesis*, the Moon is said to be as great as all the stars, but since the stars are known to be larger than the Moon, the passage was interpreted by Thomas Aquinas as indicating the size of the Moon as it is relative to us and not as it is in itself.

Pascal was impressed but never convinced by Galileo's arguments for the motion of the Earth. This is partly because he was less interested in cosmological speculation than in the question of our eternal destiny. Rather than dwell on Copernicus' theory, he would have us enquire about the nature of our soul. He may have felt that Galileo had got his priorities wrong, and that while we might know more and more about the universe, we might come to understand less and less and less about the place we are meant to occupy in it. For Pascal, this would be but an instance of the *divertissement* that shields us from biblical truth, and keeps us asking who we really are. In order to know this, we have to come to know God, but unless we are gifted with the kind of illumination that Pascal was vouchsafed on the night of the 23 of November 1654, we must follow our reason and develop a rigorous method of reasoning. Pascal attempted to show us how to do this in his *On the Geometrical Mind* and *On the Art of Persuasion*, to which we turn in the next chapter.

---

[52]Ibid., p. 267.
[53]Ibid., p. 265.
[54]Ibid., p. 266.

CHAPTER 9

The Use of Logic and the Role of Experiments

Pascal had a long-standing interest in logic and philosophy of science. As early as 1647, he had reminded Father Noël that a proposition can be recognized as true only if it is self-evident (as is the case with the axiom, *if equals are added to equals, the sums are equal*), or if it can be deduced from something that is self-evident (for instance, the statement that *three angles of a triangle are equal to two right angles*, which is not obvious in itself, can be shown to follow from self-evident axioms). Any other proposition awaits confirmation from evidence.[1] Pascal returned to this question in two essays, *On the Geometrical Mind* and *On the Art of Persuasion*, which he drafted between 1655 and 1660, and that we shall examine in this chapter.

From the two geometrical examples that he had used as illustrations in his letter to Noël, we will not be surprised to learn that Pascal was convinced that among equally sharp persons, "those who know geometry always have the edge," and that we should not waste our time on formal logic when we can do mathematics.[2] This does not mean, however, that geom-

---

[1] Letter of Pascal to Father Noël, 29 October 1647, *Œuvres de Pascal*, II, p. 519; Popkin trans., pp. 49–50.

[2] *On the Geometrical Mind*, *Œuvres de Pascal*, III, p. 391; Popkin trans., p. 174. Pascal ends *On the Art of Persuasion* by recalling that "it is not *barbara* and *baralipton* (two figures of syllogism) that teach us how to reason" (*Œuvres de Pascal*, III, p. 428; trans. p. 193). Galileo made the same point in the First Day of the *Dialogue on the Two Chief World Systems*, "The art of demonstration is learnt by reading works which contain demonstrations.

etry has the "perfect" method, which would require that all terms be defined and all propositions proven. This is impossible because the terms from which mathematical reasoning starts presuppose other terms, which presuppose others in turn, until we arrive at terms that can no longer be defined by anything clearer, and that we take for granted without proof.

Fortunately, rational discourse does not need perfect definitions. It can also start from *nominal definitions* or *arbitrary postulates*. For instance, we call numbers that are exactly divisible by two *even numbers*, "which shows," says Pascal, "that definitions are freely chosen, for nothing is more permissible than to give a clearly designated thing any name we wish."[3] It is important to bear in mind, however, that to define something is not to affirm its existence. Once chosen, a definition must retain its meaning and exclude all other ones. Should the same word happen to be used for different things, we must each time take care to mentally substitute the appropriate definition.

In geometry, all definitions are nominal and are coined within the context of a given set of problems. For instance, Euclid excluded the *unit* from his definition of *number*, which he called "*a multitude composed of units*,"[4] but this does not mean that *units* cannot be counted since they can be added like any other numbers. Euclid introduced his nominal definition because he had in mind several properties that are common to all numbers except unity, and he wanted to avoid having to say over and over again, "for all numbers save unity, such and such a condition is to be found." In other contexts, it is more convenient to include the unit in the definition, as was actually done by Pascal because he wanted to apply to all numbers, which Euclid treated as discontinuous quantities, what he had said of dimensions (lines, surfaces, solids), which he considered as continuous quantities.

These are mathematical treatises, not books on logic" (*Opere di Galileo*, VII, p. 60). His disciple, Torricelli, extolled geometry as "the only liberal art that sharpens the intellect and renders it capable of enhancing the state in time of peace and defending it in time of war. All things being equal, a mind trained in geometry has a special and manly strength that is revealed and excels in architecture, warfare and naval matters" (*Dimensions of the Parabola* in Torricelli, *Opere*, I, p. 95, quoted in Dominique Descotes, *L'argumentation chez Pascal*. Paris: PUF, 1993, p. 31, n. 1). Pascal avoided the flowery language of Torricelli.

[3]*On the Geometrical Mind*, *Œuvres de Pascal*, III, p. 394; Popkin trans., p. 174.

[4]*Ibid.*, p. 408; Popkin trans., p. 183. In Book VII of his *Elements*, Euclid defines a *unit* (*monas*) as "that by virtue of which each of the things that exist is called one" (Def. 1) and a *number* (*arithmos*) as "a multitude composed of units" (Def. 2).

We must not assume that a particular usage of a word, for instance the definition of time as "the measure of motion," is the only *correct* meaning of that word. "If I were not convinced of how necessary it is to understand this clearly," stresses Pascal, "I would not insist so much, but my experience of verbal confusion teaches me that we cannot ask for too much clarity, and it is mainly for this reason that I write this treatise."[5] Pascal applied the method that he recommended in an *Introduction to Geometry* that he wrote at the time he was drafting *On the Geometrical Mind*. Euclid had not attempted a general definition of geometry but had opened his *Elements* with limited and specific definitions of point, line, and so on. Pascal tried to replace some of Euclid's demonstrations by clearer and more natural ones. First, he defined geometry as the science of space. "The object of pure geometry," he declares, "is space which extends in three different directions that are called dimensions and are known as length, width and depth, which are self-evident notions."[6] Second, he saw that we can distinguish lines not only by their size but also by their *position*, thereby rendering possible a formulation of geometrical propositions that does not involve measurement. This was later developed by Leibniz and gave rise to topology in the nineteenth century. Third, Pascal saw the importance of the notion of *direction* that was to have such a great future in mathematics. In this he was probably influenced by Roberval, who defined the direction of a curve at a given point as its tangent, thereby allowing any curve to be characterized by the successive directions of its tangent at each of its points.

Primitive Terms

As we have seen, Pascal held that two kinds of *principles* are naturally known: (a) *primitive terms* and (b) *self-evident axioms*. Proper reasoning does not consist in attempting to define everything, or in refusing to define anything, but in defining what is not self-evident. This is why geometers do not define space, time, motion, or number, which are so obvious that any attempt to clarify them would merely render them obscure. Descartes had already made a similar point in his *Principles of Philosophy*,

---

[5]*Ibid.*, pp. 398–399; Popkin trans., pp. 177–178.
[6]*Introduction to Geometry*, Œuvres de Pascal, III, p. 435.

Part I, art. 10, where he wrote: "I have often noticed that philosophers make the mistake of employing logical definitions in order to explain what is already very simple and self-evident. The result is that they only make matters more obscure." Descartes and Pascal both saw geometrical reasoning as proceeding from the simple to the complex rather than from the general to the particular, as is the case in syllogistic deduction. But whereas Descartes made an existential claim for his first principles, Pascal considered them simply as postulates. If Descartes was mainly concerned with the intuitive clarity that warrants moving from one link in the deductive chain to the next, Pascal was more mindful of semantic slippage, namely the danger of changing the original meaning of definitions. For Descartes, logic is grounded in metaphysics; for Pascal, it is the means of constructing formally valid deductive systems. This difference in outlook can best be seen in their stated objectives: Descartes offers a method of discovery, and his aim is well expressed in the title of his most popular work, *A Discourse on the Method of Rightly Conducting One's Reason and Seeking the Truth in the Sciences*. Pascal's goal is more low-keyed. He does not teach how to make discoveries but how to structure available knowledge. The reason is that Pascal does not believe that physics can be deduced from a set of *a priori* principles, however clear and obvious. As he puts it in the *Conclusion* to his hydrostatical treatises, "Experiments are the true masters we must follow in physics," something Descartes could not have brought himself to write.

Pascal had no faith in metaphysical proofs, not only because they are removed from everyday understanding, but because they are open to doubt as soon as we cease to attend to them. For instance, he described the philosophical proofs of the existence of God as "so remote from human reasoning and so involved that they make little impact, and even if they did help some people it would only be for the moment during which they see the demonstration, because an hour later they would be afraid they made a mistake."[7] Descartes was conscious of this difficulty under a different guise: he worried about remembering each step in a sequential relationship of the type $A = B$, $B = C$, $C = D$, $D = E$, hence $A = E$. His advice was to go over the entire process in one continuous and uninterrupted sweep in such a way that memory had practically no role to play and the whole could be intuited at a glance. Pascal would have agreed with him that a principle is apprehended in a flash of intensely personal insight, but

---

[7]*Pensées*, fragment 190.

he would have added that it must be corroborated by shared experiments and active communication among scientists.

Real and Fanciful Definitions

In *On the Geometrical Mind*, Pascal offers three examples of the danger of unwarranted or slipshod definitions. The first is the notion of *being*, which we cannot begin to define without saying, "It is," and thus employing the word to be defined in the definition. The second is the concept of *man*, which Plato was said to have defined as a *featherless biped*. This invited Pascal's mock indignation: "As if the idea of man that I naturally have but cannot express was not clearer and more certain than the ridiculous and useless explanation that he gives! A man does not lose his humanity by losing his legs, just as a capon does not become a man upon losing his feathers." Pascal's source was Montaigne, who declared in his *Essais*: "The same Plato, who defines man as a chicken, says elsewhere, after Socrates, that he does not know really what man is."[8] But there is a significant difference between Pascal and Montaigne. For Pascal, the *naturalness* of the idea allows us to know *well enough* what a man is, whereas for Montaigne the definition merely illustrates the *fatuousness* and *vacuousness* of all attempts at defining concepts.

Pascal 's third example is a tautology passing muster for a definition. "There are some," he writes, "who go to the ridiculous length of explaining a word by the same word, and I know people who define light as *the luminary motion of luminous bodies*, as if the words *luminary* and *luminous* could be understood without that of light."[9] Pascal was thinking of Father Noël, who had described light as "a luminary movement of rays composed of lucid bodies that fill transparent bodies and are luminarily moved by other lucid bodies, just as steel filings are magnetically moved by a loadstone." In his reply, Pascal had chided the Jesuit for using the very word to be defined in the definition of the word: "The sentence that precedes your concluding courtesies defines light in these words: Light is a luminary movement of rays composed of lucid bodies, *that is to say, luminous ones*. Now it seems to me that it would first have been neces-

---

[8]Montaigne, *Essais*, book II, chapter 12, which is entitled *Apology of Raymond Sebon*, in the Villey-Saulnier edition, p. 545.

[9]*On the Geometrical Mind*, *Œuvres de Pascal*, III, p. 396; Popkin trans., p. 176.

sary to define *luminary, lucid* and *luminous body*, for until that is done I will not be able to understand what light is."[10] Note that Pascal sharpened Noël's definition to make it appear more ludicrous by adding *"that is to say, luminous ones."* He was careful, however, not to claim that light is a self-evident notion like the idea of being, and he made no attempt to say what light is because experiments did not allow him to discriminate between competing theories. Where experiments are ambiguous, definitions are no more than convenient labels.

Descartes also thought that the study of light was an essential part of physics and he gave his cosmological treatise, *The World*, the subtitle, *Treatise on Light*. But Descartes was willing to rush ahead of experiments and he boldly declared that light was the pressure exerted by minute spheres of matter that strive to move away from their center of rotation. Pascal called this wild speculation and, in the *Pensées*, he amuses himself at Descartes' expense by commenting on his treatment of heat and light:

> When they declare that heat is just the movement of some globules, and light the *conatus recedendi* [tendency to move away] that we feel, we are amazed. What! Is pleasure nothing but a dance of spirits? Yet we have such a different conception of it! Our sensations seem so far removed from what they are compared to and are said to be. Heat affects us in a completely different way from what we touch, hear or see. All this is mysterious, but is nonetheless said to be as simple as throwing a stone![11]

Not all words can be defined, and yet we use these words with the same assurance as if they had been explained in a completely unambiguous way:

> We assume that everyone thinks alike, a gratuitous assumption for which we have no proof. I see that some words are used in the same circumstances, and that every time two men see a body change position they both use the same word to express what they have seen, each of them saying that the body has moved. Such conformity of application provides a strong presumption of conformity of thought, but it is not absolutely convincing. We are willing to bet that it is so

---

[10]Letter of Pascal to Noël, 29 October 1647, *Œuvres de Pascal*, II, p. 527; Popkin trans., p. 55.

[11]*Pensées*, fragment 686. In his *Principles of Philosophy*, Part IV, art. 29, Descartes had defined heat as "the agitation of terrestrial particles in reference to the sense of touch."

but we know that identical conclusions are often drawn from different assumptions. That is enough to cloud the issue, to say the least, but it does not completely extinguish the natural light which provides us with certainty in such matters.[12]

We have to distinguish, therefore, between the ordinary use of reason, namely the ability to deduce the consequences implied by clearly perceived principles, and the knowledge of these first principles. These are grasped through what Pascal called the heart:

> Knowledge of first principles, such as space, time, motion, and number, is as solid as any derived through reason, and it is on such knowledge, coming from the heart and instinct, that reason has to depend and base all its argument. *The heart feels* that there are three spatial dimensions and an infinite series of numbers; reason goes on to demonstrate that there are no two square numbers of which one is double the other. *Principles are felt*, propositions proved, and both with certainty though by different means. It is just as pointless and silly for reason to demand proof of first principles from the heart before accepting them as it would be silly for the heart to demand an intuition of all the propositions demonstrated by reason before agreeing to them. Our inability serves only to humble reason, which would like to be the judge of everything, but not to undermine our certainty. As if reason were the only way we can learn! Would to God, on the contrary, that we never needed it and knew everything by instinct and feeling! But nature has refused us this blessing, and has given us instead very little knowledge of this kind.[13]

Time Cannot, and Motion Need Not, Be Defined

People agree on the everyday use of notions such as motion, space, and time, but this does not imply that they fully understand what the words really mean. The Aristotelians defined motion as the "actualization of what exists potentially in so far as it exists potentially" and assumed that it was the only correct one.[14] Descartes, always bent on caricature rather

---

[12]*Pensées*, fragment 109.

[13]*Pensées*, fragment 110.

[14]Pascal, *On the Geometrical Mind*, *Œuvres de Pascal*, III, pp. 397–398; Popkin trans., p. 177. Pascal had read the famous passage in the *Confessions* where St. Augustine voiced

than sympathetic understanding, dismissed their definition as an instance of "seeking a knot in a bulrush,"[15] but Pascal realized that it was not ridiculous. When Aristotle described change (by which he meant *movement in general*) as the process of actually realizing "what exists potentially in so far as it exists potentially," he explained the qualification "in so far as it exists potentially" with the example of building a house. The *act* of building is not the actualization of the whole character of the bricks that are used but only of that aspect of their character that can serve as building material. It is clear that before we start building it is essential to know whether the bricks have the potential of becoming suitable material for construction.

If Descartes dismissed the Aristotelian definition of motion as *strange* and passing comprehension, he offered the following as perfectly natural: "motion is the transfer of one part of matter, or one body, from the vicinity of those bodies that are in immediate contact with it, and which are regarded as being at rest, to the vicinity of others."[16] This did not seem much of an improvement to Pascal, who thought that it required just as much proof as Aristotle's.

We may not be able to define motion, space, and time, but we can know some of their properties. The most significant, according to Pascal, is that they can be increased or decreased without end because "they are suspended between nothing and infinity, and always infinitely removed from these extremes."[17] This appeared so clear to Pascal that he was be-

---

his perplexity, "What is time? If no one asks me, I know, if someone asks, I don't know" (*Confessions*, bk XI, ch. 13). St. Augustine considered both the hypothesis that time is the motion of bodies and the one that it is that by which the motion of bodies is measured. Aristotle, in his *Physics*, had defined time as "the measure of motion" (bk IV, ch. XI, 220a 25–26).

[15]Descartes, *Rules for the Direction of the Mind*, in *Œuvres de Descartes*, XI, p. 426. In a letter to Mersenne of 16 October 1639, Descartes writes: "If we attempt to define things that are very simple and naturally known such as shape, size, motion, place, time, etc., we obscure them and just get them confused. For instance, a person walking up and down a room conveys a better idea of what motion is than one who says: *est actus entis in potentia prout in potentia*" (*Œuvres de Descartes*, vol. II, p. 597).

[16]Descartes, *Principles of Philosophy*, part II, art. 25, *Œuvres de Descartes*, vol. VIII-I, p. 53.

[17]*On the Geometrical Mind*, *Œuvres de Pascal*, III, pp. 402–403; Popkin trans., p. 180. We also know that they are connected because "God made everything according to weight, number and measure," writes Pascal, paraphrasing the biblical verse, "Thou has arranged all things by measure, number and weight" (*Book of Wisdom* in the Catholic Vulgate edition, chapter 11, verse 21). This is one of the rare instances where a biblical passage is adduced by Pascal in a discussion of scientific method. Descartes quoted the same verse in *The World*, chapter 7 (*Œuvres de Descartes*, vol. XI, p. 47).

mused when "otherwise very capable persons" of his acquaintance objected that they could not imagine how we could go on dividing space without eventually arriving at extended and physically separated parts. One of these was the Chevalier de Méré whom Pascal describes in a letter to Pierre Fermat as "a very able man but no geometer. He does not even understand that a mathematical line can be divided *ad infinitum*, but believes that it is made up of a finite number of points. I have not been able to rid him of this notion. If you could do so, you would make him perfect."[18]

Méré may have been a poor mathematician but he had a high opinion of himself, and in a collection of letters that he published after Pascal's death we find one intended for Pascal in which he offers to cure him of the illusion "that the small bodies that we discussed the other day can be divided *ad infinitum*."[19] It is doubtful whether Méré actually sent this conceited letter to Pascal (there are no traces of it in the surviving correspondence), but if he did Pascal could have replied as he does in *On the Geometrical Mind*:

> No one can be a geometer without believing that space is infinitely divisible any more than he can be a man without a soul. Nevertheless, no one understands infinite division. Our only assurance of this inconceivable truth is the sole reason, which is certainly sufficient, that we understand perfectly that it is false to say that by dividing a given space we can arrive at an indivisible part, that is to say, one which would have no extension. For what is more absurd than to claim that by continually dividing a space, we finally reach a division such that after dividing it in two, each of the halves remains indivisible and without any extension, and that thus two nothings of extension together make an extension?[20]

Infinity can be known but we cannot comprehend its nature. We can grasp that numbers are not finite, and therefore that there is an infinite number, but we do not know what that number is. It cannot be even, and it cannot be odd, for adding a unit does not change its nature. Yet it is a number, and every number is even or odd!

---

[18] Letter of Pascal to Fermat, 29 July 1654, *Œuvres de Pascal*, II, p. 1142.
[19] Meré, *Œuvres de Pascal*, III, pp. 353–354.
[20] *On the Geometrical Mind*, *Œuvres de Pascal*, III, pp. 404–405; Popkin trans., p. 181.

Another Use of Reason

Reason usually goes through a sequence of logical steps, but many valuable insights are arrived at through an indirect route by showing that a proposition is true because its contradictory leads to an absurdity. The procedure is known as a *reductio ad absurdum*, and Pascal uses it to determine whether space is infinitely divisible, a proposition that might initially seem undecidable. The proof runs as follows: if two indivisible parts could be joined together they would either touch everywhere or in one place only. If they touched everywhere, they would be the same thing and, consequently, indivisible. If they touched in one part only, then they would have parts, but this is impossible because an indivisible, by definition, cannot have parts. Hence the notion that two indivisible parts could make up an extension is a manifest contradiction. It follows that space is always composed of divisible parts.

A second argument runs along similar lines: if space was composed of indivisible points and we were given two squares, one of which was double the other, then the larger would contain twice as many indivisible points as the smaller. Let those who believe this, says Pascal, arrange dots in squares until they have two squares such that one has twice the number of dots in the other! "When they achieve this," he promises, "I will see that all the geometers in the world acknowledge their feat."[21] Why even Méré would soon discover that trying to build a square with twice the number of units of a given square is doomed to failure! But what if Méré shifted his ground and asked, How can we run through an infinite number of parts in a finite time? To this Pascal answers that an infinite number of spatial parts can be paired with an infinite number of instants into which any finite period of time can be divided. The reader will remember some such reply from his introductory course to the calculus, but we can doubt whether Méré would have been convinced. This is why Pascal went on to provide a simpler illustration to show that a small extension can have as many parts as a large one.

If we observe the heavens through a telescope, he writes, we find a one-to-one relation between each small part of the surface of the lens and each huge region of the sky. Now let us imagine that an object is enlarged with more powerful lenses until it is the very size of the firmament. "Can

---

[21]*On the Geometrical Mind, Œuvres de Pascal*, III, p. 405; Popkin trans., p. 181.

we know," muses Pascal, "whether the lenses have actually changed the natural size of the objects or merely re-established the correct size that the shape of our eye had altered and reduced, as do lenses that make things appear smaller?"[22] There is a real puzzle here, according to Pascal, because we are "located between infinite extension and no extension, infinite number and no number, infinite motion and no motion, infinite time and no time. This teaches us our true worth and that we should think about things that are more important than the whole of geometry."[23]

Two Types of Mind

Pascal returns to the difficult problem of *instinctive* or *intuitive* knowledge in two famous fragments of the *Pensées*. In the first, entitled "The difference between the mathematical mind (*esprit de géométrie*) and the intuitive mind (*esprit de finesse*)," he stresses that mathematical truths are not the only ones to which we have access. Through intuition we grasp highly complex matters whose first principles are so numerous and involved that they cannot be analyzed and laid out like the first principles of mathematics. An intuitive person will grasp such matters not by proceeding deductively, step by step, but at a glance and as a whole. This kind of knowledge is very important but has little to do with solving problems in mathematics where the principles are not numerous and can be clearly stated.

Mathematical principles are abstract and removed from common usage, and it takes a rigorous, logical mind, rather than an intuitive one, to spell out their implications. Someone who has a great talent for mathematics may be lacking in intuitive insight. The mathematical mind often goes astray in practical matters because it is trained to attend only to a few clearly perceived principles and neglect others that an intuitive person takes into account without knowing that he does. As Pascal puts it, the intuitive mind "feels" these principles, rather than sees them, and uses them "tacitly, naturally and without plan."[24] The ideal mind would, of

---

[22]*Ibid.*, III, p. 406; Popkin trans., p. 182. In *Pensées*, fragment 809. The last antinomy is crucial to the case Pascal is trying to make for Christianity. He insists that we cannot have any knowledge of ourselves without "the mystery that is most removed from our knowledge," namely original sin.

[23]*Ibid.*, III, p. 411; Popkin trans., p. 185.

[24]*Pensées*, fragment 512.

course, have both kinds of understanding, but this is rare, and Pascal may be one of the few who combined deep mathematical insight with an equally profound understanding of the human heart.

As was his custom, Pascal toyed with this idea in a variety of ways. In a second fragment, it is the *mathematical mind* ("*l'esprit de géométrie*") that handles many principles without confusing them, and the *intuitive mind* (here called "*l'esprit de justesse*") that deals with those few that can be "penetrated keenly and deeply."[25] For instance, the principles required to understand the transmission of pressure through water are not numerous (there are essentially only two: the principle of virtual displacement and the principle of uniform distribution) but it is not always easy to see how they should be applied. The kind of mind that can do this is sharp and accurate but somewhat narrow, whereas the mathematical mind has greater breadth. It is clear that in this fragment Pascal is not distinguishing between the scientific and the non-scientific frame of mind but between two kinds of scientific minds. The first can follow to the last detail a deduction from very few principles, but is at a loss when the principles are too numerous or involved. The second kind of mind, called mathematical in this context, is able to attend to several principles and to discern the more significant relations between them, thereby arriving at deductions that could never be attained by the mind that is exact but narrow.

In the first fragment, where Pascal spoke of the *intuitive mind* (*l'esprit de finesse*), he was mainly concerned with the type of intelligence that is revealed in social and political matters, whether it be the choice of a friend or the criticism of a new party platform. In matters such as these, a complete analysis eludes us, and our conclusions are not supported by a rigorous chain of reasoning. In the second Fragment, Pascal is thinking of the way the *mathematical mind* arrives at conclusions that are clearly demonstrated and valid for anyone who accepts the rules of mathematical inference. There is a problem here, of course, since more than mere logical rigor is needed to arrive at these conclusions. It also takes the ability to see the significant relations between the principles, and to choose the most fruitful line of deduction from among many possibilities. Pascal, who was setting himself and others a number of novel problems in probability theory and in what came to be known as integration, understood

---

[25]*Pensées*, fragment 511.

these requirements. He forged new tools of analysis that anticipated the calculus of Newton and Leibniz. In comparison with these tasks, applying the principles of hydrostatics to atmospheric pressure was straightforward. He saw this work as belonging to the accurate kind of mind ("*l'esprit de justesse*") but even here something more than the ability to forge a logical chain of thought is necessary. Pascal had to arrange the material of the two treatises, *On the Equilibrium of Liquids* and *On the Weight of the Mass of Air*, in a way that was not merely consistent but convincing. A simple list of logical steps would have left many readers cold. Indeed, Pascal may have had no readers at all.

In the hard sciences as well as in the humanities, there would be no new knowledge if we only relied on the disclosure of our senses and the mechanical application of rules of logic. More is needed, as Pascal recognized. First of all, it is necessary to sift the evidence and choose those elements that are worth attending to, something that cannot be done by appealing exclusively to past experience and accepted rules of inference. Next, a hypothesis must be framed about the nature of the relations between these elements, and it must be formulated in a way that is open to experimental confirmation. Pascal is insistent that once the hypothesis is clearly stated it must be interpreted in the light of existing laws, and here much depends on the way these laws are selected.

Let us examine a specific instance, namely the three principles that Pascal highlighted in his *Treatise on the Equilibrium of Liquids*: (a) Torricelli's principle on the lowering of the center of gravity, (b) the principle of virtual displacements, and (c) Pascal's own principle that a change of pressure in a confined liquid is transmitted equally throughout the liquid. It took more than logical accuracy to choose these particular principles to explain the phenomena of hydrostatics. Omitting Torricelli's principle, which Pascal recognized as an instance of logical reasoning within the classical field of statics, we can briefly examine the status of the other two principles. The principle of virtual displacement is considered the kind of knowledge that an accurate but narrow mind can deduce from the few principles involved are absolutely clear. But if we look at the third principle matters are different. The uniform transmission of pressure through a liquid enclosed in a vessel is a discovery that rests on the observation of the behavior of water that is fluid and continuous. Pascal does not distinguish as sharply as we might wish between this principle, which is a physical law based upon experiments, and mathematical or logical principles that we must accept to make valid inferences from experiment, but that we cannot justify by appeals to experiments.

Feeling and Custom

Although Pascal uses the terms "heart" and "sentiment" to denote intuition, he does not for a moment identify it with "feeling" in the sense of emotional likes and dislikes. He was aware of the extent to which our emotional demands do, as a matter of fact, control our beliefs, but he regards feelings in this sense as a source of error. To accept a proposition because it is pleasant he calls unworthy of a rational mind. But how are we to distinguish between the feeling for truth, which Pascal calls intuition, and mere emotional prejudice? This is by no means an easy question to answer. If we are dealing with something that is open to scientific investigation, we can suspend judgement until it has been tested. In other cases, in that of social relations, for example, the problem is more difficult. *Fancy* often passes muster for *feeling*, and we cannot always distinguish between them.

What we call natural principles are often customs in disguise. Pascal puts this forcefully: "Custom is a second nature that destroys the first. But what is nature? Why is custom not natural? I am very much afraid that nature itself is only a first custom, just as custom is a second nature."[26] In the context of the *Pensées* where this passage occurs, Pascal wants to emphasize the lowly status of the human intellect. We cannot know, apart from faith, he says, whether man was created by a good God, an evil demon, or just by chance. Hence it is a matter of doubt whether innate principles are true, false, or merely uncertain. Indeed, our so-called first principles may be as flimsy as our dreams. A worker who dreamt every night that he was king would be as happy as a king who dreamt every night that he was a worker. And if we expected to dream every night that we were being pursued by terrorists, we would dread going to sleep as much as we dread waking up when we are afraid of what the day will bring. We appeal to common sense to dismiss such fantasies, but this can never carry full conviction because when we are asleep we are just as firmly convinced that we are awake as we are now. It even happens that we dream that we are dreaming. "Is it not possible," asks Pascal, "that this half of our

---

[26]*Pensées*, fragments 125–126.

life is itself but a dream onto which other dreams are grafted, and from which we shall awake when we die"?[27] If we dreamt in the company of others and our dreams happened to be similar, we would imagine that things had been turned upside-down if we suddenly woke up and found ourselves alone.

Order and Status

The realm of make-believe extends beyond the world of dreams. Around 1660, Pascal was asked by a great nobleman, the Duc de Luynes, to give his twenty-year-old son a short course of three lectures on the meaning to be attached to rank. This was not an uncommon practice among Christian aristocrats in the seventeenth century, and other people were often invited to attend. Although we do not have the original lectures, Pierre Nicole, a friend of the family, took notes that he published in 1670, eight years after Pascal's death.

In the first lecture, Pascal illustrates the arbitrariness of fortune with the parable of a poor young man who is shipwrecked on an island where he is taken for the lost king, whom he resembles. He enjoys his unexpected status and in public plays his part, but in private he acknowledges the fact that pure chance has put him on the throne.[28] The moral is clear: hereditary privileges are merely the result of fortuitous circumstances and the whim of the law. Another *"turn of the imagination"* of the lawgivers could have left the young duke penniless and without cause for complaint. *Imagination* is a key word here. In Pascal's political philosophy, it is the leading faculty, the mistress of our lives and the mainstay of political order. The state would crumble if the masses did not *imagine* that rulers should be honored and obeyed. That is why judges wear distinctive dress, and medical doctors green gowns.

In a fragment of the *Pensées*, Pascal extends his analysis of the role of custom or habit in creating what we might call a conditioned reflex. The awe that we experience before the King depends on the fact that we are ac-

---

[27]*Pensées*, fragment 131. The reference to dreams may have been suggested to Pascal by Descartes' *Meditations*, which was translated into French by Louis Charles marquis d'Albert, duc de Luynes (1620–1690), a friend of the Jansenist leader, Antoine Arnauld, whom he sheltered in his château of Vaumurier, where Pascal was also made welcome.
[28]Pascal, *Discourse on the Condition of the Great*, *Œuvres de Pascal*, IV, p. 1029; Popkin trans., p. 74.

customed to seeing him surrounded by guards, drums, flags and all the paraphernalia that create automatic responses of respect and fear. Even when he is unaccompanied, the King impresses his subjects with a feeling of reverence, "and the world, which does not know that this is the effect of habit (*coutume* in French), believes it to be derived from some natural force."[29]

In his second lecture, Pascal distinguishes between conventional greatness, such as titles of nobility, and natural greatness that can be physical, intellectual, or moral. Social hierarchy is the result of brute force validated by the imagination. But although institutional greatness depends on arbitrary choices, it cannot be spurned. It makes legitimate claims on our respect, and we are bound to address the King on bended knee, and stand bareheaded before a Prince. It is foolish and base to refuse to do this, says Pascal. Even a dishonorable duke for whom we have inward contempt is entitled to outward marks of respect. What is wrong is to demand a form of respect that is inappropriate to the greatness at hand: a famous mathematician, who expected to be treated like a prince, would only make himself ludicrous.[30]

In his third and last lecture, Pascal contrasts the realm of charity with the realm of material goods. Both are perfectly legitimate, but no amount of material goodwill can replace charitable motives. Pascal looked at human society from within a long Christian tradition, dating from St. Augustine, according to which the State is an artificial contrivance brought into being by the fact of human sin. In this he was influenced by Montaigne, from whom he borrowed many of his examples. Local custom sets the rule of law, and hereditary monarchs are in no way personally superior to other men. Private property is a convenient fiction, and war an absurd use of force. Nevertheless, political institutions of some kind are necessary to prevent men from gratifying their selfish impulses without restraint. Hence the need for a strong government, whatever its form, as long as it is effective.

Order, Relation, and Beauty

If Pascal would have us honor the facade while recognizing the emptiness behind it, he wanted the facade to be beautiful, a property that he defined

---

[29] *Pensées*, fragment 25.
[30] *The Condition of the Great*, *Œuvres de Pascal*, IV, pp. 1032–1033; Popkin trans., pp. 76–77.

in human terms and not by reference to some ideal form. Yet he was loath to give up the distinction between *good taste*, based on a recognized standard, and *bad taste*, of which the models are innumerable:

> There is a certain model of attractiveness and beauty that consists in a certain relation between our nature, weak or strong as it may be, and the thing that pleases us. Everything that conforms to this model attracts us, be it a house, a song, a speech, verse, prose, a woman, birds, rivers, trees, rooms, clothes, etc. Everything that does not conform to this model is displeasing to people of good taste. And just as there is a close connection between a song and a house based on a good model, because both resemble this single model though each in its own way, there is likewise a close connection between things based on bad models. It is not that there is only one bad model, because they are innumerable, but every bad sonnet, for example, whatever the false model it is based on, is exactly like a woman dressed according to that model.[31]

In another Fragment, Pascal explores the difficulty of comparing poetic beauty with the kind we find in mathematics or medicine. The problem is that whereas we know what mathematics and medicine are about, we are at a loss to spell out what makes poetry attractive, because we do not know what the natural model is,

> and for want of this knowledge we have invented strange terms, such as "golden age," "marvel of our age," "fatal," etc. We call this jargon poetic beauty! But let someone who imagines a woman dressed on this model (which consists in using big words for trivial things) see a pretty young lady loaded with mirrors and chains. He will burst out laughing because we know more about feminine attraction than about those of verse. But people who have no clue will admire her in this rig, and there are many villages where she would be taken for the Queen. That is why we call sonnets on this model "village queens."[32]

Pascal loved dressing up his ideas in new combinations of words, and like a skilful player he knew how to place them effectively:

---

[31]*Pensées*, fragment 585.
[32]*Pensées*, fragment 586.

Let no one say that I have said nothing new; the arrangement of the material is new. In playing tennis, both players use the same ball, but one places it better. I would just as soon be told that I have used old words. As if the same thoughts did not form a different argument by being differently arranged, just as the same words make different thoughts when arranged differently.[33]

The point is to choose the right combination. But what yardstick can be used to settle a dispute about what is *orderly* and *disorderly*? Those who lead disorderly lives believe that they are following nature and would be surprised if someone told them that they behave like passengers on board a departing ship who insist that it is the people on the shore who are moving away. "Language is the same everywhere," comments Pascal, "but we need a fixed point to judge it. From the harbor we can determine whether the ship is in motion, but where do we find a harbor in morals?"[34] There is no fixed point from which to determine who is moving in the right direction . . .

The Hierarchy of Orders

In spite of the relativistic stance that he sometimes takes, Pascal had the scholar's conviction that the life of the intellect is superior to such mundane activities as politics and finance. In the letter of presentation of his calculating machine to Christina of Sweden, the fulsome praise he bestows on the Queen is qualified by his own hierarchy of values:

There are degrees among geniuses as among people of rank, and it seems to me that the power of kings over their subjects is only an image of the power of great minds over lesser ones, over whom they exercise the right to persuade instead of the power of commanding as in politics. This second empire seems to me of a higher order because minds are above bodies, and also more fair because it can only be obtained and preserved by merit, while the other can result from birth or chance.[35]

---

[33]*Pensées*, fragment 696.
[34]*Pensées*, fragment 697.
[35]Letter to Queen Christina of Sweden, June 1652, *Œuvres de Pascal*, II, p. 924; Popkin trans., p. 31.

Pascal's list of parallel values can be displayed as follows:

Order of Bodies	Order of Minds
Sovereign	Person of learning
Rank	Degree of genius
Power over subjects	Power over inferior minds
Right to command	Right to persuade
Worldly affairs (*body*)	Scholarly pursuits (*mind*)

These two Orders are infinitely distant from a third dimension, the Order of charity, and Pascal contrasts them in the *Pensées*:

> Great geniuses have their power, their splendour, their greatness, their victory and their lustre, and do not need carnal greatness, which is irrelevant. They are recognised not with the eyes but with the mind, and that is enough. Saints have their power, their splendour, their victory, their lustre, and do not need carnal or intellectual greatness, which is irrelevant to them, for it neither adds nor takes away anything.[36]

Some people are known for their strength, others for their beauty, and each is master in his own house but nowhere else. "Sometimes they meet," says Pascal, "and the strong and the handsome contend for mastery, but this is idiotic because their mastery is of different kinds." Nothing is as silly as to say: "I am handsome, so you must fear me," or, "I am strong, so you must love me." But it is equally tyrannical to say: "He is not strong, so I will not respect him. He is not clever, so I will not fear him."[37]

The Order of bodies, the Order of minds, and the Order of charity are discontinuous and may be compared, to use Donald Adamson's analogy, to three wavelengths operating independently of each other and never coming into direct contact.[38] The Order of bodies is that of physical strength, worldly power and secular authority. The Order of minds is that

---

[36]*Pensées*, fragment 308. Elsewhere Pascal contrasts the modes of communication proper to each order: "The heart has its order, the mind has its own, which uses principles and demonstrations. The heart has a different one. We do not prove that we ought to be loved by setting out in order the causes of love; that would be absurd," *Pensées*, fragment 298.

[37]*Pensées*, fragment 58.

[38]Donald Adamson, *Blaise Pascal*. London: St. Martin's Press, 1995, p. 182.

of scientific and scholarly prestige. The Order of charity is that of the love of God and one's fellow men. Those who are physically weak (like Pascal himself) may be strong in intellect; those who are lacking in gifts of the mind may be rich in those of love. The problem is that some are only capable of admiring bodily strength, as if there were no such thing as greatness of the mind, while others can only admire greatness of the mind, as if there were no higher wisdom.

Pascal's fascination with continuity and discontinuity can be seen in a completely different context, the concluding paragraph of an early mathematical work, the *Potestatum Numericarum Summa*:

> In the case of a continuous quantity, the addition of quantities of any kind to a quantity of a higher kind does not add anything to that quantity. Thus points add nothing to lines, lines nothing to surfaces, and surfaces add nothing to solids, or, to use numerical terms in a treatise on numbers, roots add nothing to squares, squares nothing to cubes, and cubes nothing to squared squares. This is why the inferior degrees, having no value, can be neglected. I mention these things, which are familiar to those who study indivisibles, in order to show the admirable connection that nature, who loves unity, establishes between things that may seem very remote.[39]

We have here another instance of Pascal's paradoxical method. He abolishes the connection between successive degrees of continuous quantities only to re-establish it between the dimensions of continuous magnitudes and the summation of powers.

Proportion and Disproportion

The disproportion between the enquiring mind and what it seeks to comprehend is one of the central themes of the *Pensées*. The starting point is the realization that we are suspended halfway between the infinitely great and the infinitely small. "When I survey," Pascal writes,

> the whole universe in its dumbness and man left to himself with no light, as though lost in a corner of the universe, without knowing who

---

[39]Pascal, *Potestatum Numericarum Summa*, *Œuvres de Pascal*, II, pp. 1271–1272.

put him there, what he has come to do, what will become of him when he dies. Incapable of knowing anything, I am filled with dread like a man who is transported in his sleep to some terrifying desert island and wakes up completely lost and with no means of escape.[40]

We are like someone who wakes up in the middle of the night and does not know *who* he is because he does not recognize *where* he is. To know our location, we would have to know the spatial and temporal dimensions of the universe. But how is the limited space we occupy and the brief span of our life connected to the immensity of the world and the eternity of time? Why have such limits been set upon our knowledge and our life?[41] As Pascal exclaims,

> When I consider the brief span of my life absorbed into the eternity that comes before and after—*as the remembrance of a guest that tarrieth but a day*—the small space I occupy, which is swallowed up in the infinite immensity of spaces of which I know nothing and which knows nothing of me, I take fright and am amazed to see myself here rather than there. There is no reason for me to be here rather than there, now rather than then. Who put me here? By whose command and act were this time and place allotted to me?[42]

The immensity of space and time not only dwarfs the place and moment that I occupy; it robs them of all significance. Where am I? could only be answered if I managed to connect my little island to the universe as a whole, and in order to do this, I would have to grasp, however dimly, all there is. Memory can only link the present to the past, and take me from one island to another; it can never connect me to the mainland or tell why I am here rather than elsewhere, at this moment of time rather than at some other. To know who I am I have to start from wherever I happen to be and, using my imagination, travel into the depths of space in the hope of returning with the means of comparing my island with the world at large. This is the meaning of Pascal's imaginary cosmic voyage. It is not a scientific exploration, but a new kind of thought experiment:

> Let man contemplate the whole of nature in her full and lofty majesty, let him turn his gaze away from the lowly objects around him, let him

---
[40]*Pensées*, fragment 198.
[41]*Pensées*, fragment 194.
[42]*Pensées*, fragment 68.

behold the dazzling light set like an eternal lamp to illuminate the universe, let him see the earth as a mere speck compared to the vast orbit described by this star, and let him marvel at finding this vast orbit to be no more than the smallest point compared to that described by the stars revolving in the firmament.

The knowledge that is sought is not detached but intensely personal. We want to know the world in order to know ourselves. It is a very different kind of self-knowledge from the one that René Descartes arrived at when he realized that he was essentially a thinking self. For Descartes, this was prior and independent of any knowledge of the world. For Pascal, the order is reversed: knowledge of the world is a prerequisite to knowledge of the self. "Know thyself in the seclusion of thy mind," says Descartes. "No," replies Pascal, "Know thyself in the vastness of the universe."

The crucial question is whether man and nature are commensurable. "Before going on to a wider inquiry concerning nature," Pascal insists, "I want him to consider nature just once, seriously and at leisure, and to look at himself as well, and judge whether there is any proportion between himself and nature." As soon as I look at the universe expanding indefinitely in all directions, my own island begins to shrink and my prison becomes even smaller. Two movements, one centrifugal, the other centripetal, are going on at the same time. The first pushes the periphery of the universe outwards to an ever greater distance; the second squeezes the cosmos into a minute point. This twofold movement, whereby everything flies outwards or falls inwards, has no limits, not even those of my imagination. The more I inflate my conceptions to follow the expanding universe, the more it eludes me and flees towards the limits of nowhere. The more I try to focus on what is near me, the faster it vanishes below the threshold of sight. In the words of Pascal's younger contemporary, John Donne:

> The sun is lost, and the earth, and no man's wit
> Can well direct him where to look for it.
> And freely man confess that this world's spent,
> When in the planets, and the firmament
> They seek so many new; they see that this
> Is crumbled out again to his atomies.
> 'Tis all in pieces, all coherence gone;
> All just supply, and all relation.[43]

---

[43] John Donne, *An Anatomy of the World. The First Anniversary* (1611), lines 207–214.

Center and circumference no longer specify a geometrical figure that can be imagined. They mean just the opposite: the universe has no shape that can be represented, let alone measured. With Pascal, the saying, "Nature is an infinite sphere whose center is everywhere and whose circumference is nowhere,"[44] ceases to be an objective statement. It becomes a subjective appraisal of the world: the sphere stands neither for God, whom we cannot comprehend, nor for the Nature we know, but for the immensity of space whose limits cannot be conceived. The saying has been subverted. It is no longer the symbol of an omnipresent deity, who is in all and above all and can be compared to both the center and the circumference of a circle, but the symbol of the human mind that is infinitely removed from both the center and the periphery of the universe.

But although the sphere is no longer the symbol of God, it remains, paradoxically, a symbol of the symbol. The infinite sphere, the symbol of the endless divisibility of space, evokes the indivisible infinity of God because the visible world is the sensible image of the invisible one. But having recovered a function of the image of the sphere, Pascal subverts it again by going on to say, "It is the greatest perceptible mark of God's omnipotence that our imagination should lose itself in that thought."[45] We are once again cast adrift in a sea of unknowing, and the question, "What is man in the infinite?" continues to haunt us. "I see the terrifying spaces of the universe hemming me in," writes Pascal,

> and I find myself attached to one corner of this vast expanse without knowing why I have been put in this place rather than that, or why

---

[44]*Pensées*, fragment 199. Pascal probably found this famous saying in the Preface of Mademoiselle de Gournay to the 1635 edition of Montaigne's *Essais*: "Trimegistus calls God a circle whose center is everywhere and circumference nowhere." In a pseudo-Hermetic treatise, the *Book of the XXIV Philosopher* of the twelfth century, God is described as a "sphaera infinita cuius centrum est ubique, circumferentia nusquam." For the history of this saying, see Dietrich Mahnke, *Unendliche Sphäre und Allmittelpunkt. Beiträge zur Genealogie der mathematischen Mystik*. Halle/Saale: Max Niemeyer Verlag, 1937; Alexandre Koyré, *From the Closed World to the Infinite Universe*. Baltimore: Johns Hopkins University Press, 1957; Karsten Harries, "The Infinite Sphere: Comments on the History of a Metaphor," *Journal of the History of Philosophy* 13 (1975), pp. 5–15. Nicholas of Cusa altered the famous saying and transferred it to the Cosmos in his *De Docta Ignorantia* of 1440 (see the critical edition by E. Hoffmann and R. Klibansky. *De Docta ignorantia*, liber II, cap. 12, Leipzig, 1932, p. 103). The concept, in Cusanus' version, became basic for Giordano Bruno, for whom the innumerable worlds are all divine centers of the unbounded universe (see the Foreword of his *De la causa, principio et uno* (1584), edited by Giovanni Aquilecchia and published as volume III of the *Œuvres Complètes*, Paris: Les Belles Lettres, 1996, p. 27.

[45]*Pensées*, fragment 199.

the brief span of life allotted to me should be assigned to one moment rather than another of all the eternity that went before me and all that will come after me. I see only infinity on every side, hemming me in like an atom or like the shadow of a fleeting instant.[46]

The very contemplation of nature creates a vacuum. "Through space the universe grasps me and swallows me up like a speck."[47] If we try to leave our deserted island, we face more emptiness and are "annihilated in the presence of the infinite and become pure nothingness."[48] Yet man's greatness comes from knowing he is wretched:

Man is only a reed, the weakest in nature, but he is a thinking reed. There is no need for the whole universe to take up arms to crush him: a vapour, a drop of water is enough to kill him. But even if the universe were to crush him, man would still be nobler than his slayer, because he knows that he is dying and that the universe has the advantage over him. The universe knows none of this.[49]

Let us contrast, once more, Pascal's quest for the self with Descartes'. Having done away with everything that can be doubted, Descartes was left alone with his own mind, the new hub of the universe. For Pascal, the human mind cannot be the center of anything. If we look up to the heavens, we are overwhelmed, and if we look into the smallest living thing, we face another prodigy:

Let a mite show him in its minute body incomparably more minute parts, legs with joints, veins in its legs, blood in the veins, humours in the blood, drops in the humours, vapours in the drops. Let him divide these things still further until he has exhausted his powers of imagination.[50]

This exercise is not really about the size of a mite, but about the power of the human mind. After attempting to plumb the depths of space

---

[46]*Pensées*, fragment 68.

[47]*Pensées*, fragment 427. Or, in another fragment, "This is what I see and what troubles me. I look around in every direction and all I see is darkness. Nature has nothing to offer me that does not give rise to doubt and anxiety" (fragment 429), and in one of the most famous of the *Pensées*: "The eternal silence of these infinite spaces fills me with dread" (fragment 201).

[48]*Pensées*, fragment 418.

[49]*Pensées*, fragment 200.

[50]*Pensées*, fragment 199.

and the abyss of time, we embark on a new journey into the realm of the infinitely small. What we encounter first does not differ radically from ourselves, except in size, since a mite also has legs with joints, veins with blood and, in this blood, drops and vapors. When we study a mite, we are still, in a sense, examining ourselves. We see our world on a reduced scale, a world that, in turn, contains even smaller worlds, and so on. As the onion is pealed, it is always an onion that is revealed. But we can go further:

> I want to show him a new abyss. I want to depict to him not only the visible universe, but all the conceivable immensity of nature enclosed in this miniature atom. Let him see there an infinity of universes, each with its firmament, its planets, its earth, in the same proportions as in the visible world, and on that earth animals, and finally mites, in which he will find again the same results as in the first.

We spiral downward in a series of ever-decreasing concentric circles and each time we are enlarged beyond recognition:

> For who will not marvel that our body, a moment ago imperceptible in a universe, itself imperceptible in the bosom of the whole, should now be a colossus, a world, or rather a whole, compared to the nothingness beyond our reach?

The speck has become a world. It has been transformed into a circumference that enfolds an infinite number of spheres that can neither be counted nor measured. If we near the center, the circumference vanishes; if we approach the circumference, the center disappears. Nowhere do we find ourselves. The only way out of this predicament would be to occupy both the center and circumference. But this is impossible! We are once again suspended between infinity and nothingness.

It is not merely that we lack the means of determining the center or the circumference. The problem is much more dramatic: there is no such thing as a center or a circumference in an infinite sphere. This is why Pascal sees the difference between the geocentric and the heliocentric systems in astronomy as trivial, for both have a well-defined center in a finite universe whose boundaries are the sphere of fixed stars. Our ideas crumble in the crucible of infinity: "We are floating in a vast medium, always drifting uncertainly, blown to and fro; whenever we think we have a fixed point to which we can cling and make fast, it shifts and leaves us behind; if we follow it, it eludes our grasp, slips away, and flees eternally before us."[51] The harmonious and symbolic universe of medieval Christianity has disappeared. We are left with a feeling of vertigo. According to the vantage point chosen, we are either so puny that we cannot be seen or so large that we seem giants. Some may be less daunted by the infinitely small, and think they are more capable of reaching the center of things than of embracing their circumference, but this is an illusion:

> It takes no less capacity to reach nothingness than the whole. In either case it takes an infinite capacity, and it seems to me that anyone who could understand the ultimate principles of things might also succeed in knowing infinity. One depends on the other, and one leads to the other. These extremes eventually touch and come together by going in opposite directions, and meet in God, and God alone.

In the half-way house that we inhabit, we are limited to a very narrow range of sensations:

> Too much noise deafens us, and we are dazzled by too much light. If we are too far or too close we cannot see properly. An argument is obscured by being too long or too short, and too much truth bewilders us. I know people who cannot understand that if you remove 4 from 0 you are left with 0. First principles are too obvious for us; too much pleasure causes discomfort; too much harmony in music is displeasing; too much kindness annoys us: we want to be able to pay back our debts with something over.

The root of the problem for Pascal is that physical objects are simple, while we are composed of two opposing natures: a material body and an immaterial soul. If we were simply material, we would know nothing at all, but because we are composed of mind and matter, we cannot have adequate knowledge of things that are simply spiritual or simply corporeal. This is why we describe corporeal things as though they were animated and say that bodies *tend* to fall, that they *flee* from destruction, or that they *fear* a void. We make a similar category mistake when we speak of minds as occupying a given place, or moving from one location to another. We stamp our own composite being on the simple things we try to understand. If only we could grasp our own nature! "But this is the thing we understand

---

[51]*Ibid.*

least," laments Pascal, "man is to himself the greatest prodigy in nature, for he cannot conceive what body and, still less, what mind is and, least of all, how a body can be joined to a mind."[52] A corporeal mind is as inconceivable as the idea that matter that could know itself.

The Art of Persuasion and the Grammar of Assent

We have seen earlier that, according to Pascal, the most rigorous method of reasoning is found in geometry, which starts from self-evident principles and applies rigorous rules of inference to arrive at demonstrations. But what is the relevance of such a method when we want to persuade someone to do something in the context of his everyday life? Here, in order to convince, we have to please, and this is not easy, because what people know to be right is often at variance with what they want. We can demonstrate a new theorem by following the rules of geometry, but we cannot endear ourselves to everyone in the same way.

The art of persuasion begins from the expressed or tacit desires of a person. This is why the "method of pleasing" is so difficult and why Pascal was not certain he was up to the task. Nonetheless, he thought that there are rules of pleasing, and that if someone applied them correctly, he would succeed just as certainly in making people fond of him as in demonstrating a theorem. *The Art of Persuasion* is Pascal's attempt to show how *principles of pleasure* can replace *self-evident principles* as premises that lead to valid conclusions.[53]

"Everyone declares that he only believes or loves what he knows to be worthy," observes Pascal. So much for the ideal. In practice, our will does not always follow our reason. In order to effectively persuade someone, we would have to know what makes him tick, something that he seldom knows himself. If the underlying motives of human conduct were sufficiently stable, rules for pleasing could be devised, and we could make ourselves loved by anyone. Alas, human motives keep changing, and "no one

---

[52]*Ibid.*

[53]Revealed truths are excluded from consideration, for God "has ordained that they should enter from the heart into the mind and not from the mind into the heart." Hence, divine things must be loved in order to be known, whereas human beings must be known before they can be loved. *On the Art of Persuasion, Œuvres de Pascal*, III, pp. 413–414. The work is translated by Richard H. Popkin in *Pascal: Selections*. New York: Macmillan, 1989, pp. 185–194.

differs more from other people than he does from himself at different times in his life," declares Pascal.⁵⁴ "If someone says that he no longer loves the person he loved ten years ago," he writes elsewhere, "this is because she is not the same any more, nor is he. He was young and so was she; now she is quite different. Perhaps he would still love her as she used to be then."⁵⁵ This is the downside; the upside is that time can also heal because offender and offended change: "It is as if one angered a nation and came back to see them after two generations. They are still Frenchmen, but not the same ones."⁵⁶ We look at things not only from different angles, but also with different eyes. In the case of painting, rules of perspective determine the best vantage point. All other places are too near or too far, too high or too low. But in the case of truth and morality how can the right place be found?⁵⁷

Uncertainty in human affairs is so great that we sometimes doubt whether a person who makes a profound remark really understands it or is merely quoting from memory. Does the famous saying, "I think, therefore I am," mean the same thing for Descartes and for those who repeat it after him? There is a world of difference between merely uttering this clever sentence, says Pascal, and knowing how it proves the distinction between material and spiritual substances. The words, as used by Descartes, are as different from the same words repeated by rote as a dead person from someone who is alive.⁵⁸ Great ideas are too often mouthed parrot-like by people who hoard them without understanding. They are like jewellers who boast that they have diamonds of great price in their safe but are unable to pick them out, or apothecaries who made infusions out of several herbs but cannot tell which ones have medicinal value.

---

⁵⁴*On the Art of Persuasion*, *Œuvres de Pascal*, III, pp. 416–417; Popkin trans., p. 187–188.
⁵⁵*Pensées*, fragment 673.
⁵⁶*Pensées*, fragment 802.
⁵⁷*Pensées*, fragment 21.
⁵⁸*On the Art of Persuasion*, *Œuvres de Pascal*, III, p. 424; Popkin trans., p. 192. The connection with Saint Augustine was made by the theologian Antoine Arnauld (1612–1694), whom Pascal met around 1655. At Descartes' request, Arnauld formulated *Objections* that were published in the same volume as Descartes' *Meditations* in 1641. Arnauld refers to Saint Augustine's *De libero arbitrio*, bk. 2, ch. 3 (A.T., VII, pp. 197–198), but in a subsequent letter of 1648, he mentions the *De Trinitate*, bk. 10, ch. 10 (*Œuvres de Descartes*, vol. VII, pp. 197–198 and vol. III, p. 186).

William R. Shea

Happy Are Those Who Imagine They Are

Pascal never gave up entirely his quest for "rules as certain for pleasing as for demonstrating," and in the *Pensées*, we find him investigating the psychology of persuasion in the light of the fact that everyone wants to be happy:

> All men seek happiness and there are no exceptions. However different the means they may employ, they all strive towards this goal. The reason why some go to war while others do not is the same desire in both, but interpreted in two different ways. The will never takes the least step except to that end. This is the motive of every act of every person, including those who go and hang themselves.[59]

But the desire for happiness is never fulfilled, and everyone complains whatever his rank, fortune, health, or education. That should be enough to convince us that we are incapable of achieving happiness by our own efforts. Alas, experience teaches us very little, and because two cases are never exactly alike, we always hope that we will not be disappointed next time. So, although the present never satisfies us and experience deceives us, we go on craving for happiness. Just as physical experiments reveal a vacuum in nature, so psychological experience discloses a moral emptiness in our hearts.[60]

If only *reason* or *nature* could guide us, but neither are clearly apprehended or satisfactorily grounded. This is why *custom* passes muster for natural law. The most commonly received of all human maxims is that each man should follow the customs of his country, but it is a funny sort of justice whose limits are marked by a river, a chain of mountains, or some arbitrary national border. War is the perfect instance of this folly: "Could there be anything more absurd than that a man has the right to kill

---

[59]*Pensées*, fragment 148. The source is St. Augustine: "Everyone wants to be happy . . . and some say that happiness is found in the army, others on the farm" (*Sermo 306*, C. 3, n. 3). The opening paragraph of the *Confessions* contains the famous lines: "You made us for yourself and our heart is troubled until it rests with you."

[60]"What else does this craving, and this helplessness, proclaim but that there was once in man a true happiness, of which all that now remains is the empty print and trace? This he tries in vain to fill with everything around him, seeking in things that are not there the help he cannot find in those that are, though none can help, since this infinite abyss can be filled only with an infinite and immutable object; in other words by God himself." (*Pensées*, fragment 148).

me because he lives on the other side of the water, and his prince has picked a quarrel with mine, though I have none with him?"[61] What is the essence of justice? asks Pascal. The authority of the legislator? The convenience of the sovereign? The present custom? From the standpoint of reason, nothing is just in itself, everything shifts with time. Custom decides what is just for the sole reason that it is accepted. A return to the basic and primitive laws that unjust custom has abolished would achieve nothing. Revolutions throw off the yoke of one custom only to replace it by the fetters of another.

Custom exercises its grip through *the imagination*, which Pascal describes as the dominant faculty in man. It is all the more deceptive for not being invariably false, for it would be an infallible criterion of truth if it were infallibly wrong. It is the source not only of the respect we have for certain persons or institutions but of our value judgements. A lawyer who has been paid in advance finds the cause he is pleading all the more just. No one is allowed to be judge in his own cause, and there are some judges who, to avoid the danger of partiality, lean to the other side. The surest way to lose a perfectly just case with such magistrates is to get friends to recommend it to them. Consider a magistrate, says Pascal, who prides himself on judging things as they really are, without paying heed to the trivial circumstances that work on the imagination of weaker men. Let him go to church on Sunday in a spirit of genuine devotion, and eager to listen carefully to the sermon. But the preacher turns out to be badly shaven, not too clean, and to speak with an unpleasant and grating voice. Then, however profound the message, our magistrate is unlikely to be impressed.[62]

At a more primitive and instinctive level, imagination can completely swamp our rational faculties. The very thought of crossing a precipice over a narrow bridge that is perfectly safe will make some people break out into a cold sweat. Why, the sight of a rat or the scratching of a piece of chalk on a blackboard is enough to unhinge some of the most rational people!

If imagination can mislead us by the prestige of tradition, it can also lead us astray by the charms of novelty. This is why people accuse each other not only of being slaves to old ideas but of running after new ones. The vacuum is a case in point:

---

[61]*Pensées*, fragment 60.
[62]*Pensées*, fragment 44.

"Because," they say, "you have believed since you were a child that a box was empty when you could not see anything in it, you believed that a vacuum could exist. This is just an illusion of your senses that was strengthened by habit, and it must be corrected by science." Others say: "When you were taught at school that there is no such thing as a vacuum, your common sense was corrupted. It was quite clear about it before being given the wrong impression, and now it must be corrected by reverting to your original state." Who then is the deceiver, our senses or our education?[63]

From the Void to God

In a note originally intended for the *Treatise on the Vacuum* and subsequently published as a fragment of the *Pensées*, Pascal asks,

> What is more absurd than to say that inanimate bodies have passions, fears, horrors, or that lifeless bodies without feelings and incapable of life have passions that presuppose at least a sentient soul to receive them? Moreover, that the object of this horror should be the vacuum. What is there in the vacuum that could make them afraid? What could be baser or more ridiculous?[64]

The organic analogy must be abandoned. Mechanism triumphs and races ahead under the banner of experiments. Nature has become dumb. It no longer cries out in horror at the void. But for Pascal, and perhaps for modern man, the silence of nature is not an unqualified blessing. It is experienced as a burden, even as a threat: "The eternal silence of these infinite spaces fills me with dread," writes Pascal.[65] For Fr. Noël and the Aristotelians, nature repelled any attack on the great chain of being; for the author of the *Pensées*, nature has turned against man himself: "Man is only a reed, the weakest in nature . . . There is no need for the whole universe to take up arms to crush him: a vapor, a drop of water is enough to kill him."[66] If the outside void need no longer be feared, the inner void

---

[63]*Ibid.*
[64]*Pensées*, fragment 958.
[65]*Pensées*, fragment 201; see also fragment 113.
[66]*Pensées*, fragment 200.

remains a constant menace. Indeed, "we are full of things that impel us outwards. Our instinct makes us feel that our happiness must be sought outside ourselves. Our passions drive us outwards, even without objects to excite them. External objects tempt us in themselves and entice us even when we do not think about them. Thus, it is no good philosophers telling us: Withdraw into yourselves and there you will find your good. We do not believe them, and those who do believe them are the most empty and silly of all."[67] Our capacity for goodness "is empty." We try to fill this emptiness, says Pascal, by running out of ourselves when we should be returning to the source of plenitude: "There was once in man a true happiness, of which all that now remains is the empty print and trace that he tries in vain to fill with whatever surrounds him" for "this infinite abyss can only be filled with an infinite and immutable object; in other words by God himself."[68] The language of plenitude, once favored by Aristotelian physics, has been transposed to another key, but it remains in the service of harmony, this time a harmony that is as necessary as it is paradoxical. It is human nature that abhors a void . . .

---

[67]*Pensées*, fragment 143.
[68]*Pensées*, fragment 148.

CHAPTER 10

The Marvelous Arithmetical Triangle

Pascal was fond of mathematical topics that lent themselves to visualization and could be displayed in the form of patterns. We know this from the early work on projective geometry that he carried out when he was still a teenager, and this may be the reason why he was attracted by the Arithmetical Triangle, a way of representing numbers that had become popular among mathematicians.[1] Most readers will be familiar with Pascal's Triangle in the form that is now commonly used:

---

[1] A forerunner of Pascal's triangle appeared in a book written in the thirteenth century by Yang Hin, one of the mathematicians of the Suny dynasty in China. It was also known to the Persians, and Jamshid Al-Kashi, who died in Samarkand in 1429, discusses its properties in the *Key to Arithmetic* of 1425. In the West, versions of the Arithmetical Triangle are found in Peter Apian's *Arithmetic*, published in German in 1527, and Michael Stifel's *Arithmetica Integra* of 1544. In Italy, Tartaglia reinvented the triangle afresh and published it in his *General Trattato di Numeri et Misure* of 1556. Pascal does not seem to have known these works, and his source is probably Pierre Hérigone who published the first four volumes of his *Cursus Mathematicus* in Paris between 1634 and 1642. A striking feature of the work is the division of the pages into two columns, with the Latin text on one side and a French translation on the other. Hérigone was acquainted with the Pascal family and served with Pascal's father on the Committee that Cardinal Richelieu created to assess a method of determining longitudes that had been submitted by the astronomer Jean-Baptiste Morin. None of these mathematicians saw the full implications of their discovery, however, and it was left to Pascal to show in a clear fashion the relevance of the Arithmetical Triangle to figurate numbers, the theory of combinations, the expansion of binomial expressions, and the theory of probability. That the Arithmetical Triangle should bear Pascal's name cannot be disputed although the expression *Triangulum Arithmeticum Pascalianum* is posthumous and seems to have been first used by Abraham de Moivre in his *Miscellanea Analytica* of 1730.

DESIGNING EXPERIMENTS & GAMES OF CHANCE

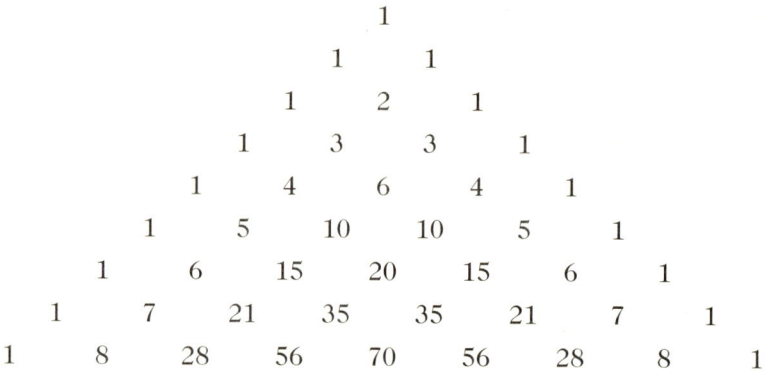

FIGURE 1. *The modern form of the Arithmetical Triangle*

All sorts of patterns greet the eye when we first glance at this triangle, but the underlying structure is simple enough: each number is the sum of the two numbers to the right and to the left in the row above. For instance, 4 is 1 + 3, and 6 is 3 + 3.

The background to the Arithmetical Triangle is the representation of numbers by points or dots that make up geometrical shapes instead of the abstract symbols 1, 2, 3, . . . , 9 which we normally use. One dot stands for 1; two dots placed apart represent 2; three dots, which represent 3, mark out the first rectilinear plane figure, a triangle, and so on. Some sequences of numbers (1, 3, 6, 10, . . .) have a *triangular shape*:

(1)   (3 = 1 + 2)   (6 = 1 + 2 + 3)   (10 = 1 + 2 + 3 + 4)   (15 = 1 + 2 + 3 + 4 + 5)

Other numbers (1, 4, 9, 16, 25, . . .) have a square shape (and are called *squares* to the present day):

(1)   (4 = 2 × 2 = 1 + 3)   (9 = 3 × 3 = 1 + 3 + 5)   (16 = 4 × 4 = 1 + 3 + 5 + 7)

242

Once displayed in this way, it is easy to see that triangular numbers are built up by adding the natural numbers, such that the nth triangular number is the sum of the first n numbers. For instance, the fourth triangular number, 10, is the sum of the first four natural numbers $(1 + 2 + 3 + 4)$. The square numbers are constructed by adding the odd numbers so that the nth square number is the sum of the first n odd numbers. Because of their shape or figure, triangular and square numbers were called *Figurate Numbers*. This may seem trivial, but it was a first step towards a systematic investigation of the properties of numbers. For instance, two successive triangular numbers taken together make a square:

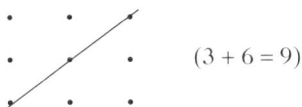

$(3 + 6 = 9)$

If we start with the simple triangular numbers $(1, 3, 6, 10, 15, \ldots)$, we can build a higher order of triangular numbers by simple addition:

simple triangular numbers:	1	3	6	10	15
next order of triangular numbers:	1	4	10	20	35
	1	$(1+3)$	$(1+3+6)$	$(1+3+6+10)$	$(1+3+6+10+15)$

The second row of numbers represents solid figures, namely tetrahedra (pyramids with a three-sided base). Using them as our building blocks, we can go on making higher and higher triangular numbers on the same plan, in the following way:

Simple Units (First Order)

```
                        1     =   1
                    1   1     =   2
                1   1   1     =   3
            1   1   1   1     =   4
        1   1   1   1   1     =   5
    1   1   1   1   1   1     =   6
1   1   1   1   1   1   1     =   7
```

Natural numbers (Second Order)

						1	=	1
					1	2	=	3
				1	2	3	=	6
			1	2	3	4	=	10
		1	2	3	4	5	=	15
	1	2	3	4	5	6	=	21
1	2	3	4	5	6	7	=	28

Simple triangular numbers (Third Order)

						1	=	1
					1	3	=	4
				1	3	6	=	10
			1	3	6	10	=	20
		1	3	6	10	15	=	35
	1	3	6	10	15	21	=	56
1	3	6	10	15	21	28	=	84

Pyramidal numbers (Fourth Order)

						1	=	1
					1	4	=	5
				1	4	10	=	15
			1	4	10	20	=	35
		1	4	10	20	35	=	70
	1	4	10	20	35	56	=	126
1	4	10	20	35	56	84	=	210

Triangulo-triangular numbers (Fifth Order)

				1	=	1
			1	5	=	6
		1	5	15	=	21
	1	5	15	35	=	56
1	5	15	35	70	=	126

Next triangular numbers (Sixth Order)

				1	=	1
			1	6	=	7
		1	6	21	=	28
	1	6	21	56	=	84
1	6	21	56	126	=	210

William R. Shea

Figurate Numbers and Pascal's Triangle

Pascal had the brilliant idea that the Arithmetical Triangle could be used to express these properties of numbers, as we can see if we look back at Figure 1. Reading diagonally downwards starting from the top, we have a row of ones, next the natural numbers, then the simple triangular numbers, and successive higher orders of triangular numbers. Pascal worked out these relationships in detail and the outcome was the epoch-making *Treatise on the Arithmetical Triangle* that he printed in 1654. It was not widely distributed, and after Pascal's death it was bound together with three related sets of papers and published in 1665. This edition contains a fold-out page with a figure of the Triangle drawn from a draft by Pascal and which we shall now consider (see Figure 2).

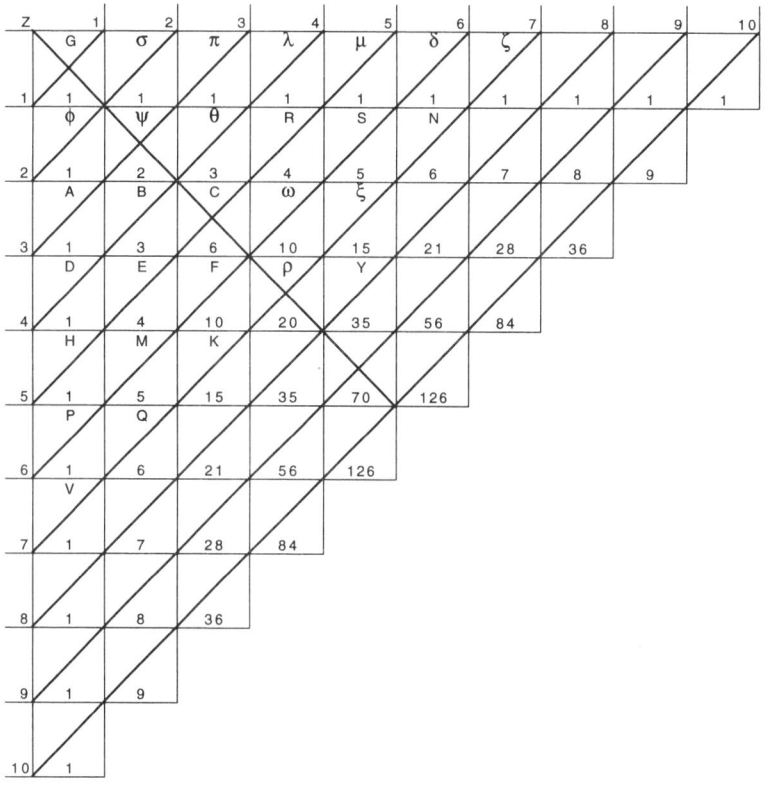

FIGURE 2   *Pascal's Arithmetical Triangle.*

Reading horizontally we have rows of units, one in each square, which Pascal calls a *cell*, and identifies by a Roman or Greek letter. The second row contains the natural numbers, which Pascal refers to as *indices*, and the rows that follow are successive higher orders of triangular numbers that can be expanded indefinitely to the right. Pascal calls positions in the same vertical column "cells of the same perpendicular rank," and those in the same horizontal row "cells of the same parallel rank." The diagonal lines that bisect the cells he terms the *bases* of the triangle, and the cells that are cut in two "cells of the same base."

Pascal delights in pointing out a number of properties of this arrangement. For instance, every cell is equal to the cell above it in its perpendicular rank plus the cell that precedes it in its parallel rank: for example, on the fourth line, 10 of cell F is the sum of 6 (in cell C just above) and 4 (in cell E to the left). Or, again, every cell is equal to the sum of the cells of the preceding parallel rank up to the one just above the cell: for example, 10 in cell F is the sum of numbers, 1, 3, 6 in cells A, B, C.

The Mathematical Justification of the Arithmetical Triangle

Pascal did not merely use the Arithmetical Triangle, he also justified it mathematically, and the argument he formulated is of great interest because it contains the first satisfactory statement of *the principle of complete induction*. Pascal begins by pointing out that of two adjacent numbers on the same base, the upper is to the lower as the number of cells from the upper to the top of the base is to the number of those from the lower to the bottom, inclusive. The wording is a bit heavy but the meaning is clear. For instance, consider cells E and C on the diagonal (called *base*) that extends from the left-hand corner of the fifth horizontal line to the top of the fifth vertical column:

E	is to	C	as	2	is to	3
(the lower number)		(the upper number)		(there are two cells between E and the first, namely E and H)		(there are three cells between C and the top, namely C, R and µ)

The proof consists in pointing out that the proposition holds for the first triangle whose base is 1 (namely the line that runs from 1 on the vertical axis to 1 on the horizontal axis) since the only cell, 1, is equal to the number of ways 1 thing can be taken from 1 thing. It clearly holds for the next parallel base (from 2 to 2), and for all the following bases, a procedure that embodies the principle of complete induction, which Pascal expresses as follows: "If this proportion is found in any base, it will necessarily be true in the next base. From which it follows that it will necessarily be true in all bases, for it is true in the second base because of the first lemma, and hence, by means of the second lemma, it is true in the third base, and hence in the fourth base, and so on to infinity."[2]

Pascal sent his proposition to Fermat, a famous lawyer from Toulouse and a mathematician of rare power, about whom we shall have more to say in the next chapter. In acknowledging receipt, Fermat says that it had came through the post from Paris to Toulouse just when his own proposition on the figurate numbers, "which in effect is the same thing, was traveling from Toulouse to Paris."[3] The letter conveying Fermat's proposition is unfortunately lost but the passage in the *Treatise on the Arithmetical Triangle* where Pascal celebrates his extraordinary agreement with Fermat deserves to be quoted in full:

> A thing can be examined in innumerable ways, and here is a remarkable instance that is most flattering for me. The very proposition that I have been developing in different ways also occurred to the famous magistrate of Toulouse, M. de Fermat. What is admirable is that without there having been any communication between us, he was writing in his home town what I was discovering in Paris, as our letters written and received at the same time testify. Happy am I to have concurred on this occasion (as I have done at other times in a way just as strange) with this great man, who excels in the highest reaches of geometry.[4]

---

[2]*Œuvres de Pascal*, II, p. 1294, trans. in D. J. Struik (ed.), *A Source Book in Mathematics 1200–1800*. Princeton: Princeton University Press, 1969, pp. 24–25. The proposition is Corollary 12 of the *Treatise on the Arithmetical Triangle*. On the significance of Pascal's statement, see H. Freudenthal, "Zur Geschichte der vollständigen Induktion," *Archives Internationales d'Histoire des Sciences* 22 (1953), pp. 17–37.

[3]Letter of Fermat to Pascal, 29 August 1654, *Œuvres de Pascal*, II, p. 1154, trans. in F.N. David, *Games, Gods and Gambling*. London: Charles Griffin and Co, 1962, p. 246.

[4]*Œuvres de Pascal*, II, p. 1328.

Fermat was a modest man and he did not bother to point out to Pascal that he had found the result eighteen years earlier.[5] He may also have realized that Pascal's proposition reached beyond the context of figurate numbers and had implications for the development of reasoning by recurrence, or mathematical induction as it has been called since the eighteenth century.[6]

The Arithmetic Triangle and Games of Chance

Pascal saw the relevance of the Arithmetical Triangle to the determination of odds when gambling or playing games of chance. We shall show how he went about this in the next chapter, but for the convenience of those readers who may wish to refresh their knowledge of probability, let us first recall some of the basic notions that anyone should acquire before placing a bet.[7]

Flipping Coins and Throwing Dice

When we toss a coin, the probability of getting heads is the same as getting tails: 1/2 or 0.5. Whatever the outcome of the first toss, the second has the same probability because the first result exerts no influence on the second. The two events are completely independent. A coin has no memory and no conscience! This is obvious, but people sometimes assume that if heads have appeared several times in a row, their chance of getting tails on the next toss is increased or, at least, becomes greater than 1/2. But the odds do not change, and the likelihood of getting tails remains one half each and every time the coin is tossed. Why then this feeling that the odds should have improved? Is it merely that "hope springs eternal in the human breast"? No, it rests on the observation that, on the long run, if we toss an unbiased coin, the number of heads and tails will even out. Let us investigate this by considering the outcome of a series of four games. Writing H for heads and T for tails, we can represent the result of the first toss as follows:

---

[5]See Fermat, *Oeuvres*, edited by P. Tannery and C. Henry. Paris: Gauthier-Villars, 4 vols. & Supplement, 1891–1922, vol. II, pp. 70, 84–85 for letters to Mersenne and Roberval in 1636.
[6]The name "mathematical induction" originated much later in De Morgan's article on "Induction (Mathematics)" in the *Penny Cyclopaedia* of 1838 (see Carl B. Boyer, *A History of Mathematics*, p. 398).
[7]We refer the reader to the excellent introductory book on probability by Warren Weaver, *Lady Luck*. New York: Dover, 1982.

$$\begin{matrix} H \\ T \end{matrix}$$

After the second toss, the result, when combined with the outcome of the first, is:

$$H \begin{bmatrix} H \\ T \end{bmatrix}$$
$$T \begin{bmatrix} H \\ T \end{bmatrix}$$

Here the first column corresponds to the first game and the second column to the second game. We see that after two games there are four possible combinations, which all have the same likelihood, namely 1/4 or 0.25. This can be represented as:

$$\begin{array}{ll} \text{HH} : & 1/4 \\ \text{HT} : & 1/4 \\ \text{TH} : & 1/4 \\ \text{TT} : & 1/4 \end{array} \Bigg\} \; 2/4 = 1/2$$

H T and T H have been bracketed together since the outcome is identical because the order in which H or T occurs does not matter.

Let us now move on to a third game and tabulate the results in three columns, with the combined outcomes of the three games appearing in the third:

$$H \begin{bmatrix} H \begin{bmatrix} H \\ T \end{bmatrix} \\ T \begin{bmatrix} H \\ T \end{bmatrix} \end{bmatrix}$$
$$T \begin{bmatrix} H \begin{bmatrix} H \\ T \end{bmatrix} \\ T \begin{bmatrix} H \\ T \end{bmatrix} \end{bmatrix}$$

There are eight equiprobable outcomes, which can be grouped as follows:

$$
\begin{array}{ll}
\text{H H H:} & 1/8 \\
\left.\begin{array}{l}\text{H H T:} \ \ 1/8 \\ \text{H T H:} \ \ 1/8 \\ \text{T H H:} \ \ 1/8\end{array}\right\} & 3/8 \\
\left.\begin{array}{l}\text{H T T:} \ \ 1/8 \\ \text{T H T:} \ \ 1/8 \\ \text{T T H:} \ \ 1/8\end{array}\right\} & 3/8 \\
\text{T T T:} & 1/8
\end{array}
$$

The gambler who bets heads has one chance out of eight of winning 3 games; 3 chances out of 8 of winning 2 games; 3 chances out of eight of winning one game; and one chance out of 8 of winning none.

The cumulative outcome of the four games is:

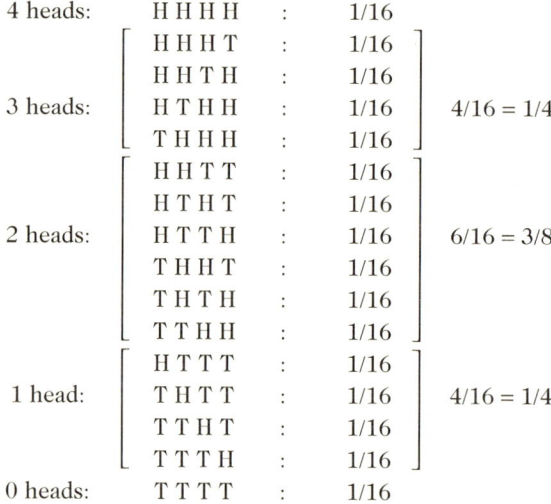

A player who bets heads has one chance out of 16 of winning 4 games; 4 chances out of 16 (or 1 on 4) of winning 3 games; 6 chances out of 16 (or 3 on 8) of winning 2 games; 4 chances out of 16 (or 1 on 4) of winning one; and one chance out of 16 of not winning any. We see how, starting from identical probabilities for heads or tails in the first game, we arrive at different probabilities for combined outcomes after a few games. For instance, after four games, there are 6 combined outcomes that yield 2 heads and 2 tails. Each of these outcomes, taken individually, is not more probable than the combined result that gives 4 tails, but when we count their number we find that the odds of getting one such combination is 6 times greater than the odds of getting H H H H. An experienced gambler (unlike the occasional visitor to a casino) knows this very well. Numerically the chances are 6 times out of 16, whereas the chance of flipping four heads is only one out of 16. But note that the combination H H T T is not more probable than H H H H. If combinations in which we find 2 heads are 6 times more probable than the one with 4 heads, the reason is not that each is more probable, but that there are 6 distinct possibilities of 2 heads showing up.

The outcomes of four consecutive games can be summarized as follows:

I. After one game

| 0 heads | [T] | probability | 1 : 2 |
| 1 head | [H] | probability | 1 : 2 |

II. After two games

0 heads	[T T]	probability	1 : 4
1 head	[HT, TH]	probability	2 : 4
2 heads	[H H]	probability	1 : 4

III. After three games

0 heads	[T T T]	probability	1 : 8
1 head	[TTH, THT, HTT]	probability	3 : 8
2 heads	[THH, HTH, HHT]	probability	3 : 8
3 heads	[H H H]	probability	1 : 8

IV. After four games

0 heads	[T T T T]	probability	1 : 1
1 head	[HTTT, THTT, TTHT, TTTH]	probability	4 : 16
2 heads	[HHTT, HTHT, HTTH, THHT, THTH, TTHH]	probability	6 : 16

DESIGNING EXPERIMENTS & GAMES OF CHANCE

3 heads   [HHHT, HHTH   probability   4 : 16
            HTHH, THHH]

4 heads   [H H H H]   probability   1 : 16

We can display these probabilities in a different way and, for our purposes, more interesting way:

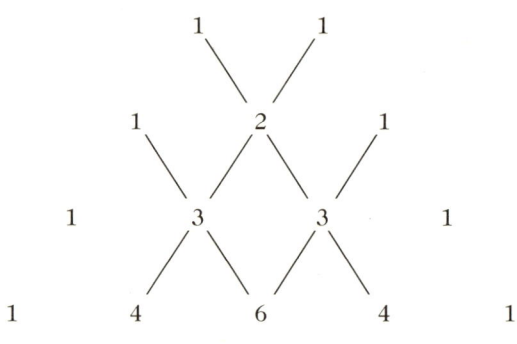

What is interesting about this pattern is that between the 1's that occur at the ends of a row, each number is the sum of the two numbers in the row above it between which it falls. When this pattern is extended, we have Pascal's Arithmetical Triangle:

```
                            1
                       1         1
                   1        2        1
               1       3        3        1
           1       4        6        4        1
       1       5       10       10        5        1
   1       6       15       20       15        6        1
1      7       21       35       35       21        7        1
   8       28       56       70       56       28        8        1
```

Permutations and Combinations

The sequence of triangular orders tells us something important about the mathematical probability when tossing a coin. It is also relevant to the way objects can be selected or combined in games such as throwing dice or playing cards. It has the great advantage of making abstract relations visible, and Pascal was quick to see this. But before we examine his procedure let us refresh our memory about permutations and combinations. A *permutation* is simply an ordered arrangement of objects. Suppose that we have the first three letters of the alphabet, A, B, and C. In how many different orders can we write them down? ABC is one way, and BCA is another, and CBA is a third. Suppose we have three boxes in a row, and we must put a different letter in each box. We can fill the first box in three ways, for all three letters are available for the first choice. We can then fill the second box in two ways, since there are now only two letters left to choose between, and we can fill the third box in only one way, because there is only one letter left over. So we can fill the three boxes in $3 \times 2 \times 1 = 6$ ways. A modern symbol for the number of permutations is $_nP_r$, where r represents the number of objects to be arranged in order, and n indicates that these objects are chosen from an original set of n objects.

The general rule is obvious. If we have n objects, all n are available for choice for position number one, so we can fill this position in n ways. Then $n - 1$ are left as candidates for position number 2, which we can fill in $n - 1$ ways. We can fill the third position in $n - 2$ ways, and so on. Clearly, if $_nP_n$ denotes the total number of orders of n objects, in other words, the number of *permutations* of n objects, then we must multiply n, the number of objects being permuted, by each successively smaller integer until we have finally run down to unity, this corresponding to the fact that there is only *one* way to fill the final box.

$$_nP_n = n(n-1)(n-2)(n-3)\ldots 3 \times 2 \times 1$$

Combinations

When the *order of selection* of r objects from a set of n objects is considered irrelevant, this is called a *combination* (for instance, when choosing AB is considered the same as choosing BA). This is the meaning of the word for Pascal, who explains it with a suitably simple example. If we take the first fours letters of the alphabet, A, B, C and D, in how many ways can we select 1, 2, 3, or 4 of these letters at a time? We agree to disregard the order of the selection, namely we make no difference between choosing A first or B first, so that AB = BC. We find, by experience, that we can combine these letters in six ways if we take them two at a time, namely AB, AC, AD, BC, BD, CD. If we take three letters at a time, we find four possible combinations: ABC, ABD, ACD, and BCD. If we take one letter at a time, we can make four selections, and if we take all the letters together, we can only have one. Then

> Number of ways of taking 1 thing from 4 things: 4
> Number of ways of taking 2 things from 4 things: 6
> Number of ways of taking 3 things from 4 things: 4
> Number of ways of taking 4 things from 4 things: 1

So the total number of possible combinations is 15.

The convenient modern short-hand symbol that was introduced after Pascal for the number of combinations is $_nC_r$, also written $\binom{n}{r}$, where n stands for the number of things from which a selection is made, and r the number of things that are selected at a time. Thus $_4C_2$ means the number of different ways in which we can select two objects from a set of four different objects. Pascal's results can be expressed as:

$$_4C_1 = 4$$
$$_4C_2 = 6$$
$$_4C_3 = 4$$
$$_4C_4 = 1$$

The series of numbers representing the ways we can take two objects at a time from 1, 2, 3, 4, 5, 6, 7, 8, . . . different objects is simply

$_1C_2$	$_2C_2$	$_3C_2$	$_4C_2$	$_5C_2$	$_6C_2$	$_7C_2$	$_8C_2$
0	1	3	6	10	15	21	28

These numbers turn out to be the simple triangular numbers and we can find them with the help of the Arithmetical Triangle. Let us see how Pascal does this using his own diagram (see Figure 2 above). Consider, says Pascal, any triangle inside the Arithmetical Triangle, for instance the one whose summit is Z and whose base runs from 4 on the horizontal axis to 4 on the vertical axis. In this case, the sum of the cells of the *second* parallel rank (1 + 2 + 3 = 6) is equal to the number of ways 2 things can be combined out of 4 things. The procedure is altogether general, and if we take another triangle, say the one whose base runs from 10 on the horizontal axis to 10 on the vertical axis, we find that the sum of the cells of the 8th parallel rank (1 + 8 + 36 = 45) is equal to the number of ways 8 things can be combined out of 10 things. The General Proposition runs as follows: "In any Arithmetical Triangle, the sum of the cells of any parallel rank is equal to the number of combinations of the number of the index of the base taken as many times as the number of that parallel rank."[8] Pascal's found a number of applications for his Arithmetical Triangle but not all of them dawned immediately to his mind, as we shall see in the next chapter when we examine the treatises that he was drafting at the time of his discovery.

---

[8] *Use of the Arithmetical Triangle for Combinations*, Œuvres de Pascal, II, p. 1305. Pascal first wrote it in Latin (*ibid.*, pp. 1232–1259). It was shortened in the French version because once the central idea had been clearly grasped the application became straightforward.

CHAPTER 11

Mastering Games of Chance

Antoine Gombauld, Chevalier de Méré and Sieur de Baussay, lived the life of a man of leisure, dividing his time between his estate in Poitou and the court at Paris. He had money to spare and he clearly enjoyed rolling dice. He also prided himself on his ability to figure the odds at the casinos, and he apparently knew that the probability of throwing a 6 with one die rises above 1/2 with four throws. This is indeed the case, for the gambler, as we shall show later in this chapter, will win on the average 671 times to every 625 times he loses.

The game was popular in a slightly more complicated form known as *sonnez* (the French term for a double-six) that is played with a pair of dice, instead of with a single die. Because in a game with *one* die the odds become favorable after four throws, we might expect in a game with *two* dice that the odds of throwing at least one *double six* (that is, a six showing on both dice of the pair) will be six times four, or twenty-four, since two dice can come up in six times as many ways as one die can. Any given face on the first die can be associated with any one of the six faces on the second, and it would seem that a gambler should feel safe to bet even odds that he will throw at least one double six in twenty-four throws. But this is not the case and de Méré knew this, also. He may have found out by sad experience at the gambling table rather than by computation but, in any event, he realized that there was "something wrong" with the theory of numbers.

He told Pascal, whom he had met a short time before in 1654 at a house party arranged by another nobleman.[1]

Pascal easily proved that the odds were slightly *against* one double six on twenty-four throws, but slightly *favorable* for twenty-five throws, but whether this convinced de Méré that arithmetic rests on a sound basis is not clear. Nonetheless, he went on to raise a second and much more difficult problem that had often been debated but never correctly solved. This was the *problem of the points* or the *division problem*, which can be stated in the form of a question: How should the stakes be shared between two players if for some reason they have to stop before the game is completed? Pascal realized that the share of the stakes a player should receive depends on the *probability* that he would win the game were it carried through to its conclusion. He worked out in detail, for several examples, how the probability of winning could be calculated from a knowledge of the nature of the game and the partial score of each contestant.

The two problems raised by de Méré and the solutions furnished by Pascal proved to be very stimulating and significant, and we shall examine them in turn.[2]

Méré's First Problem: The Odds at Dice

Before tackling the probability of getting a double six, let us work out the probability of getting one six in four throws. We go about this indirectly, attacking it from the rear, as it were. We therefore start not by calculating the probability of obtaining at least one six in four throws of a die, but by doing the easier job of calculating the probability of *not getting any* sixes in four throws of a die.

Not getting a six in four throws of a die is a compound event composed of four independent events: "not getting a six on the first throw," "not getting a six on the second throw," "not getting a six on the third throw," and "not getting a six on the fourth throw." The probability of each one of these four component events is clearly 5/6, since among the six equally probable outcomes, there are five, namely the throwing of 1, 2, 3, 4, or 5, that result in no sixes. Since what happens on the second throw is completely

---

[1] Letter to Fermat, 29 July 1654, *Œuvres de Pascal*, II, p. 1142; trans., pp. 235–236.
[2] See Warren Weaver, *Lady Luck*. New York: Dover, 1982, pp. 115–123. This lucid introduction to probability has been used throughout this chapter.

independent of what happened on the first, etc., the probability of no *sixes* in four throws is:

$$\frac{5}{6} \times \frac{5}{6} \times \frac{5}{6} \times \frac{5}{6} = \left(\frac{5}{6}\right)^4$$

A mathematical probability equal to 1 corresponds to certainty. If we toss a coin, for example, the probability of getting heads is the same as getting tails, so H + T = 1. In the case we are considering:

Probability of one six + probability of no sixes = 1
Probability of one six = 1 − probability of no sixes

$$\text{Probability of one six} = 1 - \left(\frac{5}{6}\right)^4$$

If we multiply out:

Probability of one six = 1 − 625/1296
= 671/1296

Now since half of 1296 is 648, the fraction 671/1296 is clearly greater than 1/2. Of the 1296 equally probable outcomes when throwing a die four times, 671 involve throwing at least one six, whereas the remaining 1296 − 671 = 625 outcomes do not involve any six at all. This means that the odds are 671 to 625 in favor of the gambler who bets even money that he will throw at least one six in four throws. In other words, when throwing a die four times, there is a slightly higher chance of throwing at least one six than of not doing so. Méré knew this but he did not know why.

We now move to the slightly more complicated determination of the chance of at least one double six in twenty-four throws of two dice. Again we go at it backwards. The probability of not getting a double six in one throw is 35/36. The probability of not getting a double six in twenty-four throws is 35/36 multiplied by itself 24 times. Subtracting this from 1 (which corresponds to certainty) gives the desired probability of the complementary event, namely throwing at least one double six, namely

$1 - \left(\frac{35}{36}\right)^{24}$ gives, to four decimals, 0.4914, just a little under 1/2.

Pascal made the same calculation for twenty-five throws instead of twenty-four, and he found that the probability of at least one double six then turns out to be 0.5054, which is very slightly more than 1/2, so that the odds become very slightly favorable to the gambler. The Chevalier de Méré had satisfied himself of this in some way, but he would have had to be a very assiduous gambler to be able to distinguish empirically between a probability of 0.4914 and 0.5, that is, a difference of 0.0086!

Méré's Second Problem: Dividing the Stakes

The second problem, How should the stakes be divided among two players if for some reason the game is interrupted before either has won, and when each has only *partial scores*? had been raised at the end of the fifteenth century by Luca Paccioli, a famous Franciscan Friar and mathematician. Paccioli had considered a ball game that is normally played in 6 innings but is stopped after one player has won five and the other three. The correct answer to the question of the fair division of the stakes in this case is that they should be divided in the ratio 7:1, but Paccioli argued 5:3,[3] and the matter remained controversial until Pascal appeared on the scene. He was anxious to discuss the problem with someone and he turned to Fermat, whom we have mentioned earlier as the person who discovered independently the proposition about incomplete induction that Pascal had sent him by post. Pascal had never met the famous lawyer from Toulouse but a mutual friend, Pierre de Carcavy, put them in touch, In today's world, an aspiring mathematician would present his or her ideas at a poster session at an annual meeting of one of the learned societies, but there was no such forum in Pascal's day. A small group of people might gather around someone like Father Marin Mersenne, but it was much more common to start up a private correspondence with an expert who might be able to contribute something to the investigation.

Fermat had an amazing reputation. He was an assiduous commentator on Greek and Latin literature. He spoke several European languages, and even wrote poems in some of them. He carried out research on optics and the refraction of light, and he contributed to the development of geometry and number theory. In the course of his correspondence with Pascal, he was to achieve a significant breakthrough in the theory of probability. Lady Luck may have been born in the slightly disreputable atmosphere of the gambling room, but she was given her titles of nobility in the fascinating exchange of letters between Pascal and Fermat that ex-

---

[3]F.N. David, *Games, Gods and Gambling*. London: Griffin, 1962, pp. 37–38.

tended over four months in the summer and autumn of 1654. The results were such an advance over any previous thinking about games of chance that this episode can properly be regarded as the real start of the mathematical theory of probability.[4]

Not one but three correct solutions to the problem of the points were put forward in the correspondence. The first solution, which was arrived at by the *Combinatorial Method*, was discovered independently by both Fermat and Pascal. It consists in finding the maximum number of throws that are needed to determine the winner, listing the possible sequences of heads and tails in this number of throws, and determining by inspection who is the winner in each case. The second solution, which is due entirely to Pascal, rests on the *Method of Expectations* that takes what might be expected to have occurred and works backwards. Finally, a third solution, due to Fermat but involving no new principle, simply enumerates the possibilities working forwards, and can be called the *Direct Probability Method*. We shall follow the astonishingly rapid progress of the two friends.

Fermat's Letter to Pascal in July 1654: Partial Scores and Fair Division

Not all the correspondence has survived, and the first letter from Pascal to Fermat is missing. But it is clear from Fermat's reply that Pascal was concerned with the case of a gambler who bets that he will throw a six with a die in eight tosses. After making three throws without success, the gambler has to leave for some reason or other, and the game is called off. What proportion of the stake should he be allowed to take as he departs? Pascal had proposed a solution that Fermat, in the reply that he wrote in July 1654, interpreted on the assumption that Pascal had in mind a game that is stopped, and then resumed after regular interruptions in which 1/6 of the stake is removed each time. In this case, if the game is interrupted before the first throw, the gambler has a right to 1/6 of the stake; if it is interrupted before the second throw, he has the same right and can claim

---

[4]The letters are available in English translation in F.N. David, *Games, Gods and Gambling*, appendix 4, and in D.E. Smith, *A Source Book in Mathematics*. New York and London: McGraw-Hill, 1929). The extant correspondence consists of three letters from Fermat to Pascal (July, 29 August, and 25 September), and three from Pascal to Fermat (29 July, 24 August, and 27 October). A further letter from Fermat to Carcavy (9 August) refers to their discussion.

1/6 of what remains, namely 5/36 of the total; if the game is interrupted before the third throw, he can take 25/216; and, before the fourth throw, 125/1296, and so on. The series of fractions

$$\frac{1}{6}, \frac{5}{36}, \frac{25}{216}, \ldots$$

can be written as

$$\frac{1}{6}, \frac{1}{6} \times \frac{5}{6}, \frac{1}{6} \times \left(\frac{5}{6}\right)^2, \ldots$$

which express the probability of getting a double-six with two dice after one, two, three . . . tosses.

Although Fermat understood Pascal's mathematical reasoning, he felt uncomfortable with this way of proceeding, and he raised the following objection. Since the stake remains intact, the probability of each toss is 1/6 whatever the number of unsuccessful tosses, so that the chance of winning on the next toss is always 1/6.[5] Here Fermat missed the crucial point. What has to be considered is not the stake as it is found after one or after several throws, but the *probability* that a particular player will win if the game is carried to its conclusion. This probability can be calculated from a knowledge of the rules of the game and the partial score of each contestant. It is misleading to insist on the stake, and assume that because it has not been changed or removed, the odds remain unchanged. But let us not be too hard on Fermat; almost everyone did that before Pascal, and the initial focus of the exchange of letters was on how to divide the stake and not directly on the theory of probability, as would be the case today. Once Fermat went over the problem, he saw Pascal's point and they reached full agreement.

Pascal's Reply of 29 July 1654: The Combinatorial Method

In another letter that was written in the same month of July but is lost, Fermat explained his own method of solving *the problem of the points*, something that Pascal had not done in his first letter. It turned out to be the very one that had also occurred to Pascal, who wrote back in jubilation on 29

---

[5]*Œuvres de Pascal*, II, pp. 1136–1137; trans., pp. 229–230.

July 1654: "I no longer doubt that I am right after the wonderful agreement in which I find myself with you." If he had made "something of a mystery" of his method, he adds, "it was just to be sure that I had not made a mistake."[6] In plain language: he was worried that he might have been wrong and relieved to find that this was not the case.

Fermat and Pascal had struck on the *Combinatorial Method*, and although it is only treated in specific instances, the general solution can easily be reconstructed from the examples that Pascal gives in this letter to Fermat in which he considers a game in which a coin is tossed. Two players, A and B, each stake 32 *pistoles*, a substantial sum, enough to pay for a year's rent in a fashionable apartment in Paris. The game is won by the first player to get three points, but it is interrupted when A has two points and B one. Fermat's solution "which is the first one that comes to mind," writes Pascal, consists in finding the maximum number of tosses that are needed to determine the winner, listing all the possible sequences of heads and tails in this number of tosses and, from this list, determining by inspection who is the winner. It is clear that the game will be over in at the most two tosses, for if B wins the next toss, A and B will have the same number of points, and the next toss will decide the game. The *Combinatorial Method* consists therefore in determining how two games can be combined when there are two players. If a stands for a win for player A, and b for a win for player B, we have four possible combinations: (a, a), (a, b), (b, a), (b, b). Since A only needs one favorable toss, all the combinations where a appears gives him the game. This happens three times out of four, so that the probability is 3/4.

The solution amounts to no more than an exhaustive enumeration of all the possibilities, but it does have two interesting features. First, it implicitly assumes that the total length of the game as originally intended is irrelevant: what matters are the possibilities that remain. Secondly, although the enumeration of the equiprobable endings is straightforward, it is to be noted that had the game continued it might not have gone beyond one toss if the first outcome was a. Someone with less of a feel for probability than Pascal or Fermat might object that their *Method of Combination* includes, in its enumeration, tosses that would under no circumstance have been made. As we shall see, Roberval was to make just this point when he discussed the method with Pascal some time later.

---

[6]*Œuvres de Pascal*, II, pp. 1137–1139; trans., pp. 231–232.

DESIGNING EXPERIMENTS & GAMES OF CHANCE

The Method of Expectations

It was not this difficulty, however, that inspired Pascal to seek an alternative solution, but the tedium of enumerating all possibilities. As he wrote to Fermat, "the labor of the combination is excessive." Fortunately, Lady Luck came to his rescue and he found "a short cut and indeed another method that is much quicker and neater." This is the *Method of Expectations*, which consists in asking what might be expected to have occurred. "Here," Pascal writes, "is more or less how I go about to determine the fair division of stakes when, for instance, two players have agreed to play until one has three points."[7] Pascal considers three cases in which each has put in 32 *pistoles*. He begins with one where the first player has won two throws, and needs one more point to win, while the second player has won one game and therefore needs two more points to win. At the next throw, the first player either wins, and takes all the money (64 *pistoles*), or his opponent wins, in which case, if the game is interrupted at that point, there is a draw and each takes back his original stake of 32 *pistoles*. If they agree not to play this fourth throw, the first player can say: "I am certain to get 32 *pistoles*, for I get them even if I lose, but as for the other 32, perhaps I will get them, perhaps you will get them; the odds are even. Let us then divide these 32 *pistoles* in two, and you give me one half plus the 32 of which I am certain." So he takes 48 *pistoles* and the other 16. In modern terms, the solution involves analyzing the tree of possible games and, working backward from the tips, using recursively the idea that, if expectations of gains of X and Y units are equally probable, the expectation of gain is 1/2 (X + Y) units. The stakes are to be divided according to the expectation of gain, that is, the value of a gamble is equal to its expectation.

Pascal next considers the case where the first player has won two throws, the other none. At the next throw either the first wins the full stake (64 *pistoles*), or his opponent wins, in which case, we are back to the first game situation where, as we have seen, the first could claim 48 *pistoles*. If the game is interrupted at this point, the first player can now say: "If I win, I win all the money (64 *pistoles*), and if I loose, I have a right to 48 *pistoles*. So give me these 48 *pistoles* that I can claim even if I lose the next throw, and let us share equally the remaining 16 *pistoles* since you have

---

[7]*Œuvres de Pascal*, II, p. 1138; trans., p. 231.

as much chance as I do of getting them." So the first gets 48 + 8 = 56 *pistoles*, and his opponent 8.

Finally, Pascal considers a third case where the first player has won one throw, and the second, none. If the first player wins on the next throw, he will have won two throws and his opponent none, the situation we encountered in the second game. Hence he would have a right to 56 *pistoles*. If his opponent wins, they will each have won one throw and, therefore, if the game were interrupted, each could withdraw with half the stake, or 32 *pistoles*. The first is therefore entitled to argue thus: "If we agree not to play, give me 32 *pistoles*, the amount I am sure to receive, and let us share equally the remainder of 56 *pistoles*."[8] This remainder is 56 − 32 = 24, and half of 24 is 12. So the first player receives 32 + 12 = 44 *pistoles*.

Putting the Arithmetical Triangle to Work

In the *Treatise on the Arithmetical Triangle*, which he was drafting at the time he wrote to Fermat, Pascal shows how these calculations can be made with the aid of his Triangle for a number of problems.[9] For instance, in the case where two players need a certain number of points to win a game that is interrupted, it can be used to determine the amount that is due to each, by using the following procedure. Choose the triangle (see Figure 1) whose base contains as many cells as the sum of the points that are missing for both players. Take on this base as many cells as there are points missing for the first player, and add their numerical values. The remaining cells on the base are as numerous as the points the second player is missing. Add their values in turn. The two sums are in the ratio of the chances of the two players. A concrete example illustrates the procedure: the first player needs 2 more games, the second 4. We add 2 + 4 and get 6, which tells us to select the triangle whose base runs from cell P to cell δ because this base contains 6 cells, P, M, F, ω, S, δ. The ratio of the chances of the first player to the chances of the second player are as:

$$\frac{F + \omega + S + \delta}{P + M} = \frac{10 + 10 + 5 + 1}{1 + 6} = \frac{26}{7}$$

---

[8]*Œuvres de Pascal*, II, p. 1138; trans., pp. 231–232.
[9]*Œuvres de Pascal*, II, pp. 1317–1318.

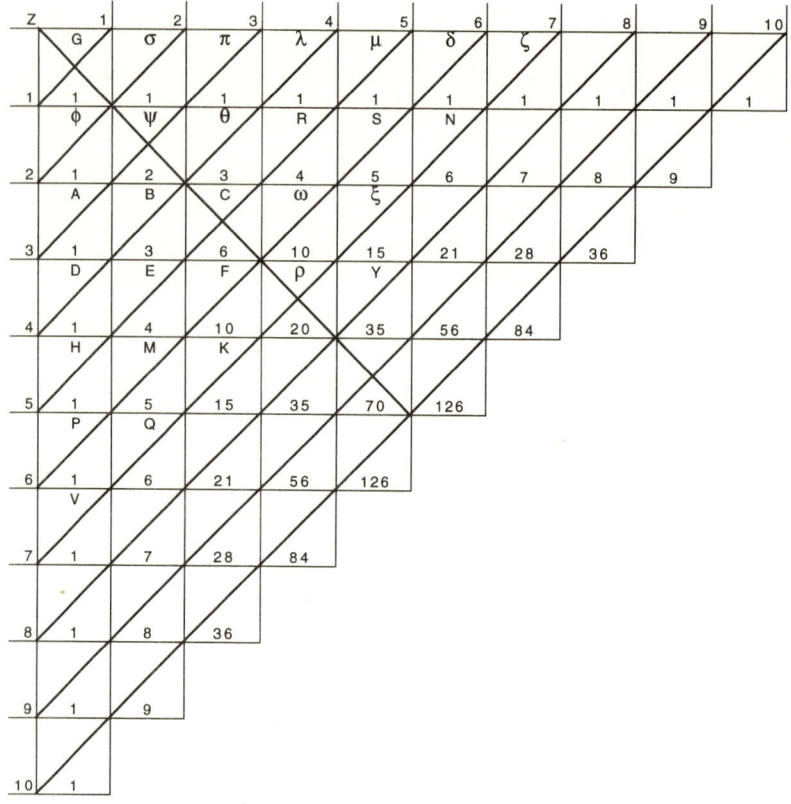

FIGURE 1 Pascal's Arithmetical Triangle.

which means that the first player's portion of the stake is:

$$\frac{F + \omega + S + \delta}{P + M + F + \omega + S + \delta} = \frac{10 + 10 + 5 + 1}{1 + 6 + 10 + 10 + 5 + 1} = \frac{26}{33}$$

The proof rests on a property of the Arithmetical Triangle that can be expressed as follows: the sum of the cells of each base is double the sum of the cells of the base just above. In other words, each number in base $n - 1$ is represented twice in base n by virtue of the nature of the triangle. For instance, the sum of the cells along the base D, B, θ, λ is 8 (namely 1 + 3 + 3 + 1 = 8), which is double the sum of the cells along base A ψ π (namely 1 + 2 + 1 = 4). Pascal formulates two lemmas before embarking on his demonstration. *First lemma*: the base φ σ of the second triangle contains the ratios of the players whose sum of missing points is 2. Since each player needs one more point, the ratio of their chances is as φ is to σ, namely as 1 is to 1, so that the probability of winning is for each 1/2. *Second Lemma*: If the base of a given triangle contains the ratios of the players whose sum of missing points is equal to the number of cells, the base of the next triangle will also contain the ratios of the players whose sum of missing points is equal to the number of cells of that base. For example, base D λ of the fourth triangle contains the ratios of the players that together need 4 more points. If the first player needs one more point and the second four, the fraction that belongs to the first is:

$$\frac{D + B + \theta}{D + B + \theta + \lambda} = \frac{1 + 3 + 3}{1 + 3 + 3 + 1} = \frac{7}{8}$$

What Pascal has to prove is that the next base, H μ, expresses the same ratio for the case when the sum of missing points of the two players is 5. For instance, when the first player needs 2 more points and the second player 3 points, the fraction that belongs to the first player is:

$$\frac{H + E + C}{H + E + C + R + \mu} = \frac{1 + 4 + 6}{1 + 4 + 6 + 4 + 1} = \frac{11}{16}$$

Pascal reverts to the strategy with which we are by now familiar, and examines the probability of winning or losing on the next toss. If the first player were to win, he would still need one point, and the second player 3 points as before. In this hypothesis, the sum of missing points for both players is 4, the situation we have just considered above and, hence, the fraction to be ascribed to the first player is:

$$\frac{D + B + \theta}{D + B + \theta + \lambda}, \text{ namely } \frac{7}{8}$$

Should the first player lose, he would still need 2 points, whereas the second player would now need 2, also. The first player's fraction is therefore reduced to:

$$\frac{D + B}{D + B + \theta + \lambda} = \frac{1 + 3}{1 + 3 + 3 + 1} = \frac{4}{8} = \frac{1}{2}$$

Summing these two possible outcomes and dividing by 2, we get for the first player:

$$\frac{D + B + \theta + D + B}{2(D + B + \theta + \lambda)} = \frac{1 + 3 + 3 + 1 + 3}{2(1 + 3 + 3 + 1)} = \frac{11}{18},$$

which is equal to

$$\frac{H + E + C}{H + E + C + R + \mu}$$

If the Loser Had Walked Off

Pascal illustrates the use of the Arithmetical Triangle with another example in which two players wager the same amount and we have to determine the amount that the loser could have taken before the last toss. He arrives at the answer by first seeking the amount (expressed as a fraction of the loser's wager) that the winner has won before the last throw is made. For instance, two players each wager 3 *pistoles* in a game of four tosses. If the first player has three points and the second none, when the game is interrupted the first player is missing one point and the second four, so the fraction that belongs to the first can be expressed, as we have seen, as:

$$\frac{H + E + C}{H + E + C + R + \mu} = \frac{1 + 4 + 6 + 4}{1 + 4 + 6 + 4 + 1} = \frac{15}{16}$$

and the fraction that belongs to the second as:

$$\frac{\mu}{H + E + C + R + \mu} = \frac{1}{1 + 4 + 6 + 4 + 1} = \frac{1}{16} = \frac{1}{2(D + B + \theta + \lambda)}$$

Since the total sum of the stake is: $2(D + B + \theta + \lambda)$, the fraction expressed in terms of the amount actually wagered by the loser is half this amount or:

$$\frac{1}{D + B + \theta + \lambda} = \frac{1}{1 + 3 + 3 + 1} = \frac{1}{8}$$

Therefore, the amount that the loser could have taken is expressed as a ratio whose numerator is one, and whose denominator is the sum of the cells of the base that contains as many cells as there are tosses in the game.

Putting All This into Words

In his letter of 29 July 1654, Pascal sums up his general argument in a way that he believed was obvious:

Since you see it all clearly, I shall no longer make a mystery of it as I did only to be sure that I had made no mistake: the value—by which I mean only the value of the opponent's money—of the last game of *two* is double that of the *last* game of three, and *four* times that of the last game of four, and eight times the last game of *five*, etc.[10]

Fermat may be excused for not finding it as clear as Pascal intended, and in order to follow Pascal's reasoning with more ease, we recall that there are two players, A and B, and that the first to score three points wins the game. Let us tabulate the results, assuming that A would win each game if it were not interrupted. A lacks one point, B lacks two, three, or four, so that the game will be decided in at most two, three, or four tosses.

Points that A lacks	Points that B lacks	Amount of the stake of 64 pistoles that A takes if the game is interrupted
1	2	$\frac{64+32}{2} = 48$ pistoles, i.e., $32 + \frac{32}{2}$
1	3	$\frac{64+48}{2} = 56$ pistoles, i.e., $32 + \frac{32}{2} + \frac{32}{4}$
1	4	$\frac{64+56}{2} = 60$ pistoles, i.e., $32 + \frac{32}{2} + \frac{32}{4} + \frac{32}{8}$

The fraction $32/2^n$ will therefore represent the value of each new game, namely $32/2$, $32/2^2$, $32/2^3$, and this is the rationale behind Pascal's rule: "The value—by which I mean only the value of the opponent's money—of the last game of *two* is double that of the last game of *three*, and four times the last game of *four*, and eight times the last game of *five*, etc." But rather than an application of the *Method of Expectations*, this can be considered a return to the *Method of Combinations*. Some of the possibilities are merely hypothetical, namely those that refer to games that could be played *after* A won three games and the competition is over. We can illustrate this with a diagram in which we follow the progress of the player who won the first game and circle the 3 that indicate a win:

---

[10] *Œuvres de Pascal*, II, p. 1139; trans., p. 232.

DESIGNING EXPERIMENTS & GAMES OF CHANCE

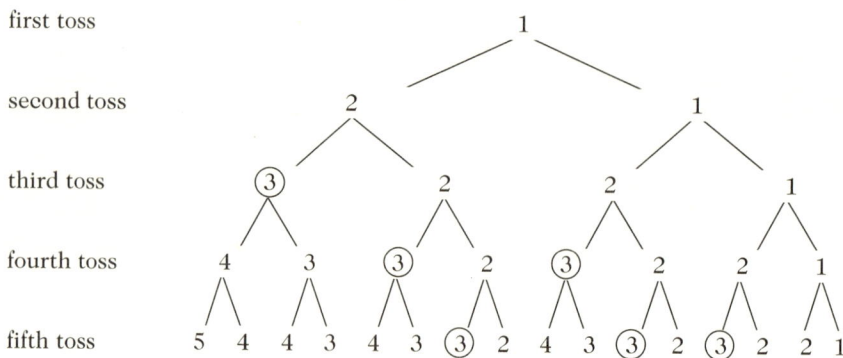

We see that the probability that A will win at the third toss is 1/4, at the fourth 2/8, and at the fifth 3/16. The cumulative probability of winning in five tosses is: 1/4 + 2/8 + 3/16 = 11/16. This is the result arrived at by Fermat, who assigned 64 × 11/16 = 44 *pistoles* to A if the game was interrupted when A needed 2 points and B 3 points. This supposes that we distinguish between the *real* and the merely *hypothetical* possibilities.

Pascal's *Method of Expectations* has one curious drawback that he recognized: although it allows us to move "backwards," it does not provide a general rule for moving "forwards" from the first game. "The proportion of the stake for the first game is not so easy to find," he writes to Fermat, and "not wishing to conceal anything," he gives his procedure or rather his formula to compute the value of the first of any number of games:

$$\frac{1 \times 3 \times 5 \times 7 \times \ldots \times (2n-1)}{2 \times 4 \times 6 \times 8 \times \ldots \times 2n}$$

For instance, if the number of games is eight ($n = 8$):

$$\frac{1 \times 3 \times 5 \times 7 \times 9 \times 11 \times 13 \times 15}{2 \times 4 \times 6 \times 8 \times 10 \times 12 \times 14 \times 16}$$

Pascal does not work out the multiplication and the division (the result is 2,027,025 divided by 10,321,920, which gives 0.196380615). It is the *Combinatorial Method* that is invoked, however, when he turns to the first eight letters of the alphabet, A, B, C, D, E, F, G, H, as in the following example:

> Take all the possible combinations of 4 letters, then all the possible combinations of 5 letters, and then those of 6, of 7 and of 8, etc., namely all the possible combinations starting from the number which

is half the total. The sum of half the combinations of 4 (i.e. 35, half of 70) with each of the combinations of the numbers above will be the nth term of the fourth progression which begins at 2 (i.e. 2^7, the fourth in the progression 2, 2^3, 2^5, 2^7), where n is half the total.[11]

The wording is slightly confusing and Pascal goes on to express the proposition in Latin "because French," he says, "is useless." Modern notation is even better than Latin, and the general theorem can be expressed as:

$$(1/2\ _8C_4) +\ _8C_5 +\ _8C_6 +\ _8C_7 +\ _8C_8 = 2^7$$
$$35 \qquad\quad + 56\ \ + 28\ \ + 8\ \ \ \ + 1\ \ = 128$$

Pascal does not offer a proof of this theorem and, presumably, he saw it as an immediate consequence of the symmetry of the Arithmetical Triangle that he was in the process of investigating. If we look at the cells to the right and the left of the straight line that is dropped from Z and divides the triangle into two equal parts, we note that the corresponding cells are equal, for instance, E = R. Furthermore, the sum of the cells that are cut across diagonally by the base of the triangle is always twice the sum of the cells in the row just above, for example, the sum of the cells along the base that runs from 9 on the x-axis to 9 on the y-axis, namely: 1 + 8 + 28 + 56 + 70 + 56 + 28 + 8 + 1 = 256, is double the cells 1 + 7 + 21 + 35 + 35 + 21 + 7 + 1 = 128 along the base just above that runs from 8 on the x-axis to 8 on the y-axis.

Reverting to French, Pascal tackles a specific instance:

> For example, in a game of 5 points if someone has one point and still needs 4, the match will be decided for certain in 8, which is twice 4. The value, in terms of the opponent's stake, of the first game in a game of 5 points, is a fraction whose numerator is half the combinations of 4 out of 8 (I take 4 because it is equal to the number of games required, and 8 because it is twice 4) and whose denominator is this same numerator plus all the combinations of higher numbers. Thus, if I have won the first game out of 5, the value of my opponent's stake due to me is 35/128; that is to say, if he has staked 128 *pistoles*, I take 35 and leave him the remainder, 93.

---

[11]*Œuvres de Pascal*, II, pp. 1139–1140; trans., p. 233.

Pascal may have thought of his mathematical formula when he noticed that 35/128 = 105/384, as he says in the next sentence of his letter, "Now this fraction 35/128 is the same as 105/384, which is made by taking the product of even numbers as denominator and the product of odd numbers as numerator" (i.e., $3 \times 5 \times 7 = 105$, and $2 \times 4 \times 6 \times 8 = 384$). The explanation ends here with a tribute that is generous to Fermat but not very helpful to the modern reader: "You will undoubtedly understand all this well, if you take a little trouble: that is why I think it unnecessary to go on with it any longer."[12]

Pascal worked with a particular example. Taking $b = 5$ (and hence $2b - 2 = 8$), he found the division to be $(2^8 + 70) : (2^8 - 70)$. For instance, if each player wagers 256 *pistoles*, the first gets $256 + 70 = 326$ *pistoles* and the second gets $256 - 70 = 186$ *pistoles*. "Thus," he writes, "if I have won the first point out of 5, 35/128 of my opponent's stake is due to me," for $_8C_4 = 70$ and $2^8 = 256$. This is explained in the *Treatise on the Arithmetical Triangle*,[13] where Pascal shows how the value of the last of two, three, and four games is determined by considering the cells of base 2, 3, and 4, namely

$$1 : (1 + 1); \; 1 : (1 + 2 + 1); \; 1 : (1 + 3 + 3 + 1)$$

The proof that is offered by Pascal rests on a Proposition in which he had shown that if one player needs one game and the other four, and the game is interrupted, the fraction of the stake that belongs to each is determined by considering the cells in the fifth base, such that A gets $(H + E + C + R) : (H + E + C + R + \mu)$, namely 15 : 16 of the stake, and B gets $\mu : (H + E + C + R + \mu)$, namely 1 : 16. Now

$$\frac{\mu}{H + E + C + R + \mu} = \frac{1}{2D + 2B + 2\theta + 2\lambda}$$

Since $2D + 2B + 2\theta + 2\lambda$ is the total stake, the value of the opponent's money is half of this:

$$\frac{1}{D + B + \theta + \lambda} = \frac{1}{8}$$

---

[12]*Œuvres de Pascal*, II, p. 1141; trans., p. 234.
[13]*Œuvres de Pascal*, II, p. 1317.

Fermat's reply to Pascal's letter of 29 July is missing, but the letter he wrote to Carcavy in August 1654, which is extant, reveals how much he valued Pascal's approval:

> I am delighted to be in agreement with Pascal, whose talent I greatly admire and whom I believe capable of solving any problem he might attempt. The friendship he offers is so dear and precious to me that I shall not scruple to take advantage of it in preparing the publication of my Treatises.[14]

Fermat goes on to express the hope that Carcavy and Pascal will become the editors of his papers on the theory of numbers, a proposal that neither of them seem to have been willing or able to entertain.

Pascal's Letter to Fermat of 24 August 1654

Meanwhile, Pascal forged ahead with the problem of points, but he encountered difficulties in extending the *Combinatorial Method* to cases where there are not only two but three players. In his letter to Fermat of 24 August 1654, he ventures to query the value of the method itself, albeit with considerable diffidence: "Allow me to present my arguments, and to ask you to let me know whether I am right or where I went wrong. I ask this in earnest for I feel safe only when you agree with me." He then makes his point, "When there are only *two* players, your combinatorial method is most reliable, but when there are *three*, I think I can prove that it is not applicable, unless you proceed in a way that I have not grasped."[15]

Pascal goes on to give a lucid summary of Fermat's method as it applies to a game with two players that is interrupted when the first player needs two points to win, and the second player needs three. The game will clearly be decided in four throws, and "from this you conclude," he writes,

> that it is necessary to determine in how many ways four games can be arranged between two players, and how many combinations would make the first win or the second win, and then share out the stakes in this proportion. I would have found it difficult to understand this if I had not arrived at the result already, and you wrote knowing that

---

[14]*Œuvres de Pascal*, II, p. 1146; trans., p. 238.
[15]*Œuvres de Pascal*, II, p. 1147; trans., p. 239.

I did. Thus, to find the number of combinations of four games between two players, we assume that they play with a die of two faces (since there are only two players), say heads and tails, and that they throw four of these dice (because they play four games). Now one has to see in how many different ways these dice can turn up. This is easy to calculate, and it is sixteen altogether, which is the square of four. Now let us suppose that one of the faces is marked a, and is favourable to the first player, and the other b, which is favourable to the second player. These four dice can turn up in one of the following sixteen ways:

$aaaa$	$abaa$	$baaa$	$bbaa$
$aaab$	$abab$	$baab$	$bbab$
$aaba$	$abba$	$baba$	$bbba$
$aabb$	$abbb$	$babb$	$bbbb$

Because the first player needs two games, he must win whenever there are two a's, and there are 11 such cases. Likewise because the second player needs 3, he must win whenever there are three b's, and there are 5 such cases. Thus the stakes must be shared in the ratio of 11 to 5.[16]

For ease of discussion let us call the players A and B. Now given that the game must finish in 4 throws, all the 16 possible outcomes for these 4 throws are equally likely. Each case where A has 2, 3, or 4 a's counts as a case favorable to A, and each case where B has 3 or 4 b's represents a case favorable to B. There are therefore 11 for A and 5 for B, so the odds are 11:5 in favor of A. It must be noted that several cases have been included where the game would have been determined in less than 4 throws, for instance, 2 throws would have settled the outcome in favor of A in the cases $aaaa, aaab, aaba, aabb$.

Why Play When You Have Won?

Pascal communicated Fermat's solution to mathematicians in Paris, and Roberval raised an interesting objection that Pascal immediately reported to Fermat. Why work out the division of the stakes on the assumption that *four* games are played when the game ceases as soon as one player wins

---

[16]*Œuvres de Pascal*, II, pp. 1147–1148; trans., pp. 239–240.

two games?[17] Pascal answered Roberval on Fermat's behalf as follows: although the game could be over in 2 throws, let us suppose that the players agree to play 4 games just for the fun of it since this cannot change the outcome, as is easily shown for cases where the first player wins in 2, 3, or 4 games:

Probability that A wins in 2 games $= \frac{1}{4}(\underline{aa}, ba, bb, ab)$

Probability that A wins in 3 games $=$ probability that he gets 1 point in 2 games × probability that he wins the third game

$= \frac{1}{2} \times \frac{1}{2}$

$= \frac{1}{4}\left(aaa, aab, abb, bbb, \underline{aba}, bab, \underline{baa}, bba\right)$

Probability that A wins in 4 games $=$ probability that he gets 1 point in 3 games × probability that he wins the fourth game

$= \frac{3}{8} \times \frac{1}{2} = \frac{3}{16}$

So the probability that A wins is:

$\frac{1}{4} + \frac{1}{4} + \frac{3}{16} = \frac{11}{16}$, as previously deduced

Roberval agreed that Pascal had got his mathematics right, but he argued that it was irrelevant unless the players were compelled to play *four* games. To which Pascal replied that although in real life the game would end when one player had the minimum necessary to win, a logically complete solution was impossible without considering all the mathematical possibilities. In a sense, the mathematical laws stand above the wishes of the players:

> For, if the first man wins the first two games out of *four* and thus has won the set, why should he refuse to play another two games, seeing that, if he wins them, he cannot win more and if he loses them he cannot win less? Should his opponent win the last two, this would not be enough for him since he needs three, and the players could not both

---
[17]*Œuvres de Pascal*, II, pp. 1148–1149; trans., pp. 239–240.

get the number they need in four games. It is easy to see that nothing is changed whether they let the game take its normal course, which means ceasing to play when one has won, or play the whole four games. Since these two procedures are equal and immaterial, the result is the same in both cases.[18]

Pascal's solution is entirely correct: the fair division of stakes is the same whether the four games are played or not. It is with some surprise therefore that we find that when he proceeded to the case with three players, he failed to apply his own solution consistently.

The Case of Three Players

Pascal considers three players A, B, and C such that A wants 1 point to win while B and C each need 2 points. Under these circumstances the game must be finished in 3 games. The 27 possible outcomes of these 3 games are enumerated, and tabulated to be read downwards:

a a a	a a a	a a a	b b b	b b b	b b b	c c c	c c c	c c c
a a a	b b b	c c c	a a a	b b b	c c c	a a a	b b b	c c c
a b c	a b c	a b c	a b c	a b c	a b c	a b c	a b c	a b c
111	111	111	111	1	1	111	1	1
	2		2	222	2		2	
		3			3	3	3	333

The numbers 1, 2, or 3 under the 27 possibilities indicate that A has won (1), B has won (2) or C has won (3). It would seem therefore that all Pascal had to do was note the winners as:

| a a a | a a a | a a a | b b b | b b b | a b c | a a a | a b c | c c c |

and give $A:B:C$ as 17:5:5. This is, of course, the correct answer, but he does not do this. To get a feeling for the problem as Pascal saw it, let us quote him at length before analyzing his procedure:

> The first player needs just *one* game to win: thus all the throws giving one a are favorable to him: there are 19. The second player needs

---

[18]*Œuvres de Pascal*, II, p. 1149; trans., pp. 241–242.

two games: thus all the throws giving two *b*'s are his: there are 7. The third player needs *two* games: thus all throws with two *c*'s are his: there are 7.

It would be a gross error to conclude that the stakes must be divided amongst the three players in the proportion 19:7:7, and I cannot believe you would do this: for some of the throws are favourable to both the first and second player, such as *a b b*, for the first player has one *a* which he needs and the second has two *b*'s which he needs; and in the same way *a c c* is favourable to the first and third player.

Thus we must not add up the throws common to two players as being worth the whole amount to each, but only half. For suppose *a c c* were thrown, the first and the third player would have the same right to the stake, each having made his point, thus they would each take the money; but if *a a b* were thrown the first player would win alone. Hence the calculations must be made in this way: There are 13 throws favourable to the first player only and 6 which give him a half share and 8 which give him nothing: thus if the stake is one *pistole*, there are 6 throws each worth 1/2 *pistole* to him and 8 worth nothing.

Thus, to share out the stakes, we must multiply

13	by one pistole which makes	13
6	by a half, which makes	3
8	by zero, which makes	0
Total 27	Total	16

and divide the sum of the products, 16, by the sum of the throws, 27, which gives the fraction 16/27; this gives the amount due to the first player, when the stakes are shared out, that is 16 *pistoles* out of 27.

The shares for the second and the third player will be found in the same way:

There are	4	throws worth 1 *pistole* to him, multiplied together	4
There are	3	throws worth ½ pistole to him, multiplied together	1½
And	20	throws worth nothing to him	0
Total	27		Total 5½

Thus, 5½ *pistoles* out of 27 are due to second player, and the same amount to the third; these three totals, 5½, 5½ and 16 added together 27. That seems to me to be the way we must solve the problem of points following your combinatorial method, unless you proceed in some other way which I do not know.[19]

---

[19] *Œuvres de Pascal*, II, pp. 1151–1152; trans., pp. 243–244.

The combinational method, applied in this way, leads to a division of stakes that is not fair because of "the false assumption" that *three* games are always played when, in fact, the play normally ceases when one player has got the number of games he needs. Now as we saw above when considering a game with two players, Roberval had objected that he could not see how "a fair division of the stakes could be made under the erroneous assumption that four games are played," and Pascal had answered him effectively. Why does he share his misgivings in the case of three players. Was he in a muddle, as F.N. David suggests?[20] The matter is both more complex and more interesting, and the answer hinges on the difference between the case when there are two players and the case when there are three.

Three Instead of Two

Let us retrace our steps a bit. Pascal had extended what *he assumed* was Fermat's *Combinatorial Method* to the case of three players and had obtained an answer that he knew was wrong because it did not correspond to the solution obtained by his *Method of Expectations*. "He had realized," as A. W. F. Edwards points out, "that in the hypothetical game in which the real game is imbedded in the method of combinations the order of heads and tails is only irrelevant with two players, and not, in general, with three or more."[21]

Pascal supposed that ignoring the order in which heads and tails occurred was an intrinsic feature of Fermat's combinatorial method (after all, this is what we mean in our modern terminology when we speak of a combination as opposed to a permutation). As we have seen, Pascal carefully worked through the original example for two players, listing all the permutations for each combination, but then assigning each to *A* or *B* according to the numbers of heads or tails, as is justifiable with only two players. When he went on to apply the identical argument to three players, he noted that some *combinations* are favorable to two players, but that to count each corresponding permutation as a win for both would be to make "a gross error." He supposed, therefore, that the method of combi-

---

[20]Games, Gods and Gambling, p. 94.
[21]A. W. F. Edwards, *Pascal's Mathematical Triangle*. London: Charles Griffin, 1987, p. 145. Edwards offers an excellent account of the discussion between Pascal and Fermat.

nations must proceed by dividing each such permutation equally between the two players, counting it as half a win for each. He was struck by the fact that in the actual conditions of the game with three players, only one person can win, for play ceases as soon as one has the required number of points. But in the hypothetical conditions, two persons can get the number of games they need. In other words, in a game with three players the problem cannot be solved by merely counting the number of points won by each player in the hypothetical game. Pascal concluded that the *Combinatorial Method* cannot in general solve the problem of points for three players, although it might work in particular cases. Indeed it works for three players but only "*by accident*":

> I hope I have made you see that the combinatorial method which is suitable for two players by accident, is sometimes suitable for three players, for example when one player needs *one* game, the second *one* and the third *two*, because in this case, no two players can have enough games needed to both win.[22]

Fermat's Letters to Pascal of 29 August and 25 September 1654

In a short letter dated 29 August and written before he had received Pascal's letter of 24 August, Fermat remarks briefly that in the game involving three players the correct answer is 17:5:5 and that this comes out of the combinatorial method. There was clearly no ambiguity in his mind.

The rejoinder to Pascal's letter of 24 August is dated 25 September. With his usual friendliness and modesty, Fermat begins on just the right note: "Do not fear that our agreement is ending. You have strengthened it in trying to destroy it, and in answering Roberval you have replied for me also." Pascal would not have been misled about the *Combinatorial Method* if he has worked out the implications of his own reply to Roberval:

> I find that there are only 17 combinations for the first player and 5 for each of the other two: for, when you say that the combination *a c c* is favourable to the first man and to the third, *it appears that you*

---

[22]*Œuvres de Pascal*, II, p. 1153; trans., p. 245.

forget that everything that happens after one of the players has won is irrelevant . . . This fiction of lengthening the match to a particular number of games serves only to simplify the rules and (in my opinion) to make all the chances equal or, to state it more clearly, to reduce all the fractions to the same denominator.

Fermat makes this clear by showing that 17:5:5 is the correct ratio whether 4 or 5 games instead of 3 are played:

So that there should be no doubt, if instead of *three* games you lengthen the hypothetical match to *four*, there would be 81 instead of 27 combinations, and we have to see how many combinations would give one game to the first player before either of the two other players got two games, and how many would give two games to each of the other players before the first player got one. You will find that there are 51 combinations favourable to the first man, and 15 to each of the other two, which comes to the same ratio as before. If you take 5 games or any other number you please, you will always get the same proportion 17:5:5.

And thus I am right in saying that the combination *a c c* is favourable only to the first player and not to the third, and *c c a* is favourable only to the third and not to the first, and therefore my combinatorial rule is the same for three players as for two and, in general, for any number of players.[23]

The difficulty was caused by the ambiguity of the word *combination* and the absence of a word for what we call *permutation*. In the *Treatise on the Arithmetical Triangle*, Pascal had felt the need to remind his readers that *combination* had been used in many different senses, and that in order to avoid ambiguity he would use the word in such a way that *order in which things are combined are not taken into account*.[24] As we have seen, this did not lead to the correct answer in the case of three or more players.

But the misunderstanding between Fermat and Pascal is happily at an end. Ever helpful, and perhaps not a little amused, Fermat adds, "for the benefit of M. Roberval," the direct probability solution for a game with three players in which one lacks one point and the others two each, sim-

---

[23]*Œuvres de Pascal*, II, pp. 1155–1156; trans., pp. 247–248.
[24]*Œuvres de Pascal*, II, p. 1302.

ply working his way forward assigning probabilities along what we would call the event-tree.[25] This *Direct Probability Method* is the third one used in the correspondence. It does not involve any new principle, but it shows Fermat as a clear expositor of difficult concepts:

> The first player can win in a single game, in two or in three. If he wins in a single game, he must, with one die of three faces, win with the first throw. Since a single die has three possibilities, his chance of winning, when one game only is played, is 1/3.
>
> If two games are played, he can win in two ways, either when the second player wins the first game and he wins the second, or when the third player wins the first game and he wins the second. Now, with two dice we have 9 possibilities. Thus the first player has a chance of 2/9 of winning when they play two games.
>
> If three games are played, he can only win in two ways, either when the second player wins the first game, the third the second and he the third, or when the third man wins the first game, the second wins the second and he wins the third; for, if the second or third player were to win the first two games, he would have won the match and not the first player. Now, three dice have 27 possibilities: thus the first player has a chance of 2/27 of winning when they play three games. The sum of the chances that the first player will win is therefore $1/3 + 2/9 + 2/27 = 17/27$.
>
> This rule is sound and applicable to all cases, so that without recourse to any artifice, the actual combinations in each number of games give the solution and show what I said in the first place, that the extension to a particular number of games is merely a reduction of the several fractions to a common denominator. There in a few words is the whole mystery, which puts us in good terms again since we are only interested in finding correct arguments.[26]

In his letter of 29 August, Fermat had tried to get Pascal interested in some of the theorems he had discovered, and he wrote that numbers of the form $2^{2n} + 1$ are always prime numbers. For instance, $2^2 + 1 = 5$; $2^4 + 1 = 17$; $2^8 + 1 = 256$; $2^{16} + 1 = 65,537$, and so on. "The proof is very difficult," he writes, "and I confess that I have not yet been able to find it fully. Had I been successful, I would not suggest that you look for it."[27] It was only a

---

[25] See the helpful figure in Edwards (fig. A1), p. 146.
[26] *Œuvres de Pascal*, II, pp. 1156–1157; trans., p. 249.
[27] *Œuvres de Pascal*, II, p. 1155; trans., p. 247.

century later that Euler proved the theorem false by showing that $2^{25} + 1$ is a composite, and hence it was just as well that Pascal did not look for the proof.[28] In his letter of 25 September, however, Fermat did send Pascal the genuine article in the form of what is one of his most beautiful theorems, and one that was to remain unproved until the nineteenth century: "Every integer is composed of one, two, or three triangles; of one, two, three, or four squares; of one, two, three, four, or five pentagons; of one, two, three, four, five, or six hexagons, and thus to infinity."[29]

Pascal's Letter to Fermat, 27 October 1654: Friendly Parting

Pascal was pleased with Fermat's solution of the problem of the points with the *Direct Probability Method*, which, he realized, was entirely Fermat's own, and the correspondence ends on a note of mutual admiration and respect. But Pascal declines to examine Fermat's theory of numbers which, he says, is beyond his mathematical ability: "Look elsewhere for someone who can follow your discoveries concerning numbers ... I confess that they are a beyond my comprehension. I can only admire them and very humbly beg you to take the first opportunity of completing them."[30]

Although no one in the seventeenth century was to share Fermat's fascination with number theory, Fermat had not been wrong in calling on Pascal. In his letter of 29 July, Pascal had demonstrated his versatility in handling numbers by proving that the difference of the cubes of any two consecutive natural numbers, when unity is subtracted, is six times the sum of all the numbers contained in the smaller one.[31] Pascal also considered another problem of number theory much discussed at the time, a formula for the sum of the mth power of the first n consecutive integers, which he related to the arithmetical triangle, to reasoning by recurrence,

---

[28]It is known that $2^{2n} + 1$ is not prime for n between 5 and 16 inclusive. See Carl B. Boyer, *A History of Mathematics*. Princeton: Princeton University Press, 1985, p. 388.

[29]Letter of Fermat to Pascal, 25 September 1654, *Œuvres de Pascal*, II, p. 1157; trans., p. 249.

[30]*Œuvres de Pascal*, II, p. 1158; trans., p. 251.

[31]*Œuvres de Pascal*, II, pp. 1142–1143; trans., pp. 236–237. Pascal treats this matter in more general terms in his *Potestatum Numericarum Summa* (*Œuvres de Pascal*, II, pp. 1261–1263). See next chapter.

and to infinitesimal calculus in an essay. He consigned the results in an essay, *The Summing of the Powers of Integers (Postestatum Numericarum Summa)*, which was published as an appendix to his *Treatise on the Arithmetical Triangle*. We shall examine this work later in Chapter 12.

Pascal was not impervious to the beauties of number theory and his unwillingness to pursue the mathematical problem raised by Fermat had as much to do with his poor health as with his growing concern with a very different kind of enquiry. Early in October, he had moved from the Right Bank of the Seine, where he had spent most of his life, to a house on the Left Bank next to the Luxembourg Gardens and near the Convent of Port Royal, where his younger sister, Jacqueline, had become a nun in 1652. As he became increasingly interested in religion, he had less time for mathematics. On 23 November 1654, he underwent his "night of fire," the mystical experience that confirmed him in his faith and led him to commit himself heart and soul to a life of Christian self-sacrifice. It is around this time that Pascal began to visit the semi-monastic retreat of Port-Royal des Champs, some thirty-five kilometers south-west of Paris and to meet with a number of prominent Jansenists. He drafted several religious works, of which the most important is the one he wrote at the request of his newly chosen spiritual director, Isaac Le Maître de Saci, in which he discussed and refuted the stoicism of Epictetus and the skepticism of Montaigne. The Jansenists, who emphasized the crucial importance of grace and were suspected of Protestant leanings, were coming under increasing fire from Catholic authorities. Pascal came to their aid by publishing between January 1656 and March 1657 eighteen *Provincial Letters* under the pseudonym of Louis de Montalte.

Comparing the Combinatorial Method and the Method of Expectations

Pascal was never again to give his undivided attention to the study of probability, but the *Combinatorial Method* and *The Method of Expectations* are impressive achievements in the history of mathematics. Let us compare them once more.

Pascal's *Method of Expectations* consists in gradually reducing the element of chance and showing that the solution is found at the end of a series of successive acquisitions. In a sense, the problem is not to determine what *will happen* or even what *might happen* but *what has already happened and is decided*. "I am sure of getting 32 *pistoles*," says the first player, "for they are mine even if I lose the next toss." His assurance does not depend on a future event but on an irreversible gain. This may strike us as strange in a game of chance where nothing is decided until the game is over, but the interruption of the game alters the situation entirely and allows for Pascal's apparently paradoxical way of reasoning where the future has been made to collapse into the present.

Pascal does not appeal immediately to the notion of probability but to the prior problem of recognizing that the money that a gambler wagers is no longer his own: "In order to understand the rules of the division, the first thing to consider is that the money that the players have put into the game no longer belongs to them. They have exchanged it for the right to expect what chance may bring them under the conditions agreed upon."[32] But players, who freely enter the game, are also free to leave at any stage. They then give up what they could have expected of chance for a part of the stake. How this part is to be determined, namely how the money is to be fairly divided, is the *division problem* or the *problem of points*, and Pascal's solution provides a way of acting in circumstances where a decision must be taken in the absence of certainty about the outcome of a series of events that was cut short, be they moves in a game or a chain reaction in chemistry. This is why Pascal is one of the founder's of the theory of probability and a forerunner of Decision Theory, although he does not use the word *probability* in the mathematical sense but in the more general one of likelihood.

The problem of dividing the stakes only arises if a game is interrupted. Since gambling on all kinds of games, from soccer to card games, was very common in the sixteenth and seventeenth centuries, interruptions due to a wide variety of factors were not rare, and several mathematicians had tried to work out a fair distribution of stakes when this occurred. The question that the Chevalier de Méré asked Pascal had become a genuine mathematical teaser. The emphasis was laid on the *fairness* to be observed and the context had legal overtones. Avoiding all moral considerations about gambling (a pet theme of contemporary preachers), Pascal presents games of chance as a simple contract that is freely entered upon by two or more persons who give up a certain sum of money in exchange for "the right to expect what chance can give." They are at liberty to terminate the agreement and leave the game, whereby "they give up what they could expect

---

[32]*On the Usage of the Arithmetical Triangle*, II, p. 1308.

from chance and recover something."³³ What is the fair value of this "something"? A hundred years earlier, in 1556, Niccolò Tartaglia despaired of finding a mathematical solution and declared: "This is a legal matter and whatever the solution it will always be opened to litigation."³⁴ Pascal was more sanguine and felt we can compare expectations and actual events by converting an *uncertain* outcome into a series of successive *certain* gains. A fair division makes it indifferent to choose to go on playing or to leave with a compensation. A player who stays in the game cannot claim to be acting more reasonably than the one who withdraws and takes his fair share of the stakes. As Pascal puts it: "The settlements must be proportioned to what they had the right to expect from chance in such a way that each gambler agrees that it is perfectly fair that he should be given the choice of retiring with the allocated sum or going on with the game."³⁵ We could also say that a player who leaves the game should receive as compensation the sum that someone who took his place would have to pay to join, since the situation that results from the interruption of the game is the same as the one that would obtain if a new game, *with different odds* for each player, were to start. Pascal does not gives this alternative formulation of his "rule," but a century earlier Girolamo Cardano, who could not calculate directly the amount that each player should receive when the game was interrupted, saw that he could get his answer by asking how much someone would have to wager to join a game in process and replace a player who had to leave. Although the case is identical, we feel on more familiar ground with this alternative approach.

Pascal only considers the case of two players, and in order to work out the rules of equitable division, he appeals to two principles. First, if at a given time in the game, one of the players has the right to a certain sum whatever the outcome, this sum is entirely his own since there is no longer any probability of loss or gain. Second, if both players have equal chances of winning on the next move, they can choose to stop and withdraw, each with half of the stakes. Assume it is agreed that if the first player wins, he gets a certain sum but if he loses, he gets a smaller one. Then if

---

³³*Ibid.*

³⁴Nicolo Tartalia, *La prima parte del General trattato di numeri e misure*, quoted in Ernest Coumet, "Le problème des partis avant Pascal," *Archives internationales d'histoire des sciences* 18 (1965), p. 254.

³⁵*Œuvres de Pascal*, II, p. 1308; see E. Coumet, "Le problème des partis avant Pascal," pp. 264–265.

the game is interrupted, the first player can retire with the lesser sum (which he would have been his had be lost), plus half the difference between the larger sum (which would have been his had he won) and the lesser sum. For instance, if the first player was promised 8 *pistoles* if he won and 2 *pistoles* if he lost, when the game is interrupted he can leave with $2 + [(8 - 2) + 2] = 6$ *pistoles*. If both players are in the same situation as the one we have just considered, again we look at the outcomes of a loss or a gain, sum them, i.e., $2 + 8 = 10$, and divide the sum equally, i.e., $(2 + 8) + 2 = 5$, for each player.

In the light of these principles and corollaries, Pascal worked out in detail for several examples how the probability of winning can be calculated from a knowledge of the nature of the game and the partial score of each contestant. He then showed how the problem can be solved with the aid of the Arithmetical Triangle, and he mentioned that "combinations" would yield the same result. Whereas Fermat looked forward and made an inventory of possible outcomes, Pascal did away with chance altogether. The future only interested him for what it could disclose about the present state of affairs. His misunderstanding of Fermat's *Combinatorial Method* resulted from this difference of approach. As we have seen, Pascal's mistake was caused by the ambiguity of the notion of *combination*. Nonetheless, he would not have slipped so easily into such an error if he had been thinking along the lines that Fermat followed. A static picture of the combinations, neglecting the element of succession, had encouraged the idea that when there are three players some combinations can be favorable to two players at the same time, for instance, when player *A* wins the first toss and player *C* the next two. We realize that this cannot be the case as soon as we think of the actual progress of the game. The very moment one player has the required points, the game ceases. Common sense requires that we attend to the *order* in which the tosses are made. This is why Fermat replied to Pascal with a touch of irony: "It appears that you forgot that anything that is done after one of the players has won is perfectly useless." Pascal's mistake was linked to the difference between the actual case and the hypothetical case that is imagined by the mathematician. "The actual game might not last as long as there are possible throws," was what Roberval had objected to the *Combinatorial Method*, and Pascal had rightly replied that this did not matter in the case of a game with two players, because the winner could go on playing if he so chose since it would make no change to the outcome. This is correct and applies equally well to the case where there are three players. Why did Pascal not see this? Over and beyond the ambiguity of the word *combination*, Pascal focused too exclusively on his own *Method of Expectations*, in which a chance event is transformed into a partial gain (determined by the risk involved). If two players have an equal claim to a given sum, it is fitting that each should get half, but this is only true if we forget about the real game in which there is one and only one winner.

Pascal was interested in working out the *partial gain secured*, as is clear by the way he went on to ask: "What is the value for the winner of the first game expressed as a fraction of the amount wagered by the loser"? The solution, not only for the first but for the second, the third, or the last toss, is easily found with the Arithmetical Triangle. The assumption is that the amount wagered by the loser is gradually eroded and progressively becomes the property of the winner. We approach the problem somewhat differently in our modern textbooks, in which we ask about *the probability that an event will occur*. Pascal's question and our own are related, of course, but it is interesting to contrast them. Let us examine a particular case where A and B have wagered 32 *pistoles* each. A has one point, B none, and three are required to win. *What is the amount that A can claim to have taken from B after this first toss?* Pascal shows in the relevant section of the *Arithmetical Triangle* how this can be readily computed with his triangle. The answer is 6/16th of the amount wagered by B and, since A and B put in the same amount, this is 3/16 of the total. Were the game interrupted, A could leave the table with his half plus 3/16, namely 1/2 + 3/16 = 11/16 of the stakes. In our case, this means (64:2) + (64:3/16) = 44 *pistoles*. Today we would ask, What is the probability that A will win the game after winning the first throw? The answer is 11/16 since for 2 players there are 16 possible outcomes of 4 tosses (the maximum possible to settle this game): 11 are favorable to A, and 5 to B. Pascal proceeds by what we might call a transfer of property that removes all uncertainty when the game is interrupted, and the *latent* or *virtual ownership* is actualized. Pascal's method is like an instantaneous photograph that *captures* the motion by *freezing* it at a given moment in time. The modern approach to probability looks to the future, and the more confidently the longer the time available. For instance, the greater the number of tosses of a coin, the greater the assurance that the odds will turn out to be exactly one half for heads and one half for tails. This consideration is not one that seems to have occurred to Pascal.

Conclusion

For Pascal, pure chance becomes amenable to mathematical reasoning when a game is interrupted. What seemed totally fortuitous enters the realm of computation and rationality. Pascal was fully aware of the novelty and importance of this idea, as we can see from an important passage in his Letter to the Parisian Academy in 1654, in which he calls his method the *Geometry of Chance*, and describes it as

> a completely new and unexplored subject, namely the combination of events in games of chances, what is called *faire les partis des jeux* (the problem of points) in French. Here the uncertainty of fortune is so controlled by the fairness of reason that each of two players is always assigned exactly what belongs to him by right. This is the more to be sought by reasoning, the lesser it can be investigated by experiment. The ambiguous outcome of fortune is rightly ascribed to chance rather than natural necessity. This is why the issue has remained uncertain to this day. But if it has proved refractory to experience, it can no longer escape rational enquiry. We have turned it into an assured form of knowledge with the help of geometry whose certainty it shares. It combines mathematical demonstration with the uncertainty of chance, and having shown that these are not contraries, it borrows its name from both and proudly calls itself the *Geometry of Chance*.[36]

The correspondence with Fermat highlighted not one but three correct solutions to the problem of points. The first solution is the *Combinatorial Method*, which consists in finding the maximum number of throws that are needed to determine the winner, listing the possible sequences of heads and tails in this number of throws, and determining by inspection who is the winner in each case. Fermat and Pascal both arrived at this solution, but Pascal initially misunderstood Fermat's procedure and assumed, erroneously, that he had intended it to be applied to three players, just as it had been applied to two, by ignoring the order of occurrence of the tosses. The second solution, which is due entirely to Pascal, is *The Method of Expectations*, which works backwards. The third, *Direct Probability Method*, due to Fermat but involving no new principle (and exem-

---

[36] *Œuvres de Pascal*, II, pp. 1034–1035.

plified in his reply to Roberval in his letter to Pascal of 25 September), consists in enumerating the possibilities working forwards. An important tool in all of this was Pascal's Arithmetical Triangle and we turn, in the next chapter, to some of the other important applications that Pascal discovered.

CHAPTER 12

Putting the Arithmetical Triangle to Work

The *Treatise on the Arithmetical Triangle*, as it was published posthumously in 1665, begins with a description of the triangle, and is followed by three sets of essays. The first two sets show how the Arithmetical Triangle can be used: (1) in the theory of figurate numbers, (2) in the theory of combinations, (3) in dividing the stake in games of chance, and (4) in finding the powers of binomial expressions. The third set of papers consists of one paper in French and six in Latin. The last two of these, *The Summing of the Powers of Integers* and *The Recognition of Multiple Numbers by the Mere Sum of their Digits*, were written first. Their typography is different and the Arithmetical Triangle is not mentioned. They were completed by the Spring of 1654 and a Latin version of the *Arithmetical Triangle* was drafted soon thereafter, followed by a French version in which Pascal applied it to the problem of points, as we have seen in Chapter 11. How it can be used for other purposes will be outlined in this chapter. We first consider *The Summing of the Powers of Integers*, in which Pascal made an important contribution to the problem of finding the sum of the powers of the first natural numbers.[1]

Finding the Powers of Binomial Expressions

A striking feature of the Arithmetical Triangle is the way it can be used to raise binomials to different powers. This had been discovered earlier in the

---

[1] *The Summing of the Powers of Integers. Œuvres de Pascal*, II, pp. 1259–1272. See Dominique Descotes, "Arithmétique et littérature: le *Potestatum Numericarum Summa*," *Revue des Sciences Humaines*, no. 244, octobre–décembre 1996, pp. 53–79.

seventeenth century by Pierre Hérigone, and Pascal gave him full credit. "I do not give a demonstration of all of this," he writes, "because others have treated it such as Hérigone. Besides the matter is self-evident."[2] Pascal illustrates the procedure by showing how to compute the value of $(a + 1)^4$.

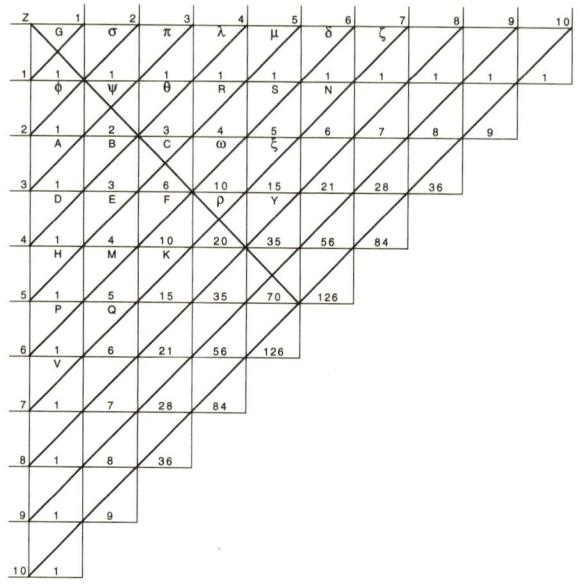

FIGURE 1

We start by taking the base of the Arithmetical Triangle (Figure 1) whose number is *one more* than the degree to which the binomial is raised, namely the diagonal line that runs from 5 on the *x*-axis to 5 on the *y*-axis, and whose cells or boxes are numbered 1, 4, 6, 4, 1. So that

$$(a + 1)^4 = 1a^4 + 4a^3 + 6a^2 + 4a + 1$$

When $a = 1$, the result is $1 + 4 + 6 + 4 + 1 = 16$, the value of $(1 + 1)^4$. Pascal then went on to find more interesting results, for instance, 5^4, expressed as $(a + 1)^4$ where $a = 4$. The equation now reads:

$$(4+1)^5 = (1 \times 4^4) + (4 \times 4^3) + (6 \times 4^2) + (4 \times 4) + 1 = 625$$

---

[2]*On the Use of the Arithmetical Triangle to Find the Power of Binominal Expressions*, Œuvres de Pascal, II, p. 1323.

Once algebra, which was just getting underway in the first half of the seventeenth century, became common, the link between the now familiar binomial theorem and the Arithmetical Triangle became even clearer. In practical terms, Pascal's triangle shows how to write out fully the expression $(x+y)^n$ without multiplying out. Although Pascal did not express his discovery in algebraic form, his results are easily rendered in this way, and if we carry out the multiplication of $(x+y)^n$ for $n = 1, 2, 3, 4, 5$, we find:

$$(x + y)^1 = x + y$$
$$(x + y)^2 = x^2 + 2xy + y^2$$
$$(x + y)^3 = x^3 + 3x^2y + 3xy^2 + y^3$$
$$(x + y)^4 = x^4 + 4x^3y + 6x^2y^2 + 4xy^3 + y^4$$
$$(x + y)^5 = x^5 + 5x^4y + 10x^3y^2 + 10x^2y^3 + 5xy^4 + y^5$$

The *coefficients* or numbers in front of each term in these expressions are those that we find in the rows of Pascal's triangle. Thus $(x+y)^6$ is:

$$x^6 + 6x^5y + 15x^4y^2 + 20x^3y^3 + 15x^2y^4 + 6xy^5 + y^6$$

The process leads to the formula that is familiar under the name of the Binomial Theorem:

$$(x + y)^n = x^n + nx^{n-1}y + \frac{n(n-1)}{2}x^{n-2}y^2 + \frac{n(n-1)(n-2)}{2.3}x^{n-3}y^3 + \ldots + y^n$$

We now return to Pascal and his gambling friends. How could they gain assistance from the Arithmetical Triangle? Suppose that Pascal was asked whether his chances of getting three heads out of four tosses were greater than those of getting five heads out of eight tosses (meaning at least three heads in the first case and at least five in the second). He would look at the fifth diagonal line and note that the sum of the cells is $1 + 4 + 6 + 4 + 1 = 16$. Hence the possibility of getting four or three heads in three games is equal to $1 + 4 = 5$ out of 16 possible outcomes, namely 5/16. Moving down four horizontal lines in the Arithmetical Triangle, he would ask the same question for the likelihood of getting eight, seven, six or five heads in eight games. This time the probabilities are $1 + 8 + 28 + 56 = 93$ out of 256 possible outcomes, hence 93/256, which is slightly higher than 5/16. So Pascal could bet on getting 5 heads out of 8 tosses rather than 3 heads out of 4 tosses. The reason why this is the case can be made clearer by examining the *coefficients* in the Binomial Theorem, and rewriting them in modern form:

$$(x+y)^n = x^n + \binom{n}{1}x^{n-1}y + \binom{n}{2}x^{n-2}y^2 + \ldots \binom{n}{3}x^{n-3}y^3 + \ldots + \binom{n}{0}y^n$$

where

$$\binom{n}{1} = n$$

$$\binom{n}{2} = \frac{n(n-1)}{2!}$$

$$\binom{n}{3} = \frac{n(n-1)(n-2)}{3!}$$

$$\binom{n}{0} = 1$$

Coin Tossing and the Binomial Expansion

The problems that Pascal usually considered were series of trials confined to two outcomes that were either *gain* or *loss*. Let us designate the probability of winning as p and the probability of losing as q. Since there are only these two mutually exclusive outcomes, it follows that $p + q = 1$, and $q = 1 - p$.

We suppose that the successive trials are completely independent one of another, and that p, and hence q also, does not change from trial to trial. The game might be a long series of coin-tossings, in which case we would have $p = q = 1/2$. Or it might be the rolling of a die in the hope of obtaining the face 2, say, in which p would be 1/6 and q would be 5/6. Since $p + q = 1$,

$$(p+q)^n = (1)^n = 1 = p^n + \binom{n}{1}p^{n-1}q + \ldots + q^n$$

On the right, we have the sum of the probabilities of all the mutually exclusive outcomes in n trials. Some one of these *must* happen, so the total probability of all the cases properly is 1, or certainty. If $p = q = 1/2$, as it is with coin-tossing, then the binomial expansion simplifies to the form:

$$(p+q)^n = \left(\frac{1}{2} + \frac{1}{2}\right)^n = \left(\frac{1}{2}\right)^n \times \left[1 + \binom{n}{1} + \binom{n}{2} + \binom{n}{3} + \ldots + 1\right]$$

and the results correspond to what we read as we move down the rows of the Arithmetical Triangle:

$$\left(\frac{1}{2}+\frac{1}{2}\right)^2 = \frac{1}{4}(1+2+1)$$

$$\left(\frac{1}{2}+\frac{1}{2}\right)^3 = \frac{1}{8}(1+3+3+1)$$

$$\left(\frac{1}{2}+\frac{1}{2}\right)^4 = \frac{1}{16}(1+4+6+4+1)$$

$$\left(\frac{1}{2}+\frac{1}{2}\right)^5 = \frac{1}{32}(1+5+10+10+5+1)$$

The various terms in the parentheses build up in size, reach their maximum at the middle term (if there is a middle one, and otherwise at the *two* central terms), and then decrease. This means that when a coin is tossed a number of times, the least likely result is to get only heads or only tails, and the probability of mixtures of heads and tails steadily increases as we move away from the extreme cases, until it reaches its maximum at an even break between heads and tails (if the number of tosses permits an even break), or a shared maximum for the two symmetrical cases nearest to an even break (if the number of tosses is odd).

The Summing of the Powers of Integers

In general terms, the question that Pascal set himself was to determine *the sum of the powers of an Arithmetical progression or Arithmetical series*, namely a series in which each term is formed from that immediately preceding it by adding or subtracting a constant number. For instance, the sum of the sequence of natural numbers:

$$1+2+3+4+5+6+7+\ldots,$$

or the sum of their squares:

$$1+4+9+16+25+\ldots \text{ (i.e., } 1^2+2^2+3^2+4^2+5^2+\ldots\text{)}.$$

An Arithmetical progression can have a common difference of 2

$$1+3+5+7+\ldots$$

or 3

$$1 + 4 + 7 + 10 + \ldots$$

It can also begin with another number than 1, for example:

$$2^2 + 4^2 + 6^2 + 8^2 + \ldots$$

or

$$2^n + 4^n + 6^n + 8^n + \ldots$$

And, in general,

$$a^n + (a + b)^n + (2 + 2b)^n + (a + 3b)^n = \sum_{x=a}^{x=(a+3b)} x^n$$

The quest for the sum of an Arithmetical Series went back to Antiquity. The Pythagoreans were familiar with the rule for finding triangular numbers, that is, with the formula

$$1 + 2 + 3 + \ldots + n = \frac{n(n + 1)}{2},$$

and Archimedes had already found the sum of square numbers,[3] which can be expressed as:

$$3\sum x^2 = n(x_n)^2 + (x_n)^2 + 1\sum x$$

For instance, if we sum the squares of the five first integers, the formula yields

$$3(1^2 + 2^2 + 3^2 + 4^2 + 5^2) = 5(5)^2 + 5^2 + 1(1 + 2 + 3 + 4 + 5)$$

The sum of a series of cube numbers was also discovered in Antiquity. It was known to Roman surveyors and to Greek mathematicians

---

[3] See *On Spirals* in *The Works of Archimedes*, edited by T.L. Heath. Cambridge University Press, 1912, reprinted New York: Dover, no date, *On Spirals*, Prop. 10, p. 162, with the reference to Lemma 2 to Proposition 2 of *On Conoids and Spheroids*, pp. 107–109.

such as Nicomachus of Gerasa, who flourished about 100 A.D.[4] The quest for an algorithm for the sum of powers greater than 3 became popular at the beginning of the seventeenth century, and the German Johann Faulhaber found the sum of the powers of the first n positive integers up to the seventeenth power, but his formulas were discovered empirically from a detailed and laborious study of figurate numbers and did not suggest a scheme for extension to higher values of the powers.[5] The first recognition of such a general rule was made in 1636 by Fermat whose theorem on figurate numbers may be written in modern notation as follows:

$$\sum_1^n = \frac{n(n+1)}{1 \cdot 2}$$

$$\sum \frac{n(n+1)}{1 \cdot 2} = \frac{n(n+1)(n+2)}{1 \cdot 2 \cdot 3}$$

$$\sum \frac{n(n+1)(n+2)}{1 \cdot 2 \cdot 3} = \frac{n(n+1)(n+2)(n+3)}{1 \cdot 2 \cdot 3 \cdot 4}$$

and so on.

With this theorem, Fermat found the sums of the powers of integers and applied the results to determine the areas under paraboliform curves. Pascal, unaware of Fermat's anticipation, reached a similar conclusion some eighteen years later when he was studying the figurate number relationships in his Arithmetical Triangle, and made the key observation that the sum of the figurate numbers of any order is expressible in terms of the figurate number of the next higher order. This was in 1654, as we know from the list of works in progress that he wrote down for the *Parisian Academy of Mathematical Sciences*. The first work mentioned is closely related to *The Summing of the Powers of Integers*, and is titled, *On the Boundaries of the Powers of Integers*, where the word *boundary* or *ambitus* in Latin replaces the more common word *gnomon*. We have a clue about his procedure in the way the Ancients constructed the series of odd numbers and separated the sums 3, 5, 7, 9, 11, 13, etc., by successive gnomons or boundaries:

---

[4] See Thomas Heath, *A History of Greek Mathematics*, 2 vols. London: Constable, 1921. Reprint New York: Dover, 1981, vol. I, pp. 108–110.

[5] See Ivo Schneider, *Johannes Faulhaber 1580–1635*. Basel: Birkhäuser, 1993, pp. 135–140.

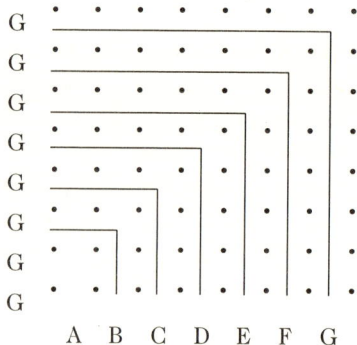

This method of formation shows that the sum of any number of successive terms of the series of odd numbers starting from one is a square number. For instance,

$$1 + 3 = 4$$
$$1 + 3 + 5 = 9$$
$$1 + 3 + 5 + 7 = 16$$

In this way we can determine the *difference* between two powers of like order in Arithmetical Progression. For instance, finding the difference between $(A + B)^2 - A^2$ (namely $(1+3)^2 - 1^2$) consists in finding the number of points that have to be added to the square of side A to obtain the square of side A + B. This leads to the summing of powers of like order in Arithmetical Progression, as Pascal saw.

Figurate Numbers

Figurate numbers can be expressed as the products of continuous numbers, but this idea, or it least its application, dawned gradually in Pascal's mind. In the early Latin version of *The Summing of the Powers of Integers*, figurate numbers are only mentioned incidentally and their relation to the Arithmetical Triangle is not made as clear as in the treatise entitled *On Figurate Numbers or Numerical Orders* that Pascal wrote some time later, where they are presented as follows:

William R. Shea

Indices	Roots					
	1	2	3	4	5	etc.
1. Units or First Order	1	1	1	1	1	etc.
2. Natural numbers or Second Order	1	2	3	4	5	etc.
3. Triangular numbers or Third Order	1	3	6	10	15	etc.
4. Pyramidal numbers or Fourth Order	1	4	10	20	35	etc.
5. Triangulo-triangular numbers or Fifth Order	1	5	15	35	70	etc.

The numbers of the sequence of numerical orders (1, 2, 3, 4, 5 on the left) are called *indices*, and the sequence of numbers within a particular order are called *roots*. Since the numerical orders are the same as the parallel rows of Pascal's Arithmetical Triangle, we can use all the properties of the Triangle. For instance, if we want to know the sum of the first three *roots* of the third numerical order (1, 3, 6), all we have to do is look at the number just under 6 in the next parallel row, namely 10, since every cell in the Arithmetical Triangle is equal to the sum of the cells of the preceding parallel row up to the one just above that cell.

Pascal acknowledges in the *Summing of the Powers of Integers* that the Ancients had found an expression for the sum of the squares or the cubes of a series of natural numbers, but he claims to have gone beyond by providing a "single and very general method" for the sum of powers of any degree of a sequence of numbers in Arithmetical Progression, whether the series begins with 1 or any other term, and regardless of the size of the common difference. He stresses that his method is valid even if the first number of the series is not part of the progression (e.g., it works for the series 5, 8, 11, 14 whose common difference is 3 but where 5 lies outside the progression). "This happy discovery," Pascal writes, "is so simple that it can be expressed in a few lines without the help of letters that are necessary in more difficult propositions."[6] Let us follow him on this happy path, and expand the binomial $(x + 3)^4$ to obtain:

$$x^4 + 12x^3 + 54x^2 + 108x + 81$$

---

[6] *The Summing of the Powers of Integers*, Œuvres de Pascal, II, p. 1260.

Pascal adds, without explanation, that the coefficients 12, 54, and 108 "are formed partly by figurate numbers and partly by the number 3, the second [of the two numbers] in the binomial."[7] Indeed,

$$12 = 4 \times 3^1$$
$$54 = 6 \times 3^2$$
$$108 = 4 \times 3^3$$

but that would have been apparent only to the trained eye of a geometer.

The coefficient 1 of x^4, the first member of the binomial expansion of $(x + 3)^4$, is invisible, so to speak. Furthermore, the figurate numbers 1, 4, 6, 4 and 1 do not belong, to use Pascal's terminology, to the same *order*, namely the same horizontal column of the Arithmetical Triangle, but to the same *base*, the line that runs diagonally across the Arithmetical Triangle from number 5 on the vertical column to the left to number 5 on the horizontal column to the right.

Going Beyond Figurate Numbers

This lack of clarity, which is unusual in Pascal, is explained by the fact that *The Summing of the Powers of Integers* was drafted before he had worked out the implications of his Arithmetical Triangle. When he realized the importance of the *base*, he saw that the *coefficients* are not only *figurate numbers* but also numbers on the *bases*. Evidence for this is an extended comment to *Proposition 7* of the *Treatise on Numerical Orders*. The proposition reads: "A number, whatever its order, is to the next larger number of the same order as the root of the smaller number is to the sum of this root and the index of that order minus one."[8] This can be represented graphically as:

---

[7]*Ibid.*, p. 1261. The source is probably Pierre Hérigone in his *Cursus Mathematicus* in the section on *Algebra* in vol. II, p. 41 (Descotes, art., p. 62, n. 11). Incidentally, Pascal refers to the numbers that precede the *x*'s by the new word *coefficient*, a term introduced by Viète, something he probably ignored.

[8]*Œuvres de Pascal*, II, p. 1326.

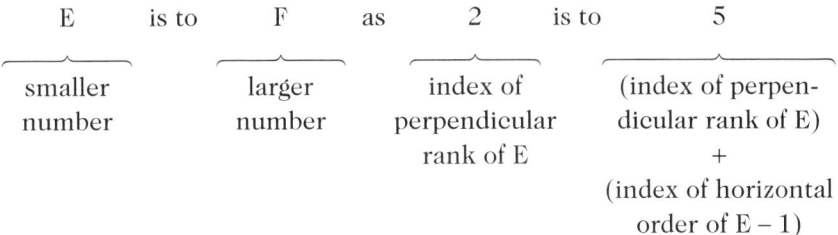

"This follows," writes Pascal, "from Corollary 14 of the Arithmetical Triangle that states that each cell is to the one that precedes it in its parallel rank (horizontal column) as the index of the base of the preceding cell is to the index of its perpendicular rank (vertical column)." Corollary 14 is illustrated with cells F and E (which have the numerical values 10 and 4):

	F	is to	E	as	5	is to	2
	larger number		smaller number		index of base E		index of perpendicular rank of E

Nonetheless, the transition from Corollary 14 to Proposition 7 is not entirely obvious, as Pascal admits. The reason is that we do not immediately see the connection between the base of the triangles and the numerical orders. "Here is how it is found," he tells us, "Instead of the *index of the base*, mentioned in Corollary 14, substitute *the index of the parallel rank plus the index of the perpendicular rank minus one*. This yields the same result with the advantage that we know the relation of these indices to the numerical orders since, in this new way of expressing ourselves, we say: *the index of the order, plus the root, minus one*. I mention this to make clear the method that renders these operations easier."[9] This wordy description is clarified with the help of the following particular case:

$$(14 + 3)^4 = 14^4 + (12 \times 14^3) + (54 \times 14^2) + (108 \times 14) + 81$$

---

[9]*Ibid.*, pp. 1326–1327.

After subtracting 14^4 from each side of the equation, Pascal obtains

$$(14 + 3)^4 - 14^4 = (12 \times 14^3) + (54 \times 14^2) + (108 \times 14) + 81$$

For the modern reader, this is a trivial operation, but for Pascal and his contemporaries it represented a new way of looking at the problem, a genuine change that made it possible not only to raise a binomial to a given degree, but to study the difference between two terms of an Arithmetical Progression of difference 3, namely 14^4 and $(14 + 3)^4$, i.e., 17^4. The new equality provided a *Rule* for handling the difference between two powers of like order, and constituted a major step towards a general solution. It is embodied in a *lemma*: "The difference between two powers of like order is equal to the difference of their roots raised to the given power, plus the smallest root raised to all the powers inferior to the given power where each is multiplied by the coefficients that precede the first number x in a binomial expansion. For instance, $14^4 - 11^4 = (12 \times 11^3) + (54 \times 11^2) + (108 \times 11) + 81$."[10]

Pascal gives a second example in which he expresses 174 as:

$$17^4 = 17^4 - 14^4 + 14^4 - 11^4 + 11^4 - 8^4 + 8^4 - 5^4 + 5^4$$

which he obtains by adding and subtracting the same numbers, namely

$$17^4 = 17^4 (- 14^4 + 14^4) + (-11^4 + 11^4) + (- 8^4 + 8^4) + (5^4 + 5^4)$$

The terms are then rearranged in a more convenient way,

$$17^4 = (17^4 - 14^4) + (14^4 - 11^4) + (11^4 - 8^4) + (8^4 - 5^4) + 5^4$$

so that the lemma can be applied to the difference of each pair of powers of like order.

The next step consists in moving from the difference to the sum of powers of like order. Pascal illustrates his procedure with the arithmetical series 5, 8, 11, 14 that has 5 as first term, and 3 as common difference. In order to find the sum of $5^3 + 8^3 + 11^3 + 14^3$ he raises the next term of the series $(14 + 3)^3$ or 17^3 to a power higher by one degree: $(14 + 3)^4 = 17^4$. Since we already know that

$$17^4 = (17^44 - 14^4) + (14^4 - 11^4) + (11^4 - 8^4) + (8^4 - 5^4) + 5^4$$

---

[10] *The Summing of the Powers of Integers*, Œuvres de Pascal, II, pp. 1262–1263.

the *Lemma* can be applied to each of the two terms inside the parentheses to obtain the following equalities:

$$17^4 - 14^4 = (12 \times 14^3) + (54 \times 14^2) + (108 \times 14) + 81$$
$$14^4 - 11^4 = (12 \times 11^3) + (54 \times 11^2) + (108 \times 11) + 81$$
$$11^4 - 8^4 = (12 \times 8^3) + (54 \times 8^2) + (108 \times 8) + 81$$
$$8^4 - 5^4 = (12 \times 5^3) + (54 \times 52) + (108 \times 5) + 81$$

We substitute these values in the equation:

$$17^4 = (17^4 - 14^4) + (14^4 - 11^4) + (11^4 - 8^4) + (8^4 - 5^4) + 5^4$$

and we move 5^4 to the left side after changing its sign:

$$17^4 - 5^4 = (12 \times 14^3) + (54 \times 14^2) + (108 \times 14) + 81$$
$$+ (12 \times 11^3) + (54 \times 11^2) + (108 \times 11) + 81$$
$$+ (12 \times 8^3) + (54 \times 8^2) + (108 \times 8) + 81$$
$$+ (12 \times 5^3) + (54 \times 5^2) + (108 \times 5) + 81$$

This operation is equivalent to adding the four differences. The terms can be rearranged once more (we would say by finding common factors) to yield:

$$17^4 = 108 \ (5 + 8 + 11 + 14)$$
$$+ 54 \ (5^2 + 8^2 + 11^2 + 14^2)$$
$$+ 12 \ (5^3 + 8^3 + 11^3 + 14^3)$$
$$+ 4 \times 81$$
$$+ 5^4$$

The third line, $12 \ (5^3 + 8^3 + 11^3 + 14^3)$, is the sum of the powers multiplied by coefficient 12. Hence we get:

$$17^4 = 108 \sum n + 54 \sum n^2 + 12 \sum n^3 + \left(4 \times 81\right) + 5^4$$

We rearrange the terms to obtain the sum of the cubes:

$$12 \sum n^3 = 17^4 - 5^4 - 54 \sum n^2 - 108 \sum n - \left(4 \times 81\right)$$

We may summarize by saying that Pascal succeeded in formulating what is now called an algorithm, namely a procedure to solve a problem by applying a rule. He prided himself on the generality of his method, which provides an expression for the sums of powers of like order of a sequence of numbers in Arithmetical Progression, whatever the first term or the number of terms. In a concluding *Remark*, he stresses the validity of his method in the case of "a natural progression beginning with any number whatsoever, where the difference between the square of the smallest term and the square of the term that follows the largest in the series, minus the number of terms in the series, is equal to the sum of the terms multiplied by 2."[11] For instance, given the series 5, 6, 7, 8, we have $9^2 - 5^2 - 4 = 2(5 + 6 + 7 + 8)$. If $n = 5$ and $p =$ the largest number in the series, we can write the progression as

$$n, n + 1, n + 2, n + 3 + \ldots + n + p$$

so that the number of terms is $p + 1$, and the number that follows the last term is $n + p + 1$. Pascal's formula or rule can then be written as

$$\left[(n + p + 1)^2 - n^2\right] - (p + 1) = 2\sum_{x=n}^{x=n+p} x$$

Pascal claims that "similar rules for the sum of other powers and other Arithmetical Progressions can easily be devised," but although this is true in a general sense, the computations involved are very lengthy, and it is doubtful whether Pascal tried to carry them out.

The Products of Consecutive Integers

Having successfully tackled the sum of the powers of an Arithmetical Progression, Pascal moved on to the more difficult problem of computing the products of consecutive integers, namely the result of the multiplication of numbers in a natural series, "something," he writes, "that has never been examined by anyone."[12] The development rests on a proposition he had formulated as a corollary in his treatise on *Figurate Numbers or Numerical Orders*: "Given any two numbers whatsoever, the product of all the numbers that precede the first is to the product of as many continuous numbers beginning with the second as the product of all the numbers that precede the second to the product of as many continuous numbers

---

[11]*Ibid.*, p. 1270.
[12]*The Products of Consecutive Integers or Numbers Produced by the Multiplication of Numbers in a Natural Series, Œuvres de Pascal*, II, p. 1215.

beginning with the first."[13] For instance, if the numbers are 5 and 7, we write:

$$\frac{1 \times 2 \times 3 \times 4}{8 \times 9 \times 10 \times 11} = \frac{1 \times 2 \times 3 \times 4 \times 5 \times 6 \times 7}{5 \times 6 \times 7 \times 8 \times 9 \times 10 \times 11}$$

Pascal's strategy becomes clear when we note that the right-hand side of the equation has been produced by adding $5 \times 6 \times 7$ to both the numerator and the denominator of the first fraction on the left-hand side. We shall not follow Pascal's laborious calculation, about whose difficulties he is candid: "The method by which I obtain the resolution of the orders is perfectly general," he writes, "but I had to look for it a long time." He dealt initially with each order separately, and to find a root of the third order he multiplied the given number by 2, and then extracted the square root. This gave him either the root or the root plus one. For the fourth order, he multiplied the given number by 6 (namely the product of $1 \times 2 \times 3$) and then extracted the cubic root. Again this gave him the root or the root plus one. For the fifth order, he multiplied the given number by 24 (the product of $1 \times 2 \times 3 \times 4$), extracted the quadratic root, and found either the root or the root plus one. "And so," he tells us,

> I sought the roots of the other orders, not by using a general construction but one suited to each. This did not displease me entirely since the way of extracting *roots* is not more general, for the square root is extracted in one way, the cubic in another, etc. Although the principle is the same, the way of proceeding is different. Since there was as yet no general method to extract roots, I could hardly hope to find a general method of extraction for the orders. But the results were better than my expectations and I found the general method that I have described. This gave pleasure to my learned friends, who are fond of general solutions, and they urged me to find a general method to extract pure roots. I complied and I was again lucky enough to succeed.[14]

The reader may have been struck by the fact that Pascal arrived at his results with a very limited use of algebra. He even prided himself on having found a general solution "without the help of letters" (such as x, y, a,

---

[13]*On Figurate Numbers or Numerical Orders*, Œuvres de Pascal, II, p. 1207.
[14]*Ibid.*, pp. 1213–1214.

b, etc.).[15] But the absence of letters as symbols imposed a limitation. Had Pascal expressed his binomial as $(x + y)$ instead of $(x + 3)$, the expansion

$$x + y = x^4 + 4x^3y + 6x^2y^2 + 4xy^3 + y^4$$

would have immediately revealed the figurate numbers of the coefficients 1, 4, 6, 4 and 1. But this advantage would have been lost as soon as Pascal moved on to display the differences between powers of like order, since he would have had to write:

$x + y$ for the term that comes after the last in the series (17 in our case)
$x + y - x$ for the last term of the series (14 in our case)
$x + y - 2y$ for the penultimate term

These formulas would have been very messy, especially since it did not occur to Pascal to use parentheses as we now do. His restricted use of letters and his preference for concrete examples was therefore justified.

In the process of applying his Arithmetical Triangle, Pascal became aware of what might be called "elegant variations," and he makes the following remark:

> These propositions (indeed any proposition) can be changed into a variety of different propositions, for instance, given that *a number is to another number as a third number is to a fourth number*, can we not infer that *the product of the first by the fourth is equal to the product of the second by the third?* Or again that *the product of two of these numbers divided by one of the two other numbers is equal to the remaining number?*

What he means becomes immediately clear when we write in symbols,

$$\frac{a}{b} = \frac{c}{d}, \text{ hence } a \times d = b \times c, \text{ and } \frac{a \times d}{b} = c$$

"In this way," concludes Pascal, "different formulations of the same proposition have different uses. This is what the study of geometry is all about, for skillfully modified formulations lead to different and important theorems by relating what seemed completely independent."[16]

---

[15]*Ibid.*, p. 1260.
[16]*Œuvres de Pascal*, II, pp. 1202–1203. The Latin version is reproduced, pp. 1196–1214 and the French, which is much shorter, on pp. 1300–1302. When he wrote the French version, Pascal realized that the appelation *figurate number* was too restrictive and he dropped it from the title, leaving only *On Numerical Orders*.

William R. Shea

The Recognition of Multiple Numbers by the Sum of their Digits

A good illustration of Pascal's skillful manipulation of mathematical relationships is his demonstration of the well-known rule to find whether a number can be divided by 9. We simply sum up the digits, and if the sum is divisible by 9, then so is the number. For instance, the sum of the digits of 1719, which is 18, can be divided by 9; therefore so can 1719. "Nothing is more trivial for mathematicians," writes Pascal, "but although it has often been observed, it has never, as far as I know, been demonstrated or extended." This is Pascal at his most characteristic, a man for whom a mathematical rule or a physical law must, first, be proved, and then generalized. Pascal also saw that once proven, the rule could be extended to other numbers "not only in the decimal progression according to our system of numbering (for the decimal system suffers from shortcomings and merely rests on human convention, not natural necessity as is commonly believed) but on any progression that is chosen as the basis of a numbering system."[17]

So how do we find whether one number can be divided by another number merely by adding the digits of the given number or, in more general terms, how do we determine whether any (non-zero) integer is a multiple of another and, if not, what remainder the divisor will produce? In order to obtain a general solution, Pascal replaces the numbers by the letter A for divisor and the letters TVNM for the four-digit dividend. The procedure is as follows:

Write the series of natural numbers, 1, 2, 3, 4, 5, 6, 7, 8, 9, 10 but in the reverse order:

10	9	8	7	6	5	4	3	2	1
K	I	H	G	F	E	D	C	B	1

Under the first column at the right, place the unit, 1. Multiply this unit by 10, and from the product subtract, as often as possible, the value of the divisor A. The remainder, B, is placed under column 2. Now multiply B by 10 and, as before, subtract the value of divisor A as often as possible. The remainder, C, is placed under column 3. Repeat this procedure as often as is necessary.

---

[17]*Œuvres de Pascal*, II, pp. 1272–1273. *The Recognition of Multiple Numbers by the Sum of their Digits* is published in *Œuvres de Pascal*, II, pp. 1272–1287.

DESIGNING EXPERIMENTS & GAMES OF CHANCE

The trick, or rather the mathematical insight, consists in constructing the second row in such a way that (10 – B), (10 B – C), etc. are multiples of A. How this works can be seen in the way Pascal deals with numbers that can be divided by 7.

We write down, as before, the natural numbers 1, 2, 3, 4, 5, 6, 7, 8, 9, 10 in the inverse order:

10	9	8	7	6	5	4	3	2	1
6	2	3	1	5	4	6	2	3	1

Following the procedure described above for the second row, we write 1 under the first column.

| Under 2, | write | $(10 \times 1) - 7$ | = | 3 | (which occupies the place of B) |
| Under 3, | write | $(10 \times 3) - (7 \times 4)$ | = | 2 | ($7 \times 4 = 28$ being the maximum of times we can subtract a multiple of 7 from 30) |

Under 4,	write	$(10 \times 2) - (7 \times 2)$	=	6
Under 5,	write	$(10 \times 6) - (7 \times 8)$	=	4
Under 6,	write	$(10 \times 4) - (7 \times 5)$	=	5
Under 7,	write	$(10 \times 5) - (7 \times 7)$	=	1
Under 8,	write	$(10 \times 1) - (7 \times 1)$	=	3
Under 9,	write	$(10 \times 3) - (7 \times 4)$	=	2
Under 10,	write	$(10 \times 2) - (7 \times 2)$	=	6

Take any number, say 287,542,178. To know whether it can be divided by 7, construct a third row under the second one, using the same procedure.

6	2	3	1	5	4	6	2	3	1
4	24	7	25	16	12	2	21	8

Under 1,	write	8×1	=	8
Under 3,	write	7×3	=	21
Under 2,	write	1×2	=	2
Under 6,	write	2×6	=	12
Under 4,	write	4×4	=	16
Under 5,	write	5×5	=	25
Under 1,	write	7×1	=	7
Under 3,	write	8×3	=	24
Under 2,	write	2×2	=	4
				119

If 119 can be divided by 7, so can 287,542,178. The division of 119 by 7 can be carried out by the standard method, but the procedure just used can be applied once more:

$$
\begin{array}{llccc}
\text{Under 1,} & \text{write} & 9 \times 1 & = & 9 \\
\text{Under 3,} & \text{write} & 1 \times 3 & = & 3 \\
\text{Under 2,} & \text{write} & 1 \times 2 & = & 2 \\
\hline
& & & & 14
\end{array}
$$

If 14 can be divided by 7, so can 119 and 287,542,178.

Towards the Calculus

Pascal was also aware of the relevance of his work to the determination of quadrature (as the area of a figure bounded by a curve was called), "something that will be clear," he writes "to anyone at all familiar with the doctrine of indivisibles," by which he meant those who were acquainted with the work of the Italian mathematician Benedetto Cavalieri, whose *Geometria indivisibilibus continuorum* showed that an area can be thought of as made up of lines or "indivisibles," and that a solid volume can be regarded similarly as composed of areas that are indivisibles or quasi-atomic volumes.

By extending to *continuous* quantities the results found for numbers, namely discontinuous quantities, Pascal is able to state several rules for a natural progression beginning with unity:

1. The sum of a certain number of lines is to the square of the greatest as 1:2
2. The sum of their squares is to the cube of the greatest as 1:3
3. The sum of their cubes is to the fourth power of the greatest as 1:4

Hence the general rule: "The sum of like powers of a certain number of lines is to the power of next higher degree of the greatest of these as unity is to the exponent of this latter power."[18] Let us illustrate what Pascal had in mind by applying the second of his three rules. Consider a num-

---

[18]*Ibid.*, p. 1271.

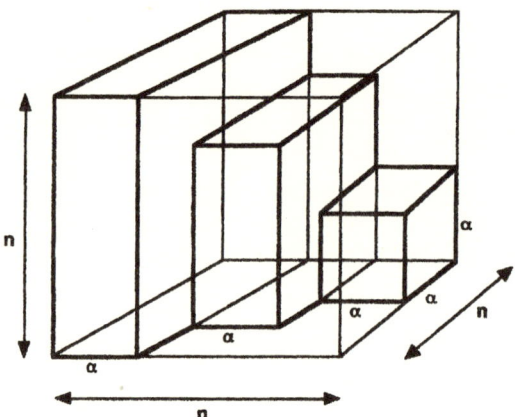

FIGURE 2 *Diagram to show the sum of squares.*

ber of lines whose lengths are in a natural progression of 1, 2, 3, ..., n, and construct a square on each of these lines. The surfaces will be as

$$1^2, 2^2, 3^2, \ldots, n^2$$

These squares, which all have a width of a (where $a = 1$) are placed in contact, as in Figure 2.[19] The result is a three-tiered pyramid-like structure whose height is n (this is arrived at by multiplying the sides, a, by the number of squares, namely $1 \times n = n$). This pyramid will fit into a cube whose sides are equal to n and whose volume is n^3. If we increase the number of squares and decrease their thickness until they are infinitesimally thin, so thin as to become "indivisibles," we can then apply the second rule: "The sum of the squares of a certain number of lines is to the cube of the greatest as 1:3." Hence, if the number of squares is increased indefinitely, we have, in the limit, a pyramid whose volume is one-third of the circumscribing cube.

A similar application can be made in the case of a parabola. Let line OC = n, and divide it into equal segments of length a where $a = 1$. At the end of each of these segments, erect lines whose heights y_1, y_2, y_3, etc. are

---

[19]Figure 2 is taken from Dominique Descotes, "Arithmétique et littérature: le *Potestatum Numericarum Summa*," p. 70.

successively as the squares of segments OA, OB, OC, and so on. The result is the sequence of the squares of natural numbers,

$$1^2, 2^2, 3^2, \ldots, n^2$$

When these segments are divided repeatedly, their size decreases accordingly until, in the limit, the curve line that joins the corner of each rectangle has the shape of a parabola. Pascal's second rule states that the area under the curve, i.e., the sum of the squares, is 1/3 of the rectangle whose base is n and whose height is n^2, namely

$$1/3n \times n^2 = 1/3n^3$$

The algebraical equation for a parabola is $y = x^2$, and since in the figure $n = x$, here this equation will read $y = n^2$. In the language of modern calculus, we would say that the value of the area, $A = 1/3n^3$, is obtained by integration. "The guiding principle," writes Pascal,

> is that in continuous quantities, when quantities of one kind are added to a quantity of a higher order nothing is effectively added to that quantity. Thus points add nothing to lines, lines add nothing to surfaces, surfaces add nothing to solids, or to use the language of numbers in a numerical treatise roots add nothing to squares, squares nothing to cubes, and cubes nothing to the fourth power. Thus, the lower degrees, being of no value, are not to be taken into consideration. This is familiar to those who study indivisibles but I mention it in order that the admirable connection that nature, which is fond of unity, establishes between what seems most remote may be seen in this case where the dimensions of continuous quantity are connected with the summing of the powers of integers.[20]

In other words, we neglect points when adding lines, lines when adding surfaces, and surfaces when adding volumes because the extension of the rule for the summing of the powers of integers to the determination of areas is possible only if we do not take into consideration the value of the lower degrees.

---

[20] *The Summing of the Powers of Integers, Œuvres de Pascal*, II, pp. 1271–1272. See Carl B. Boyer, "Pascal's Formula for the Sums of Powers of the Integers," *Scripta Mathematica* 9 (1943), pp. 240–241

Summary

Pascal treated the more general case of the powers of the terms in any arithmetical progression, not only those starting with 1. He proved the result for a particular case, and then observed that the method could be used generally. In essence, he argued as follows: Consider, as a particular example, the identity

$$(x + 1)^4 - x^4 = 4x^3 + 6x^2 + 4x + 1$$

and put $x = 1, 2, 3, \ldots, n$ in it in turn, summing the results:

$$
\begin{aligned}
2^4 - 1^4 &= 4 \times 1^3 + 6 \times 1^2 + 4 \times 1 + 1 \\
3^4 - 2^4 &= 4 \times 2^3 + 6 \times 2^2 + 4 \times 2 + 1 \\
4^4 - 3^4 &= 4 \times 3^3 + 6 \times 3^2 + 4 \times 3 + 1 \\
&\ldots \\
(n+1)^4 - n^4 &= 4 \times n^3 + 6 \times n^2 + 4 \times n + 1
\end{aligned}
$$

The general result is now easily seen by analogy, and may be expressed thus in algebraic terms:

$$(n + 1)^r - (n + 1) = r\sum n^{r-1} + \left(\frac{r}{2}\right)^{r-2} + \left(\frac{r}{3}\right)\sum n^{r-3} + \ldots + r\sum n$$

where $\sum n^r$ is written for $\sum_{i=1}^{n} i^r$

This Recurrence Formula for the sum of the powers of integers was to become the standard textbook method. Pascal also saw that the summing of the powers of the integers could be applied to the determination of the area of a figure bounded by a curve (what was called the quadrature), and using an arithmetical progression with a large number of terms and very small differences, he obtained, in the limit, what is equivalent to the well-known calculus formula

$$\int_o^a x^k dx = \frac{a^{k+1}}{k+1},$$

Pascal arrived at his result through a laborious pairing of geometric indivisibles, but he foreshadowed the development of the infinitesimal, and hence the integral calculus.

CHAPTER 13

The Brilliance and Pride of a Gambler and a Mathematician

The Greatest Gamble

Pascal's work on probability is the background against which we must read the famous passage in the *Pensées* where he argues that it is not unreasonable to wager that God exists, although from a purely rational point of view the probability may only be 50-50. Pascal frames the question, "God is, or he is not," in terms of a game of chance infinitely far in the future. "At the far end of this infinite distance," he writes, "a coin is being spun which will come down head or tails." How should we wager when what we risk is our mortal life in the hope of winning an eternal one? Because our life span is finite and will soon end, we are compelled to wager, for not bothering to do so is actually to wager that God is not.

But how can reasoning help us choose that God exists or that God does not exist? The odds seem even on both sides, and when the *likelihood* of two outcomes is the same, we usually act in the light of experience, our own or that of reliable persons. But we have no direct experience of God, and we are not willing to trust the experience of those who say they do, because they may themselves be the victims of self-deception. The way out, says Pascal, is to introduce a new element in the decision making. We must consider the *consequences* of believing in God or rejecting God. Is the outcome that God exists preferable to one where he does not exist? By all means, since he offers an infinitely happy life.

> What you are staking is finite, and that leaves no choice. Wherever there is infinity and where there is not an infinity of chances of losing against that of winning, there is no room for hesitation, you must give everything. And this, since you are obliged to play, you must be renouncing reason if you hoard your life rather than risk it for an infinite gain, just as likely to occur as a loss amounting to nothing.[1]

Now if God does not exist, it is immaterial whether you live a good or a bad life. If God exists, then you are bound to lead a moral life but the reward is eternal happiness. Furthermore, following one's moral conscience may be more conducive to happiness here below than indulging one's appetites.

Pascal anticipates the objection that there is an infinite distance between a certain risk and an uncertain gain. "There is indeed," he writes,

> an infinite distance between the certainty of winning and the certainty of losing, but the proportion between the uncertainty of winning and the certainty of what is being risked is in proportion to the chances of winning or losing. . . . The certainty of what we are risking is equal to the uncertainty of what we may win. Thus our argument carries infinite weight, when an infinite prize is at stake and what is wagered is finite in a game where the chances of winning and losing are even."[2]

We could paraphrase the objection as follows: my earthly existence is finite but certain; eternal life is infinite but uncertain. The distance between certainty and uncertainty is as great as the one between the finite and the infinite. If this is the case, there can be no wager: the game ends here. Pascal's reply is to observe: "Every gambler takes a certain risk for an uncertain gain, and yet he is taking a certain finite risk for an uncertain finite gain without sinning against reason."[3] In another fragment of the *Pensées*, Pascal develops the same idea:

---

[1] *Pensées*, fragment 418.
[2] *Ibid.*
[3] *Ibid.*

If we must never take chances, we ought to have nothing to do with religion, for it is not certain. But how many chances we do take: sea voyages, battles. Why we should do nothing at all, for nothing is certain. Now there is more certainty in religion than that we shall live to see tomorrow. For it is not certain that we shall see tomorrow but it is certainly possible that we shall not. We cannot say the same of religion. It is not certain that it is true, but who would dare to say that it is certainly possible that it is not? Now when we work for tomorrow and take chances we are behaving reasonably, for we ought to take chances, according to the rule of probability already demonstrated.[4]

But mathematical subtlety is not enough to force anyone to wager his life, and Pascal fully realized that the mind cannot risk everything without involving the heart. He believed and risked his all because he underwent an overmastering experience of the reality of God. As we have already seen, Pascal had sewn a description of this event into his coat so that he could wear it next to his heart. The concluding passage of the fragment on *The Wager* in the *Pensées* is written with this experience in mind. Pascal counsels the unbeliever who agrees with his argument but protests, "I am being forced to wager and I am not free. I may be held fast, but I am so made that I cannot believe," to act in a manner that will dispose him to believe in God. The person who lives an upright life is already wagering that God exists. This is both reassuring and challenging, but it is first and foremost a vibrant personal testimony.

According to Pascal, we are so made that we know that the infinite exists without being able to know its nature. But can we know that God exists without knowing who he is? Pascal does not eschew the difficulty: "If there is a God, he is infinitely beyond our comprehension, since, being indivisible and without limits, he bears no relation to us. We are therefore incapable of knowing either what he is or whether he is." But we are also conscious of a paradox: a double infinity is present in all things and all forms of knowledge:

> When we know better we understand that, since nature has engraved her own image and that of the author on all things, they almost all share her double infinity. Thus we see that all the sciences are infinite in the range of their researches, for who can doubt that mathe-

---
[4]*Pensées*, fragment 577.

matics, for instance, has an infinity of infinities of propositions to expound? They are infinite also in the multiplicity and subtlety of their principles, for anyone can see that those that are supposed to be ultimate do not stand by themselves, but depend on others, which depend on others again, and thus never allow of any finality. But we treat as ultimate those which seem so to our reason, as in material things we call a point indivisible when our senses can perceive nothing beyond it, although by its nature it is infinitely divisible.[5]

The Gambler's Ruin

In the interval of jotting down his *Pensées*, Pascal occasionally tackled some mathematical problem at the prompting of his friends. Claude Mylon remarked to Huygens that Pascal and Pierre Carcavy were often to be seen together at church, a place he was not himself in the habit of frequenting, "Although Pascal has withdrawn from the world to lead a religious life," he adds, "he has not lost sight of mathematics. When Carcavy manages to see him and put a question to him, he gets an answer especially in the matter of games of chance which he was the first to investigate."[6] Pierre Carcavy reported on his conversations with Pascal to Huygens and it is largely thanks to his letters that we can follow the activities of Pascal and Fermat, who had remained in touch.

The most challenging problem that Carcavy raised concerns the respective probabilities in a game of chance in which two players each has three dice, the one aiming to throw 11s, the other 14s. The winner is the first to reach twelve points, but points are not accumulated in the usual way. If player *A* throws 11 on the first toss, he gets a point, but if player *B* throws 14 on the next toss, he does not get a point. It is *A* rather who loses one point. In other words, a player gains a point only if his opponent's score is nil, otherwise it is subtracted from his opponent's score. When Huygens heard about it, he worked out a solution that he published in a book on games of chance, *De Ratiociniis in Ludo Aleae*, in 1657. Huygens has each player starting with twelve points, a win for a player involving the transfer of a point from his opponent to himself, and the overall winner being the one who bankrupts the other of points. It is in this form that the problem has come down to us as the "Gambler's Ruin."

---

[5]*Pensées*, fragment 199.
[6]Letter of Mylon to Huygens, 2 March 1657, *Œuvres de Pascal*, III, p. 867.

Pascal had worked out the solution before sending the problem to Fermat, but he had found it so "incomparably more difficult than all the others" that he confided to Carcavy, perhaps with a touch of self-complacency, that he doubted whether even Fermat would be able to solve it. His doubt was unfounded and Fermat returned a correct solution via Carcavy immediately: the odds, he said, were between 1156:1 and 1157:1. Pascal, fearing that he might be thought to have posed Fermat a problem that he had not solved himself, promptly revealed the exact solution to Carcavy:

$$\frac{150,094,635,296,999,121}{129,746,337,890,625}$$

Carcavy conveyed this to Huygens but said nothing about the way Fermat and Pascal reached the solution beyond that they seemed "to have used different methods."[7] Pascal, Fermat, and Huygens easily provided the answer to the first part of the problem and saw that the odds of scoring 11 versus 14 with three dice are 9:5. A. W. F. Edwards conjectures that Fermat asked himself, as he had done in the cases about which he had corresponded with Pascal in 1654, what would be the maximum number of throws required to complete the game. He realized that the game could continue indefinitely but that the chance of a very long game was small. He then proceeded to analyze the games, starting with the shortest. If A takes 12 points straight off (probability = p^{12}), he wins, as does B (probability = q^{12}). In this shortest possible game, the odds are therefore $p^{12}:q^{12}$. The next shortest game has 14 throws, 13 points for one player and 1 for the other, but when Fermat enumerated the permutations, he saw that two of the fourteen lead, in each case, to a win at the twelfth throw, and therefore that the odds are $12p^{13}q:12pq^{13}$, or $p^{12}:q^{12}$ again. The next shortest game has 16 throws, each permutation leading to a win for A having a probability of $p^{14}q^2$ and each one leading to a win for B having a probability of p^2q^{14}, and, however many permutations there are in each case, symmetry requires the same number for wins by A as for wins by B. Once again the odds are $p^{12}:q^{12}$. The argument is obviously quite general: games of any length command odds of $p^{12}:q^{12}$.

Knowing that the odds of scoring 11 versus 14 are 9:5, Fermat could then compute the answer for $p = 9$ and $q = 5$, using logarithms. With a table of five-figure logarithms of four-figure numbers, the most natural format

---

[7]Letter of Carcavy to Huygens, 28 September 1656, *Œuvres de Pascal*, III, pp. 681–682.

for the result is to write that it lies between 1156:1 and 1157:1. Edwards' reconstruction of Fermat's solution is faithful to Fermat's method of enumeration and explains both the speed and form of his answer.[8] How Pascal might have proceeded is more difficult to establish. Edwards suggests that Pascal extended his *Method of Expectations* to reach the same conclusion as Fermat $p^{12}:q^{12}$, but instead of substituting $p = 9$ and $q = 5$ as Fermat had done, he put $p = 27$ and $q = 15$ (because 9:5 = 27:15) and went through the long and laborious computation that leads to the exact answer quoted above.

There was one further exchange of letters between Fermat and Pascal in 1660. On July 25th, Fermat wrote from Toulouse to Pascal, who was in Clermont-Ferrand, suggesting that they meet at a place midway between the two towns. Fermat explained that his health was too poor to make the whole journey. Pascal replied, on August 10th, that he appreciated the offer but that he was himself so ill that he found it difficult to read or write. "Were I in good health," he added,

> I would have flown to Toulouse, and I would not have allowed a man such as you to take one step for someone like me. Allow me to say that although you are the greatest geometrician in the whole of Europe, I would not seek you out for that, but for your qualities of mind and character. If I may speak candidly, I grant that geometry is the highest mental exercise but, at the same time, I find it so useless that a man who is nothing but a geometrician is not much different from a clever craftsman. It may be the best craft in the world, but it is only a craft. As I have often said, it is good for testing our strength but we should not give it all our time. In other words, I would not take two steps for geometry and I feel certain that you are of the same mind. Furthermore, my studies have taken me so far from this way of thinking, that I can scarcely remember that there is such a thing. I had returned to it, a year or two ago, for a particular purpose, and having satisfied it, I may never think about it again, especially since I am not yet strong enough. I am so weak that I cannot walk without a stick nor even ride a horse. I can only manage three or four leagues in a carriage, and it took me twenty-two days to come here from Paris.... Such is the present state of my health, which I wanted to describe to explain why I must decline the honour you so kindly offered me. I

---

[8] See A.W.F. Edwards, "Pascal's Problem: The 'Gambler's Ruin,'" *International Statistical Review* 51(1983), 73–79, reprinted Edwards, pp. 151–160.

hope, with all my heart, that one day I shall be able to meet you or your children for whom I have a special regard because they bear the name of the greatest man in the world.[9]

The Affair of the Cycloid

One evening in the Summer of 1657, not long after the completion of the *Provincial Letters* and after he had begun his *Pensées*, Pascal suffered an excruciating attack of toothache. He went to bed but could not sleep, and the pain only increased. To alleviate his discomfort by thinking about something else, he tried to work out the properties of a curve that is traced out by a point on the circumference of a circle revolving at uniform speed. This is called a cycloid and its path can be visualized by imagining that a nail has been fixed to the rim of a carriage wheel that is rolling at a constant speed on a straight, even road. Although it is easy to trace, the cycloid was never investigated by the Ancients, not even by Archimedes. It was mentioned by Nicholas of Cusa in the fifteenth century, but it was only in the seventeenth that Galileo noticed that arches of bridges have the shape of a cycloid. He set about calculating its surface and he estimated it at three times that of the circle that describes it. When his results did not quite agree with this estimate, he "made an experiment," which consisted in weighing a plate in the shape of a plane bounded by a cycloid and comparing it with the weight of a plate of the same material whose shape was traced out by the generating circle. He found, upon repeated trials, that the ratio was less than 3:1. This suggested that the numbers were incommensurable and he gave up further research.[10]

Investigation of the cycloid in Pascal's day may have been stimulated by the paradox that goes by the name of the *Wheel of Aristotle*, and was often discussed at the time. When two concentric circles rotate together (think of the rim and the hub of a carriage wheel), they both trace out a path of the same length, but if they revolve separately, their paths are in the same ratio as their respective sizes. Mersenne was interested in this

---

[9]*Œuvres de Pascal*, IV, p. 923; translated in F.N. David, *Games, Gods and Gambling*. London: Charles Griffin, 1962, pp. 252–253.
[10]See the letter of Galileo to Bonaventura Cavalieri, 24 February 1640, *Opere di Galileo*, vol. XVIII, pp. 153–154, and the letter of Evangelista Torricelli to Gilles Personne de Roberval, 1 October 1643, *Correspondance de Mersenne*, vol. XII, p. 331.

paradox, and according to Pascal, he was the first to draw attention to it around 1615.[11] Mersenne did not get very far with the problem, but he was a spur to others' achievement rather than a man of achievement himself, and he made it known to Fermat and Roberval. They were better mathematicians and they set to work. Roberval made considerable progress in 1634 but he did not publish his results. This was not due to modesty on his part. He had just been named to the Chair of Mathematics at the Collège Royal and his appointment would be up for renewal, after the customary three years, on the basis of a competitive examination at which he would be invited to set the questions. So he kept his results to use them, if necessary, in 1637.[12]

Meanwhile, Mersenne had suggested that the path was elliptical, and Roberval wrote to lift the veil somewhat. He informed Mersenne that the cycloid was not an ellipse and that he could describe all its characteristics, if he had time. But he made sure he had none until the end of 1637, when his reappointment to the Chair at the Collège Royal became assured. He then allowed Mersenne to broadcast that he had found the "quadrature" of the curve, namely the surface area of the geometrical figure bounded by an arc of the cycloid, and that this area was indeed three times that of the generating circle as Galileo had surmised. In a supplement to his *Harmonie Universelle* of 1638, Mersenne christened the curve *roulette*, the name under which it became known in France.

Descartes grudgingly admitted that he had not thought of the curve before, but he played down its importance. "What Roberval says is clever enough," he wrote to Mersenne, "but I cannot see why such a fuss should be made about something so easy. Anyone who knows a little geometry cannot fail to see it as soon as he looks for it." Descartes provided his own demonstration that the area of the cycloid is three times that of its generating circle but he could not resist making yet another niggardly remark: "If I bragged about finding such a thing, it would be just as if I cut an apple

---

[11] Pascal, *History of the Cycloid*, Œuvres de Pascal, IV, p. 214. The paradox is also mentioned by Galileo in the First Day of the *Discorsi*, Opere di Galileo, VIII, p. 68.

[12] See Roberval's letter to Torricelli dated 1 January 1646 (*Correspondance de Mersenne*, XIV, p. 4). By the conventions of the time, there was nothing unusual in this behavior or in setting prize questions. In 1657, Fermat was to issue two challenges to his fellow mathematicians, one of them based on his study of the indeterminate equation $x^2 - q = my^2$ for non-square m. Likewise, Huygens later challenged Leibniz to find the sum of the infinite series of reciprocal triangular numbers. In 1674, Leibniz was challenged in turn, and defeated, by Ozanam's "six-square problem": that of finding three numbers x, y and z such that $x - y$, $y - z$, $x - z$, $x^2 - y^2$, $y^2 - z^2$ and $x^2 - z^2$ are all squares.

in half and boasted that I had seen something that no one else had ever seen."[13] In spite of his condescending manner, Descartes was nonetheless anxious to know what Roberval thought of his proof and he asked Mersenne in August: "I would like to know what he thinks of the latest version of my demonstration of his *roulette*. I have made it so clear that were he to deny it, he would be laughed at by school-children."[14]

Fermat was initially sceptical of Roberval's claim that he had determined the quadrature of the cycloid, but he soon worked it out for himself and offered a new demonstration. Meanwhile, Roberval remained silent about his own proof and withheld his new name for the cycloid, which he had christened *trochoïde*, from *trochos*, the Greek for wheel. Once the quadrature of the cycloid was known, Mersenne asked about the tangent to this curve. Roberval had found the solution, but his method was mechanical and, as such, fell short of the ideal of rigorous geometrical demonstration. Fermat and Descartes sent in their solutions in August 1638. Never to be stopped Mersenne, went on to raise a further problem, What is the volume of the solid produced by the revolution of the cycloid around its axis? Again Roberval claimed to have found the answer but he provided no demonstration. Descartes refused to consider the question, alleging that he had "given up geometry." When Mersenne tried to provoke him by interpreting his reply as an admission that the problem could not be solved, Descartes assured him that the problem was easy but that he had no time to waste on it.[15]

Mersenne publicized the curve in his correspondence with Theodore Haak and John Pell.[16] The problems that Mersenne raised in these letters concern the determination of the surface area of the cycloid, the volume of the solid produced by its rotation around the base, and the ratio of the line to the circumference of the circle. In a further letter to Pell on 1 March

---

[13]Letter of 27 May 1638 to Mersenne, *Correspondance de Mersenne* VII, pp. 226–227. Descartes' demonstration was not as straightforward as he had surmised, and he offered an extended version a couple of months later in a letter of 27 July 1638 to Mersenne, *Correspondance de Mersenne* VII, pp. 407-412.

[14]Letter of Descartes to Mersenne, 23 August 1638, A.T. II, p. 332; *Correspondance de Mersenne*, VIII, p. 60.

[15]Letter to Mersenne of 11 October 1638, *Correspondance de Mersenne*, VIII, p. 111, and 29 January 1640, *Correspondance de Mersenne8, IX, p. 88.

[16]Letters of 18 December 1639 and 20 January 1640, *Correspondance de Mersenne*, VIII, p. 692: IX, p. 51. In the first letter, he asked Haak, in London, to advertise a problem concerning the *roulette*. In the Latin of the second letter, this became *cycloid*, apparently the first appearance of the word.

1640, Mersenne added yet another question concerning the volume of the solid produced by the rotation of the cycloid around the axis.[17]

The Parisian mathematician, Jean de Beaugrand, raised the same questions about the cycloid in a letter to Bonaventura Cavalieri, whom he had met in Bologna a few years earlier. These were forwarded to Galileo, and this is probably how they reached Torricelli, who successfully determined the surface area of the cycloid, as he informed Roberval in a letter of 1 October 1643. Torricelli published his results in 1644 as an appendix to his *Opera Geometrica* and he discussed the normal cycloid as well as prolate and curtate cycloids, that is, the curves that are traced out when the point lies above or below the circumference of the generating circle. Roberval procrastinated as usual and when he replied, more than two years later, it was to insinuate that the Italians had plagiarized the French!

Torricelli appealed to Mersenne on 7 July 1646 to bear witness to the independence of his results arrived at by considering the point describing a cycloid as endowed with two simultaneous motions, one uniform and the other varying. When Torricelli died on 25 October 1647, however, the testimonial letter that he expected from Mersenne had not yet arrived. At the time, accusations of plagiarism were frequent since few realized that it is not unusual for researchers to arrive at more or less identical results independently. Thus, about 1635, Cavalieri, Roberval, Fermat, and Descartes had all been working on the geometry of indivisibles and constructing tangents to cycloids. Likewise, thirty or forty years later, Newton and Leibniz independently discovered the integral and differential calculus.

A challenge to many and probably a headache to more, the properties of the cycloid had become a current topic of discussion among mathematicians, and it is not surprising that they should have caught Pascal's attention. What is amazing is the speed with which he made his discoveries. According to his sister Gilberte, Pascal's concentration on the night made sleepless by a toothache produced a rapid succession of brilliant insights. "The first was followed by a second," she writes "and the second by a third, and finally by a multitude of ideas each succeeding one another. These revealed to him, almost involuntarily, the proof of the properties of the cycloid, so that he himself was surprised."[18] These proofs banished his pain,

---

[17]*Correspondance de Mersenne*, XIV, p. 6; the letter is dated 1 January 1646, but probably was sent later since Torricelli received it in March, *Correspondance de Mersenne*, XIV, p. 147.

[18]*Œuvres de Pascal*, I, pp. 585–586. The story is told by Gilberte Périer, *Œuvres de Pascal*, I, pp. 585–586, 623, and by her daughter Marguerite, *Œuvres de Pascal*, I, pp. 1103–1104.

which, writes Marguerite Périer, was "all that he desired of them." Whether it was, in fact, a cure, or merely a distraction from a peculiarly distressing symptom, is another matter. But without that bout of toothache, Pascal would never have returned with so much zest, intensity, and ingenuity to mathematics. The Duc de Roannez, who had been with him the previous evening and was anxious about his state of health, returned the next morning to see how he was. Instead of finding him in pain, he found that he had just triumphantly re-entered the scientific arena. Roannez was deeply impressed and urged Pascal to consider how his mathematical skill could best be used as a weapon in the religious apology upon which he had embarked. As Donald Adamson points out in his book on Pascal: "This was not bravado on the duke's part. His attitude in no way resembled that of the promoter of a prizefighter keen to take on all comers and thus to assert his own particular, if not general, superiority. Like so many of his contemporaries, Roannez almost certainly felt that no philosophical issue could properly be settled by anyone lacking mathematical ability: Leibniz has expressed the same point of view."[19]

In short bursts of fierce concentration, Pascal worked out the properties of the cycloid between June 1658 and February 1659. This was to be his last major contribution to mathematics, and it led him away from the religious concerns that had so occupied him and back to something of the aggressive worldly superiority that had characterized his earlier dispute with the Jesuit Etienne Noël. The desire to shine, to outrun one's rivals, and to establish a priority in scientific discovery brought out the worst in Pascal's character. Scientific bickering inspired to some extent by personal vanity was a shortcoming of the whole seventeenth century. It had poisoned the relations between Roberval and Descartes twenty years earlier, and it was to lead to a disgraceful dispute between the followers of Newton and Leibniz in the eighteenth century.

Convinced by the Duc de Roannez that his discoveries were not alien to his work concerning religion and would serve to combat "atheists and libertines," Pascal issued a *Challenge* to mathematicians, a practice that was common in his day. He wrote it out in Latin at breath-taking speed. "It is incredible," writes Gilberte, "how quickly he dashed that down on paper. He wrote as fast as his hand would go, and it was finished within a few days. He did not make a copy of what he wrote, but handed the sheets over as he went along. They were printing something else of his which he

---

[19]Donald Adamson, *Blaise Pascal*, London: St Martin's Press, 1995, pp. 190–191.

was handing them in the same way as he went along, and so he was supplying the printers with two different things simultaneously."[20] The *Challenge* was issued on 1 June 1658, and mailed by Pascal and his friends Carcavy and Mylon to as many mathematicians as they could think of. These included Fermat in France, John Wallis and Christopher Wren in England, Leibniz in Germany, Michelangelo Ricci in Italy, Frans Van Schooten and Huygens in the Netherlands, and the Flemish canon François de Sluse in Liège. The intended connection with religious issues was implied by the pseudonym Amos Dettonville, which is an anagram of Louis de Montalte, the name under which Pascal had issued his *Provincial Letters*, and also of Salomon de Tultie, which he had chosen for the publication of the *Pensées*. Prizes of forty and twenty *pistoles* respectively were offered to the two mathematicians who managed to find a solution to following six problems:[21] (1) the quadrature of a segment of an arch of the cycloid bounded by the curve, the axis (through the vertex perpendicular to the base) and a chord parallel to the base; (2) the centre of gravity of this segment; (3) the volume generated by revolving the segment about the chord; (4) the volume generated by revolving the segment about the axis; (5) the centers of gravity of the volumes generated; and (6) all the above for the halves of the volumes when these are sliced by a plane through the axis.

The highly competitive atmosphere thus generated could hardly serve the prestige of religion. Things began to go wrong within weeks of the *Challenge* being distributed. Pascal suddenly realized, or perhaps someone brought it to his attention, that of the six questions involved in the challenge, the first four had already been successfully solved by Roberval. It may seem surprising that Pascal should have been unaware of Roberval's work, but we must recall that it was largely unpublished. Pascal therefore instructed Carcavy, who had agreed to chair the panel of judges, not to take these four into consideration, and to judge only the answers to the last two. Another letter was issued in July 1658 to clarify yet another ambiguity: the cycloid was a simple one, not curtate or prolate. The deadline of 1 October for the submission of entries was maintained, however, but it now only applied to the results, and not to the demonstrations.

---

[20]Gilberte Périer, *Œuvres de Pascal*, I, p. 623.

[21]These sums appeared to be offered by the Anonymous Challenger himself, who deposited them on trust with Carcavy, but Roannez may have helped his friend to produce sixty *pistoles*, a handsome sum, enough to pay the yearly rent on a house in one of the fashionable parts of Paris. The deadline for the certified despatch of entries was 1 October 1658.

The competition attracted more bystanders than participants. Only two mathematicians entered the list, although many toyed with partial solutions and followed the proceedings with great curiosity. The Flemish canon René de Sluse informed Pascal that he had long since discovered how to calculate the area bounded by a cycloidal arc, but that he could not find the volumes and the centers of gravity. Huygens came up with a partial solution of the problems but did not submit an entry. Another mathematician who made some progress, but without going far enough, was Michelangelo Ricci in Rome. In Oxford, the twenty-five-year-old Christopher Wren made no headway at all with the specific problems but pursued an ancillary line of investigation and found that the length of the arc of a cycloid is four times that of the diameter of its generating circle. This was irrelevant to the competition but it delighted Pascal. "Nothing," he later wrote, "is finer than what has been sent in by Mr. Wren; for quite apart from the fine method which he provides for calculating the plane surface of the cycloid, he has provided the method of comparison between the actual curve and its proportionate relationship to the straight line."[22]

The two candidates who entered the competition, John Wallis, the professor of mathematics at Oxford, and the French Jesuit Antoine Lalouvère, came out of the contest full of grievances and, especially in the Jesuit's case, sorely bruised. Both men felt that Pascal had given them too little time. On 19 August 1658, Wallis wrote to Carcavy to complain of the closeness of the deadline, but Pascal indignantly rejected the suggestion that he should have allowed eight or even twelve months. The second contestant, Antoine Lalouvère, did not to formally compete but he sent Carcavy a solution of the first three of the cycloid problems by the end of July. Dropping the mask of anonymity, which had probably become useless, Pascal replied on 4 September 1658 that answers to three questions were not enough. Lalouvère rose to the challenge and managed, in less than a week, to solve the fourth. Pascal congratulated him on 11 September but added that the problem had already been dealt with by Roberval. There remained the fifth and sixth questions on which Lalouvère worked frantically for four days, but no sooner had he dispatched his solutions to Carcavy than he realized that they contained some errors and he wrote to withdraw from the competition. The last two problems were too difficult for him, but it was no mean feat for a college teacher to have achieved what

---

[22]Pascal, *History of the Cycloid*, *Œuvres de Pascal*, IV, p. 221.

mathematicians of the first rank had so signally failed to do: tackling and mastering the first three cycloid questions within ten days, and making inroads on the more arduous problems that remained.

Lalouvère was not a friend of the Jansenists,[23] and this may have influenced Pascal against him, but it is disconcerting to find so many disparaging references to Lalouvère in the *History of the Cycloid* that Pascal published on 10 October 1658, only ten days after the deadline for the despatch of entries to the cycloid competition. The pamphlet, which is anonymous, was probably written in collaboration with Roberval. It comments unfavorably on the submissions received from Lalouvère and Wallis, but it was not these remarks that roused Wallis to a fury and made the *History of the Cycloid* the most controversial scientific paper Pascal ever wrote, even more controversial, in a sense, than the *Provincial Letters*. What made the *History of the Cycloid* so outrageous was its treatment of Torricelli.

Pascal began by claiming that Mersenne was the first to notice the cycloid's existence, but he damned him with faint praise. "He tried to investigate its nature and properties," he writes, "but could not fathom them. He had a special gift for formulating good questions, in which, perhaps, he had no equal, but he was not equally fortunate in solving them, and this is where fame really lies. Nevertheless we are under an obligation to him for the opportunity of several fine discoveries, which might never have been made had he not provided the stimulus." Unable to find the properties of the cycloid himself, Mersenne put the problem "to all those in Europe whom he thought capable of it, including Galileo. But no one had any success, and all despaired." Finally in 1634, according to Pascal's account, Roberval proved that the area of the cycloid is three times as great as that of its generating circle. He communicated his discovery to Mersenne and asked for a competition! According to Pascal, Mersenne wrote to mathematicians throughout Europe, allowing them a year in which to determine the ratio of the area of a cycloid to that of its generating circle. Pascal, at this time, was eleven years old, and what he writes can hardly be based on personal experience. He is clearly relying on Roberval's self-serving reconstructing of the events. As a matter of fact, no such formal competition had been organized. In 1638, so the account continues, the French achievements were passed on to Galileo by Jean de

---

[23]See his letter of 7 June 1659 to a fellow Jesuit, *Œuvres de Pascal*, IV, pp. 655–656.

Beaugrand, who died in December 1640, to be followed by Galileo in January 1642. Torricelli came into possession of Galileo's papers, and found among other things "the solutions to the *roulette* (here called the cycloid) in the handwriting of de Beaugrand, who seemed to be their author. Beaugrand was dead and Torricelli thought that enough time had elapsed for people to have forgotten about him, and he decided to take advantage of this." Pascal goes on to say that Torricelli brazenly plagiarized Roberval's solution, and when Mersenne and Roberval complained, he privately admitted his guilt but never withdrew his claims publicly.[24]

Such a travesty of truth could not go without a reply. From all over Europe, mathematicians sprang to the defense of Torricelli, who had unquestionably arrived at his resolution of the quadrature of the cycloid by independent methods. How could Pascal have lent his name to such an unfair accusation? He must have been blinded by Roberval's hostility towards the Italian scientist. But over and above Pascal's willingness to believe a friend, his ruthless denunciation of someone who was no longer alive to defend himself shows a deplorable lack of decency that is compounded by his arrogant and patronizing treatment of Lalouvère.

On 24 November 1658, the panel of judges met, under Carcavy's chairmanship, and found the two entries unsatisfactory. When Lalouvère protested, he was administered one of the sharpest attacks Pascal ever wrote against the erring waywardness of human nature. In a *Sequel to the History of the Cycloid*, published on 12 December 1658, Lalouvère was denounced for gross incompetence, and turned into a jest. "Geometrical matters are so serious in themselves," Pascal writes in his opening sentence, "that I welcome the opportunity of showing that they have a lighter side."[25] The final blow, the *Letters from A. Dettonville* (i.e., Pascal) to *M. de Carcavy*, followed three weeks later with the solutions to all the cycloid problems, including three additional ones. Still smarting, the Jesuit attempted a replied in a pamphlet in January 1659, outlining conclusions similar to Pascal's own. But Pascal would not let matters rest. In a *Footnote to the Sequel to the History of the Cycloid*, published in the same month, he excoriated Lalouvère for inaccuracies, oblivious of the fact that only four months before writing his *History of the Cycloid*, he had been himself ignorant of Roberval's work on the subject.

---

[24]Pascal, *History of the Cycloid*, Œuvres de Pascal, IV, pp. 214–217.
[25]*Œuvres de Pascal*, IV, p. 238.

Lalouvère was unwise enough reply in February 1659, but he was no match for a man who handled words as effectively as he manipulated mathematical symbols. The competition, which Roannez hoped would redound to the greater glory of the religion, thus ended in the unedifying spectacle of two Christian believers at each other's throats. Nevertheless, if we abstract from the passages where Pascal vents his feelings and those of Roberval, we can only be impressed by the importance of his geometrical discoveries during the last few months of 1658. Inspired by Archimedes' method of determining the quadrature of the parabola by means of the equilibrium of the lever, Pascal developed a new approach to infinitesimal quantities that was to have enormous consequences for the development of the integral calculus. One of the geometrical essays that he drafted at that time was later to open Leibniz's eyes to a profound mathematical truth. In a deleted postscript to a letter meant for Jacob Bernoulli in 1703, Leibniz acknowledged his indebtedness to Pascal's *Treatise on the Sines of Quarter-Circles*, which he had come across during his time in Paris.[26] Leibniz states that upon glancing at one of Pascal's figures he saw in a flash that whereas the quadrature of the cycloid depends on the sum of the ordinates for infinitesimal intervals in the abscissas, the *differences* between the ordinates and the abscissas are what determines the tangent to a cycloidal curve. But what struck Leibniz so forcefully was something that the author of the treatise had failed to realize for himself. Using Figure 1, Pascal had showed that the sum of the sines of the arcs of a quadrant multiplied by the infinite numbers of arcs corresponding to the tangents is equal to the portion of the base between the extreme sines, multiplied by the radius. This is equivalent to our formula:

$$\int_{x_0}^{x_1} \sin x\, dx = \cos x_0 - \cos x_1 .$$

Pascal does not express himself in this way, but a few words of explanation will enable us to follow his geometrical way of reasoning with the aid of Figure 1.

ABC is the quadrant of a circle and D is any point on the arc from which DI is drawn to the radius AC. As was common in his day, Pascal calls DI the sine (this is quite straightforward since the sine of angle CAB is equal to DI/AD, and AD is equal to the radius AC whose value is given as

---

[26] Postscriptum to the letter of Liebniz to Jacob Bernoulli, April 1703, in G. W. Liebniz, *Mathematische Schriften* (edited by C. I. Gerhardt). Hildesheim: Georg Olms, 1971, vol. III/I, pp. 72–73.

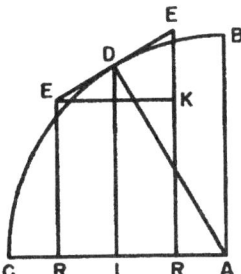

FIGURE 1 *Pascal's diagram to illustrate the sum of the sines of a quadrant.*

1). EDE is a tangent to the circle such that ED = DE. Perpendiculars, dropped from the two points marked E, meet the radius at the points marked R. From elementary geometry, Pascal knew that triangles ADI and EKE are similar because both are right-angle triangles and < EEK = < EDI. Hence the ratio of the sides is:

$$\frac{AD \ (the \ radius)}{DI \ (the \ sine)} = \frac{EE}{EK} = \frac{EE}{RR}$$

It follows that DI × EE = AD × RR (or as Pascal put it, the rectangle whose sides are DI and EE is equal to the rectangle whose sides are AD and EK). The next step is to divide the arc BC into an infinite numbers of sines (such as DI) and the radius AC into an infinite number of equal parts, each corresponding to the infinite number of sines, as the RR corresponded to the DI. All the rectangles formed by the sines multiplied by the infinitesimally small arcs can then be considered equal to the rectangles formed by all the portions RR multiplied by the radius. In other words, Pascal is equating the sums of two infinite number of products, one factor of which is an infinitesimally small quantity (here EE or RR). "Although this equality is not true when the number of sines is finite," he writes, "it is true when the number is infinite because the sum of all the equal tangents EE differs from the entire arc BC *by less than any given quantity*, and similarly the sum of all the small portions RR from the entire length RR."[27]

---

[27]The *Treatise on the Sine of Quarter-Circles* was published in the *Letters from Dettonville* (Paris, 1659), in *Œuvres de Pascal*, IV, pp. 478–487. Pages 478–480 are translated in D.J. Struik (ed.), *A Source Book in Mathematics 1200–1800*. Princeton: Princeton University Press, 1986, pp. 238–241, the quotation is from p. 241, and the figure from p. 239.

We now see how Pascal's discussion of the related concepts of the infinitely large and the infinitely small could have influenced Leibniz, but why did Pascal miss what leaped out of the page for Leibniz? It is true that Pascal was inclined to see something of a mystery in the infinitely small, but then Leibniz also left unanswered the question of the nature of the infinitesimal. The reason why Pascal's infinitesimal calculus fell short seems twofold. First, Pascal shunned the rule-of-thumb procedures of algebraic algebra. Had he been acquainted with Wallis's work, he would have known better, but he was to remain ignorant of the English mathematician's effective use of what Hobbes called the "scab of symbols." Second, Pascal's health was declining and he could only work in feverish bursts of concentration on one topic, in this case the cycloid. When he found the areas and volumes connected with cycloids, he felt that he had exhausted the interesting possibilities. He was not concerned with tangents to the curve and missed the significance of the *characteristic triangle*, which Leibniz was to see in his diagram. Pascal, Leibniz remarks, seemed scientifically "blindfolded" in his failure to discern some of the applications of his work on cycloids. But it is always easier to see a hidden truth when the light has been turned on and Leibniz had time to accustom his eyes to its dazzling brightness.

Pascal developed related insights in a number of important mathematical letters to friends. In an open letter of 10 December 1658 to a mysterious correspondent disguised under the initials A.D.D.S. (and presumed to be Antoine Arnauld), he showed that the parabolic curve and the Archimedean spiral are of the same length, using this time not the method of the indivisibles but the manner of the Ancients.[28] In an open letter to Sluse, Pascal demonstrated certain properties of the so-called *escalier* (or *staircase*), a solid consisting of spiral steps of uniform height and surface area mounting within the arc of a circle and with a quarter of that circle as its base. Pascal established that the volume of this solid was equal to one quarter of the square of the base arc multiplied by the radius.[29] Yet another open letter, this time addressed to Huygens and composed in January 1659, greatly enlarged the scope of the *Letters from Dettonville* by providing a method of determining the quadrature of *all* cycloids, not only the common ones, but also prolate and curtate cycloid curves.[30] A final letter,

---

[28] *Œuvres de Pascal*, IV, pp. 542–558.
[29] *Œuvres de Pascal*, IV, pp. 533–539.
[30] *Œuvres de Pascal*, IV, pp. 524–530.

written to Carcavy but intended for Huygens, concerned the convex dimension of the parabolic conoid, which Pascal hoped in due course to extend to the parabolic spheroid.[31] In vain, however, as his health did not permit it. With this letter to Carcavy, his mathematical research had come to an end.

---

[31]*Œuvres de Pascal*, IV, pp. 353–356

Conclusion

From the age of eighteen, Pascal was a sick man, never knowing a painless day till he died at the early age of thirty-nine. Yet he managed to conquer his abiding feeling of physical discomfort and respond with passionate interest to the intellectual stimulus provided by scientific acquaintances. When his father and a friend became interested in determining whether a vacuum existed, Pascal designed and performed a number of clever experiments to show that this was possible. He did away with the concept of Nature's horror of a vacuum and confirmed the role of atmospheric pressure by having his brother-in-law go up and down the Puy-de-Dôme with what is now known as a barometer. He also devised an ingenious experiment that consisted in creating a vacuum-in-a-vacuum. For us, this is an instance of standard laboratory procedure but it was highly original and unconventional at the time. The recognition of atmospheric pressure was the first step in the discovery of the Principle (Pascal's Principle) that hydrostatical pressure is the same in all directions about a point in a fluid. Pascal realized that it could be applied to pumps, and later generations extended it to hydraulic brakes, elevators, and cranes.

Designing Experiments

What is unusual in Pascal's scientific approach is its modest aim and its comparative freedom from metaphysical presuppositions. Most of his contemporaries, be it Descartes or the Jesuit professors, assumed that scientific knowledge was arrived at by demonstration from first principles. This meant following a long chain of consequential events without leaving gaps. Descartes found the notion of a vacuum metaphysically impossible because he could not imagine God leaving a void between a succession of causes and events. Pascal saw no difficulty in this. His belief in God did not preclude his acceptance of the existence of a vacuum, which was the result of objective experimentation. He was aware of Descartes' argument but he dismissed it in the *Pensées* on both scientific and scriptural grounds: "It is a remarkable fact," he writes, "that no canonical author ever used an argument from nature in order to prove that God exists . . . David, Solomon, etc., never said: 'The void does not exist. Therefore there is a God.'"[1]

At a time when the Bible was the cornerstone of European science and culture, this was an important point. The inspired writers did not appeal to nature's alleged horror of nothingness to prove that God is everywhere and fills everything. For Pascal, Scripture provided an insight into the purpose of creation, not its detailed structure. This was the job of science. Here Pascal found himself again at variance with Descartes for whom the universe is a clock-work mechanism presided over by a Deity who is silent about his aims. Pascal quipped that all Descartes' God ever had to do was just to snap his fingers and so set the world in motion. For Pascal, God intervenes in human history but in such a way that men and women remain free to work out their personal destiny. In this sense, the world is ultimately contingent, and the exercise of reason is subject to caution. Knowledge that flows from self-evident principles is absolutely certain but is limited to mathematics. The truths of physics cannot be established deductively even if this is sometimes a useful way of presenting them. Without experimentation, we can only tell likely stories, like astronomers who use different models to describe the paths of celestial bodies. Whether the Earth goes around the Sun was problematical for Pascal, but he considered it unwise to behave as though the absence of actual proof precluded the possibility of eventually finding one.

The task of the physicist is to examine the evidence and subject hypotheses to rigorous testing. A little looking is infinitely better than a lot of guessing. The Jesuit Etienne Noël *assumed* that light was refracted in the empty space above the mercury in an evacuated tube and promptly *deduced* that no real vacuum had been produced. Had he taken the trouble of carefully observing the actual path of the ray of light, comments Pascal, he would have noticed that it is not bent in the empty space but only when it passes through the glass of the tube. Elementary? Yes, but not for the

---

[1]*Pensées*, fragment 463.

head of a seventeenth-century Jesuit school. Experiments, for Noël, were a matter of common sense observations, not precise measurement, and he felt perfectly justified in pulling out his silk handkerchief in an overheated and dry room to produce the crackling sparks that "proved" that air contains particles of fire.

Outside the field of theology, where we find divine truths, and mathematics, where we have rigorous demonstrations, Pascal insisted that nature can only be known through experiments. He makes no attempt to provide a complete world-picture. In the *New Experiments on the Vacuum*, he stresses that one is not required to accept the hypothesis that a vacuum is possible in order to appreciate the main outcome of his experiments, which is that the force that resists the formation of a vacuum—be it apparent or real—is measurable and constant within narrow limits. In a tube it is equal to the weight of 27 inches of mercury. Pascal's approach stands in marked contrast with the style of Noël, who believed it was futile to proceed with experiments without first determining whether they were philosophically sound. Noël did not think it was of much interest to examine how light goes through a vacuum unless he first ascertained whether a vacuum was possible. Physics was a question to be debated rather than a matter to be tested, and Noël went about his business by invoking grand principles such as nature's refusal to allow a vacuum, the correspondence between the macrocosm and the microcosm, and the absence of gaps in the great chain of being. Pascal was usually low-key and spoke of the vacuum merely as a place empty of any material that we can know. In his letter to Noël he exposed himself, once, to the high winds of speculation and located the vacuum halfway between matter and nothingness. He was never to indulge in such metaphysical flights again.

What we are looking for determines what we see. Pascal and his opponents agreed that the mercury drops to a measurable height when the tube is inverted and that an apparently empty space is left at the top of the tube. But the fact could be discussed in the light of two very different questions. This first asked what was left in the apparently empty space. Something or nothing? This opened a debate on what that something could be. Some suggested rarefied air, others Descartes' unobservable subtle matter. The relevance of experimentation receded as the discussion became increasingly metaphysical. The second question did not focus on the empty space but asked what kept the column of mercury from falling all the way down. Why did it stop at a height of 27 inches? Torricelli had guessed that this was on account of the weight of the atmosphere, and Pascal proved that this was indeed the case.

Pascal saw that the ascription of such properties as "sympathies" and "antipathies" to inanimate objects was an obstacle to a clear rendering of the result of experiments. Here again he sounds perfectly modern but many of his contemporaries found his dismissal of organic analogies wrongheaded or, at least, premature. In the treatise, *On the Weight of the Mass of Air*, that he intended for publication, Pascal was careful not to pour scorn on those who, like Noël, spoke of nature as experiencing horror. He restricted himself to showing that the phenomena that were usually attributed to the horror of a void could be explained by hydrostatic principles. When the way nature reacts in the presence of a vacuum is seen to follow from the general laws of equilibrium, an appeal to horror becomes redundant. Pascal avoided getting entangled in abstruse metaphysical discussions about whether or not we can know that nature has emotions. It was enough that such a notion be recognized as superfluous.

Once the occurrence of a vacuum was understood in the light of the general laws of hydrostatics, the outcome of experiments could be anticipated and determined in quantitative terms. In his treatises on the equilibrium of liquids and the weight of air, Pascal does not claim to have performed all the experiments he describes. He imagines what will happen, and in some instances he is clearly carrying out a thought-experiment. A case in point is his description of a man sitting under 20 feet of water and holding a tube that rests on his thigh and extends above the surface of the water. Pascal wants the reader to realize that although the man would feel his flesh rise as if it were being pulled up, this really happens because the rest of his submersed body is uniformly pressed by the weight of the water. The experiment (without an oxygen tank) is impractical but the result can be predicted because the new physics tells us what will happen. There is no deception here, just the legitimate desire to provide a striking illustration of a well-confirmed physical law, and to underline that what is *experienced* as an *attraction* coming from *inside* the tube is really *pressure* from the water *outside*.

Games of Chance

Pascal was not only gifted for mathematics, he was a child prodigy. Without help or instruction from anyone he mastered the elements of Euclidean geometry. At the age of fourteen or fifteen he grasped what Desargues was attempting to achieve with his new projective geometry, and he wrote an *Essay on Conic Sections*. Anxious to apply the mathematics that he knew, Pascal devoted himself for three years to the design and production of a business machine to help his father, who was crushed by the burden of computing taxation rates. Fifty prototypes were made, and Pascal seems to have been the first person to manufacture and market a desktop mechanical calculator.

The projective geometry that Desargues had inaugurated and that Pascal had embraced proposed to examine a large number of formal relationships including but not limited to perspective. Pascal was attracted by the fact that when numbers are represented by points or dots some sequences have a triangular shape (1, 3, 6, 10, ...) while others have a square one (1, 4, 9, 16, 25, ...). The concrete representation of these numbers (called *figurate* on account of their figure or shape) led to an investigation of their properties. Triangular numbers were found to be constructed by adding the natural numbers, such that the nth triangular number is the sum of the first n numbers. Likewise square numbers were discovered to result from adding the odd numbers so that the nth square number is the sum of the first n odd numbers. Further examination showed that two successive triangular numbers taken together make a square, e.g. $3 + 6 = 9$.

Gradually more and more of such patterns were revealed, and by the time Pascal appeared on the scene they had been brought together in the form of the Arithmetical Triangle. What had not been done, and what Pascal was to achieve in a masterful way, was to show the relations between the different aspects of the numbers. Using mathematical induction Pascal first worked out the properties of the numbers in an original way, and then showed how these properties were relevant to the study of figurate numbers, the theory of combinations, the expansion of binomial expressions, and the solution of important problems in the theory of probability. His interest in the last application was aroused by a practical problem. The Chevalier de Méré, an assiduous gambler but no mathematician, asked Pascal to work out the odds of throwing a double six with a pair of dice. Pascal made the calculations and Méré went on to raise a second and much more difficult problem that had often been discussed but never properly solved. This was the problem of points or how to work out an equitable distribution of stakes between participants in a game that is interrupted before it is completed. Pascal's great insight was to grasp that the share of the stakes that a player should receive depends on the probability that he would have won the game if it had been allowed to continue. He worked out in detail, for several examples, how the probability of winning could be calculated from a knowledge of the nature of the game and the partial score of each contestant. He discussed his solution with the Toulouse mathematician Fermat, and their fascinating exchange of letters, which extended over a short period of four months in 1654, marks the real start of the mathematical theory of probability.

Pascal extended the results he had arrived at with the Arithmetical Triangle to the determination of the sum of the powers of an Arithmetical progression and to the more difficult problem of the multiplication of numbers in a natural series. He saw the relevance of his work to the determination of the areas of figures bounded by a curve, what was known as the problem of quadrature, and his results foreshadowed the development of infinitesimal calculus. Pascal's last contribution to mathematics was his determination of the properties of the cycloid in the course of which he developed a new approach to infinitesimal quantities that was to influence Leibniz and have decisive consequences for the development of the calculus. Using a nautical metaphor, we might say that although Pascal did not actually land on the new mathematical continent, he glimpsed it from afar and, more importantly, he taught others where to look.

We cannot be certain which of his numerous achievements Pascal valued most, but there is little doubt that he derived great satisfaction from the application of his theory of probability to what he considered the most important question that we face as human beings: Does God exist? From the evidence that surrounds us, the odds that God exists are matched by those that he does not. The matter seems undecidable. The way out, for Pascal, is to introduce a new element in the decision making. We must consider the consequences of our wager. If God exists we can expect "an infinity of an infinitely happy life"; if he does not, we lose nothing. This is because, according to Pascal, leading a moral life here below is more conducive to happiness than indulging one's appetites. It would be foolish not to bet when we have a win-win situation.

Objections and qualifications to Pascal's Wager are easy to find, but the fascination that the argument has exerted on the best philosophical minds for over three centuries is an indication of its originality and its paradoxical power. Is it merely a thought-provoking parable or does it carry convincing mathematical weight? Men and women continue to the present day to ask whether they should take the gamble. Some do, others don't. Very few dismiss it as trivial or uninteresting. Pascal felt they we are compelled to make the wager. Life, as he experienced it, was a preparation for something greater, and he harnessed all his immense mathematical and scientific talent to a practical purpose: getting people to make the great experiment of betting for their life.

Selected Bibliography

Pascal's Works

Pascal, Blaise. *Œuvres*, edited by L. Brunschvicg, P. Boutroux, and F. Gazier, 11 vols. Paris: Hachette, 1908–1914. Reprinted Vaduz: Kraus Reprint, 1965. The standard edition is now being replaced by the *Œuvres Complètes de Pascal* by Jean Mesnard, of which four of eight proposed volumes have appeared to date. Paris: Desclée de Brouwer, 1964–.

Pascal, Blaise. *Œuvres Complètes*, edited by Louis Lafuma. London: Macmillan, 1963.

Pascal, Blaise. *Pensées*, edited by Philippe Sellier. Paris: Mercure de France, 1976.

Translations

Pascal, Blaise. *Great Shorter Works*, translated by Emile Cailliet and John C. Blankenagel. Philadelphia: Westminster Press, 1948.

Pascal, Blaise. *Selections*, edited by Richard H. Popkin. London: Macmillan, 1989.

Pascal, Blaise. *The Physical Treatises*, translated by I.H.B. and A.G.H. Spiers. New York: Columbia University Press, 1937.

Secondary Sources

Adamson, Donald. *Blaise Pascal*. London: St. Martin's Press, 1995.

Agagi, Shozo. "Comment interpréter les expériences nouvelles touchant le vide," in Thérèse Govet (ed.), *Pascal, Port Royal, Orient, Occident*. Paris: Klincksieck, 1991, pp. 199–209.

Allard, Robert. "Pascal et la trente-deuxième proposition d'Euclide," in Thérèse Goyet (ed.), *Pascal, Port Royal, Orient, Occident*. Paris: Klincksieck, 1991, pp. 191–198.

Arnold, Keith. "Pascal's Great Experiment," *Dialogue* XXVIII (1989), pp. 401–415.

Arnold, Keith. "Pascal's Theory of Scientific Knowledge," *Journal of the History of Philosophy* 27 (1989), pp. 531–544.

Auger, L. "Les idées de Roberval sur le système du monde," *Revue d'histoire des sciences* X (1957), pp. 226–234.

Bacon, Francis. *A Critical Edition of the Major Works*, edited by Vickers. Oxford University Press, 1996.

Bacon, Francis. *Works*, edited by J. Spedding, R.L. Ellis et al., 14 vols. London, 1857–1874. Reprint, Stuttgart-Bad Cannstatt: Frommann, 1963.

Baillet, Adrien, *La vie de Monsieur Descartes*. Paris: Daniel Horthemels, 1691. Facsimile reprint, Geneva: Slatkine Reprints, 1970.

Barker, John. *Strange Contrarieties. Pascal in England during the Age of Reason*. Montreal: McGill–Queen's University Press, 1975.

Baron, M.E. *The Origins of the Infinitesimal Calculus*. Oxford: Pergamon Press, 1969.

Beeckman, Isaac. *Journal*, edited by Cornélis de Waard. 4 vols. The Hague: Martinus Nijhoff, 1939–1945.

Bernhardt, Jean. "La question du vide chez Hobbes." *Revue d'histoire des sciences* XLVII (1993), pp. 225–232.

Bouchilloux, Hélène. *Apologétique et Raison dans les Pensées de Pascal*. Paris: Klincksieck, 1995.

Boyle, Robert. *Works*, edited by Thomas Birch, 6 vols. London, 1772. Facsimile reprint Hildesheim: George Olms, 1965.

Bras, Gérard, and Cléro, Jean Pierre. *Pascal: Figures de l'imagination*. Paris: PUF, 1994.

Brun, Jean. *La philosophie de Pascal*. Paris: PUF, 1992.

Burton, Roger. *The Anatomy of Melancholy*, first published in 1621. Everyman's Library reprints the 6th edition of 1651, 3 vols. London: Dent, 1932.

Carraud, Vincent. *Pascal et la philosophie*. Paris: PUF, 1992.

Chevalley, Catherine. *Pascal contingence et probabilités*. Paris: PUF, 1995.

Clagett, Marshall. *The Science of Mechanics in the Middle Ages*. Madison: University of Wisconsin Press, 1961.

Cohen, Morris C., and Drabkin, I. E. *A Source Book in Greek Science*. Cambridge, MA, 1958.

Coleman, Francis X. J. *Neither Angel nor Beast. The Life and Work of Blaise Pascal*. London: Routlege & Kegan Paul, 1986.

Corsano, Antonio. "Pascal critico della scienza esatta," *Rivista di filosofia* 50 (1959), pp. 423–442.

Costabel, Pierre. "La physique de Pascal et son analyse structurale." *Revue d'histoire des sciences* XXIX (1976), pp. 309–324.

Cottingham, J., Stoothoff, R., and Murdoch, D. *The Philosophical Writings of Descartes*, 2 vols. Cambridge: Cambridge University Press, 1984.

Coumet, Ernest. "Le problème des partis avant Pascal." *Revue internationale d'histoire des sciences* 18(1965), pp. 245–272.
Croquette, Bernard. *Pascal et Montaigne*. Genève: Droz, 1974.
Daston, Lorraine. *Classical Probability in the Enlightenment*. Princeton: Princeton University Press, 1988.
David, F.N. *Games, Gods and Gambling*. London: Charles Griffin, 1962.
Davidson, Hugh M., and Dubé, Pierre H. *A Concordance to Pascal's Pensées*. Ithaca: Cornell University Press, 1975.
Derriulat, Jacques. *L'arithmétique de la grâce. Pascal et les carrés magiques*. Paris: Les Belles Lettres, 1994.
Descartes. *Œuvres de Descartes*, edited by Charles Adam and Paul Tannery. 13 vols. (vol. 12 contains the bibliography and index; vol. 13 a biography of Descartes by Charles Adam). Paris, 1897–1913. The first 11 vols. were revised and reprinted, Paris: Vrin, 1964–1974.
Descotes, Dominique. "Arithmétique et littérature: le *Potestatum Numericarum Summa*," *Revue des sciences humaines*, no. 244, October–December 1996, pp. 53–79.
Descotes, Dominique. *Blaise Pascal: littérature et géometrie*. Clermont-Ferrand: Presses universitaires Blaise Pascal, 2001.
Descotes, Dominique. "Genèse des Corollaires 1 et 2 de la *Lettre à Carcavy* de Blaise Pascal," *Revue d'histoire des sciences* 51 (1998), pp. 127–138.
Descotes, Dominique. *L'argumentation chez Pascal*. Paris: PUF, 1993.
Descotes, Dominique. "L'expérience du vide dans le vide à Clermont-Ferrand," *Courrier du Centre International Blaise Pascal*, no. 121 (1989), p. 24.
Drake, Stillman. *Cause, Experiment and Science*, with a translation of Galileo's *Bodies that Stay atop Water or Move in It*. Chicago: Chicago University Press, 1981.
Drake, Stillman. *Galileo at Work*. Ann Arbor: University of Michigan Press, 1970.
Duhem, Pierre. "Le principe de Pascal," *Revue Générale des Sciences Pures et Appliquées* 16(1905), pp. 599–610.
Duhem, Pierre. "Le P. Marin Mersenne et la pesanteur de l'air," *Revue Générale des Sciences Pures et Appliquées* 17(1906), pp. 769–782, 809–817.
Duhem, Pierre. *Le système du monde*, 10 vols. Paris: Hermann, 1913–1959.
Duhem, Pierre. "Roger Bacon et l'horreur du vide," in A.G. Little (ed.), *Roger Bacon*, Oxford: Clarendon Press, 1914, pp. 214–284.
Edwards, A.W.F. *Pascal's Arithmetical Triangle*. London: Charles Griffin, 1987.
Fanton d'Andon, Jean-Pierre. *L'horreur du vide*. Paris: Editions du C.N.R.S., 1978.
Franklin, James. *The Science of Conjecture: Evidence and Probability Before Pascal*. Baltimore: Johns Hopkins Press, 2001.
Galilei, Galileo. *Le Opere di Galileo Galilei*, edited by Antonio Favaro. 20 vols. Florence: Barbèra, 1899–1909.

Galluzi, Paolo, and Torrini, Maurizio. *Le Opere dei Discepoli di Galilei. Carteggio 1641–1648*. Florence: Barbèra, 1975.
Gardies, Jean-Louis. *Pascal entre Eudoxe et Cantor*. Paris: Vrin, 1984.
Gassendi, Pierre. *Opera*, 6 vols. Lyon, 1658. Facsimile, Stuttgart-Bad Canstatt: Frommann, 1964.
Gouhier, Henri. *Blaise Pascal. Commentaires*. Paris: Vrin, 1971.
Grant, Edward. *Much Ado About Nothing*. Cambrige: Cambridge University Press, 1981.
Guenancia, Pierre. *Du Vide à Dieu*. Paris: François Maspero, 1976.
Hacking, Ian. *The Emergence of Probability*. Cambridge: Cambrige University Press, 1975.
Hacking, Ian. *The Taming of Chance*. Cambridge: Cambridge University Press, 1990.
Hara, Kokiti. "L'œuvre mathématique de Pascal," *Mémoires de la Faculté des Lettres de l'Université d'Osaka*, no. 212 (1981), pp. 1–239.
Hero of Alexandria. *Pneumatics*, translated by J.G. Greenwood. London, 1851. Facsimile, London: Macdonald, 1971.
Hooykaas, Reyer. "Pascal: His Science and His Religion," *Tractrix* 1 (1989), pp. 115–130.
Humbert, Pierre. *Cet Effrayant Génie. L'oeuvre scientifique de Blaise Pascal*. Paris: Albin Michel, 1947.
Huygens, Christiaan. *Oeuvres Complètes*, 22 vols. The Hague: M. Nijhoff, 1888–1950.
Jones, Richard Foster. *Ancients and Moderns*, first published in 1936, reprinted Berkeley: University of California Press, 1965.
Keats, John. *Complete Poems*, edited by Jack Stillingfleet. Cambridge: Harvard University Press, 1978.
Knowlson, James. *Universal Language Schemes in England and France 1600–1800*. Toronto: University of Toronto Press, 1975.
Koyanagi, Kimiyo. "Date de rédaction des œuvres posthumes de Pascal traitant de physique," in Thérèse Goyet (ed.), *Pascal, Port Royal, Orient, Occident*. Paris: Klincksieck, 1991, pp. 211–219.
Koyanagi, Kimiyo. "Les expérience du vide dans le vide." *Courrier du Centre International Blaise Pascal*, no. 11 (1989), pp. 3–23.
Koyanagi, Kimiyo. "Pascal et l'expérience du vide dans le vide," *Japanese Studies in the History of Science*, no. 17 (1978), pp. 105–127.
Koyré, Alexandre. "Pascal Savant" in *Blaise Pascal. L'homme et l'Ouvre*. Paris: Editions de Minuit, 1956, pp. 259–295.
Krailsheimer, Alban. *Pascal*. Oxford: Oxford University Press, 1980.
Krüger, Lorenz (ed.). *The Probabilistic Revolution*, 2 vols. Cambridge, MA: MIT Press, 1987.
Le Guern, Michel. *L'image dans l'oeuvre de Pascal*. Paris: Klincksieck, 1983.

Le Guern, Michel. *Pascal et Descartes*. Paris: Nizet, 1971.

Leavenworth, Isabel. *A Methodological Analysis of the Physics of Pascal*. New York: Institute of French Studies, 1930.

Leibniz, Gottfried Wilhelm. *Die Philosophischen Schriften*. Edited by C.J. Gerhardt. 7 vols. Berlin, 1875–1890. Reprint Hildesheim: Olms, 1978.

Loeffel, Hans. *Blaise Pascal 1623–1662*. Basel: Birkhäuser, 1987.

Mach, Ernst. *The Science of Mechanics*, translated by J. McCormack. La Salle, Illinois: Open Court, 1960.

Magnard, Pierre. "Pascal et le sens du vide," *Baroque* 12 (1987), pp. 71–79.

Magnard, Pierre. *Pascal. La clé du chiffre*. Paris: Editions Universitaires, 1991.

Maire, Albert. *Bibliographie générale des oeuvres de Blaise Pascal*, 5 vols. Paris 1926–1927.

Mathieu, Félix. "Pascal et l'expérience du Puy-de-Dôme, *Revue de Paris*, mars–avril 1906, pp. 565–589; mai–juin 1906, pp. 179–206; mars–avril 1907, pp. 176–224, 347–378, and 385–876.

Matton, Sylvain. "La grande expérience du Puy de Dôme revisitée," *Chrysopoeia* 2 (1988), pp. 305–364.

Mazauric, Simone. *Gassendi, Pascal et la Querelle du Vide*. Paris: PUF, 1998.

Meneghelli, Ruggero. "Osservazioni critiche sopra una recente interpretazione di Pascal," *Rivista di Filosofia* 51 (1960), pp. 426–451.

Mersenne, Marin. *La Correspondance du P. Marin Mersenne*, edited by P. Tannery, C. de Waard, A. Beaulieu et al. 17 vols. Paris: Editions du CNRS, 1933–1988.

Mesnard, Jean. "Baroque, science et religion chez Pascal," *Baroque*, 7ème cahier (1974), pp. 71–83.

Mesnard, Jean. *Pascal*. Paris: Desclée de Brouwer, 1965.

Mesnard, Jean. "Pascal et Copernic," in *Avant, Avec, Après Copernic*. Paris: Blanchard, 1975, pp. 241–249.

Middleton, W. E. Knowles. *The History of the Barometer*. Baltimore: Johns Hopkins Press, 1964.

Mourlevat, Guy. *Les machines arithmétiques de Blaise Pascal*. Clermont-Ferrand: La Française d'Edition et d'imprimerie, 1988.

Noël, Etienne. *Le Plein du Vide*, Paris, 1648, reprinted in the *Œuvres de Pascal*, edited by the Abbé Bossut. The Hague, 1979, vol. IV, pp. 108–146.

Nonnoi, Giancarlo. "Galileo e Pascal: Idee e Esperienze," in Antonio Cadeddu (ed.), *Filosofia Scienza e Storia*. Milan: Franco Angeli, 1995, pp. 151–176.

Pintard, René. "Autour de Pascal: l'académie Bourdelot et le problème du vide," in *Mélanges offerts à Daniel Mornet*. Paris: Librairie Nizet, 1951.

Plantié, Jacqueline, and Mesnard, Jean. "Un écho burlesque aux expériences de Pascal: *La Folie du Vide (1648)*," *Courrier du Centre International Blaise Pascal*, no. 9 (1987), pp. 9–25.

Porta, John Baptista. *Natural Magick*, Anon. Trans. London: Thomas Young and Samuel Speed, 1658. Facsimile reprint, New York: Basic Books, 1957.

Poulet, Georges. *Les Métamorphoses du Cercle*. Paris: Flammarion, 1979.
Prager, Frank D. "Berti's Devices and Torricelli's Barometer from 1641 to 1643," *Annali dell'Istituto e Museo di Storia della Scienza* 2 (1981), pp. 35–53.
Sabra, A.I. *Theories of Light From Descartes to Newton*. London: Oldbourne, 1967.
Schaffer, Simon, and Shapin, Steven. *Leviathan and the Air-Pump*. Princeton: Princeton University Press, 1985.
Schmitt, Charles B. "Experimental Evidence for and against a Void: The Sixteenth-Century Arguments," *Isis* 58 (1967), pp. 352–366.
Schobinger, Jean-Pierre. *Kommentar zu Pascal's Reflexionen über die Geometrie im Allgemeinen*. Basel: Schwabe, 1974.
Sellier, Philippe. *Pascal et Saint Augustin*. Paris: Albin Michel, 1995.
Shea, William R. *Galileo's Intellectual Revolution*. 2nd ed. New York: Science History Publications, 1972.
Shea, William R. *The Magic of Numbers and Motion. The Scientific Career of René Descartes*. Canton, MA: Science History Publications, 1991.
Smith, D.E. *A Source Book in Mathematics*. New York: McGraw-Hill, 1929.
Sprat, Thomas. *The History of the Royal Society*. London, 1667.
Taton, René. "L'annonce de l'expérience barométrique en France," *Revue d'histoire des sciences et de leurs applications* 16 (1963), pp. 77–83.
Taton, René. "*L'Essay pour les Coniques* de Pascal," *Revue d'histoire des sciences et leurs applications* 8 (1955), pp. 1–18.
Thirion, J. "L'horreur du vide et la pression atmosphérique," *Revue des questions scientifiques* 12 (1907), pp. 384–451; 13 (1908), pp. 149–251; and 14 (1909), pp. 149–201.
Thirouin, Laurent. *Le Hasard et les Règles: Le Modèle du Jeu dans la Pensée de Pascal*. Paris: Vrin, 1991.
Todhunter, Isaac. *A History of Mathematical Theory of Probability*. London: Macmillan, 1865.
Waard, Cornélis de. *L'expérience barométrique*. Thouars: Imprimerie Nouvelle, 1936.
Weaver, Warren. *Lady Luck. The Theory of Probability*. New York: Dover Publications, 1982 (reprint of 1963 edition).
Westfall, Richard S. *Force in Newton's Physics*. New York: Elsevier, 1971.
Westfall, Richard S. "The Problems of Force in Galileo's Physics," in *Galileo Reappraised*, edited by Carlo L. Golino. Berkeley: University of California Press, 1966, pp. 67–95.
Zouckermann, R. "Air Weight and Atmospheric Pressure from Galileo to Torricelli," *Fundamenta Scientiae* 2 (1981), pp. 185–204.

Index

abhorrence, of the void: see horror, of the void
Adamson, Donald, 227, 323
Agricola, Georg, 176
Aiguillon, Duchesse d', 14
air, igneous, 67, 68, 78–80, 82, 86, 99, 159
air, weight of, 18–21, 33, 37, 39, 51, 52, 82, 85, 96, 100, 101, 106, 115, 122, 127, 134, 151, 154, 155–185
Albert of Saxony, 73
Albert the Great, 192
Aldovrandi, Ulisse, 62
Aleotti, G.B., 71
algebra, 6, 9
Al-Kashi, Jamshid, 241
alphabet, 15–16
altitude, 171–172, 175, 178, 182–185
antipathy, 116, 117, 336
antiperistasis, 79, 116, 117, 118, 119
antiquity, 189–194
Apian, Peter, 241
Apollonius of Perga, 6, 8, 9
Aquinas, Thomas, 206, 207
Archimedes, 38, 109, 130, 136, 148, 153, 296, 319, 328
Aristarchus, 89
Aristotelians, 33, 46, 47, 48, 127, 159, 215, 239
Aristotle, 19, 70, 71, 75, 85, 94, 118, 119, 139, 141, 151, 155, 176, 177, 192, 193, 216, 319

Arithmetical triangle, 241–255, 265–268, 286–287, 289, 291–306, 337–338
Arnauld, Antoine, 223, 236, 330
atmospheric pressure, 18, 33, 37, 39, 42, 60, 80, 82, 85, 101, 106, 113, 120, 126, 159, 172, 175, 185, 221, 335–336
atomism, 70, 76, 88, 94, 233
Auger, L., 89
Augustine, Saint, 191, 201, 206, 215–216, 224, 236, 237
authority, 169, 87–188
Auvergne, 1, 126
Auzoult, Adrien, 51, 103–104, 123, 165
Averroes, 192

Bacchin, Simonetta, xi
Bacon, Francis, 190–191
Baillet, Adrien, 125, 191
Baliani, Giovanni Battista, 18–21, 23, 33, 54, 156, 167
Bannier, Friar, 109
Barancy, François, 120
Barberini, Francesco, 24
barometer, 134, 172, 185, 333
Beaugrand, Jean de, 326–327
Beaulieu, Armand, 3
Beeckman, Isaac, 37, 157
Begon, Antoinette, 1
Begon, Councillor, 110
Benedetti, Giovanni Battista, 144, 145

347

Index

Bernès, Catherine, 38
Bernier, François, 62, 92, 121, 160
Bernoulli, Jacob, 328
Bernouilli, Johann, 138
Berti, Gasparo, 24, 25, 26, 28, 31, 35, 36, 41, 44
binomial theorem, 292–293
Borgatto, Graziella, xii
Bose, Georg Mathias, 35
Bouchard, Jean-Jacques, 24
Bouchilloux, Hélène, 195
Boulliau, Ismaël, 5, 120
Bourdelot, Abbé, 14, 41
Boyer, Carl B., 7, 248, 282, 311
Boyle, Robert, 29–31, 92–93, 102, 113, 114, 129, 151–152, 161, 162, 167–168
Bradwardine, Thomas, 73
Brahe, Tycho, 68, 96 192, 205
braccio, 17, 18, 20, 21, 23, 24, 33, 35, 84, 175
Bretel, Cornelius, 149
Bruno, Giordano, 71, 231
Brunschvicg, Léon, 1
Bucciantini, Massimo, 85
Buridan, John, 118
Burton, Robert, 117

calculating machine, 9–15, 95, 96, 152, 337
Calculus, 309–312, 322, 338
Calimani, Eugenio, xi
Campanella, Tommaso, 193
capillarity, 134
Capponi, Vincenzo, 24
Capra, Giovanni Paolo, 144
Carcavy, Pierre, 125, 126, 260, 261, 273, 316, 317, 324, 325, 327, 331
Cardano, Girolamo, 155, 285
Castelli, Benadetto, 172
Cavalieri, Benedetto, 309, 319, 322
Cesarini, Ferdinando, 172
chance, games of, 248–255, 257–289
Chanut, Pierre, 41, 42, 43, 53, 57, 71, 176, 182

Chapelain, Jean, 15
Chastin, Fr., 110, 111
Christina, Queen, 14, 15, 336
Clagett, Marshall, 118
Clavius, Christoph, 192
Clermont, 1, 48, 106, 107, 109, 111, 129, 121, 122, 179, 182–184
Coimbra, 75
Colbert, 5
Collège de France, 13
College of Clermont, 65, 83
College of la Flèche, 65
College of Montferrand, 84
Collège Royal, 5, 320
Commandino, Federico, 71
Condé, Prince de, 13, 14
combination, 251, 252–255, 280, 286
conics, 6–9
Copernicus, Nicolaus, 68, 96, 207
Cornelio, Tommaso, 26
Cornier, Robert, 155
Costa, Giovanni, xi
Cracow, 85
currency, French, 10, English, 10, Polish, 13
Cusanus, Nicholas, 231, 319
custom, 222–225, 237–238
cycloid, 319–331, 338

David, F.N., 247, 260, 261, 278, 319
Dati, Carlo, 32, 35
Della Porta, Giovanni Battista, 117, 148
De Morgan, Augustus, 248
Democritus, 76
Desargues, Girard, 5, 6, 7, 337
Descartes, René, 5, 6, 8, 9, 13, 14, 42, 60–62, 65, 68, 70, 74–75, 77, 78, 82, 83, 93–96, 99, 108, 116, 122, 123, 126, 130, 137, 139–141, 157–159, 191, 214–216, 223, 230, 232, 236, 320, 323, 33–335
Descotes, Dominique, 57, 170, 210, 291, 300, 310

Index

Desnoyers, Pierre, 46, 48, 49, 50, 57, 84, 88, 182
Dieppe, 42, 179
division problem: see problem of points
Dominicy, Marc-Antoine, 42, 53, 176
Donne, John, 194, 230
Drake, Stillman, 18, 20, 26, 138, 192
Duhem, Pierre, 102 135 143, 145, 156, 157
Dungarvan, Lord, 114
Du Verdus, François Bonneau, 39, 51, 52

Edward, A.W.F., 278, 281, 317–318
Effiat, Marquise d', 108
Elizabeth, Queen, 189
Epictetus, 283
Epicurus, 76
Euclid, 3–4, 210
experiment, 3, 18–23, 24–39, 41–58, 60–62, 78, 83, 88–93

Fabri, Honoré, 120
falling bodies, 70, 73, 156
Fanton d'Andon, Pierre, 62
Faulhaber, Johann, 297
Favaretto, Irene, xi
Fermat, Pierre, 203, 217, 247–248, 258, 260–288, 297, 316, 317, 318, 320, 321, 338
Florence, 31, 36, 39, 41, 49, 175
Florentine Academy, 172
Fonseca, Pedro, 75
Forton, Jacques, 188
Fracastoro, Girolamo, 117
François, Jean, 15
Frederick Henry of Orange, Prince, 13
Freudenthal, H., 247
Furetière, Antoine, 3

Galileo, x, 17–26, 32, 33, 49, 53, 68, 71–73, 79, 84, 94–95, 101, 137–139, 151, 155–156, 160, 172, 175, 176, 181, 205, 206, 207, 209–210, 319, 320, 326, 327
Galluzzi, Paolo, xi, 33
gambler's ruin,, 316–318
Gassendi, Pierre, 5, 76, 92, 120, 121, 122, 157, 160
Gellius, Aulus, 191
Genoa, 18, 20
geometry, 4–9, 210, 219, 235, 288, 337
Gesner, Conrad, 62
Giorgi, Alessandro, 72
Girard, Albert, 145, 147
Golefer, sieur de, 190
Goodman, Godfrey, 189
Gournay, Marie le Jars de, 231
Grant, Edward, 73
gravity, center of, 136, 142–145, 153
Greatorex, Ralph, 167
Guericke, Otto von, 30
Guiffart, Pierre, 46–47

Haak, Theodor, 8, 321
Hallé de Monflaines, Raoul, 108
happiness, 237, 240
Harries, Karsten, 231
Hartlib, Samuel, 92
heart, 215, 220, 222, 227, 235, 237, 315
Heath, Thomas, 297
Henri II, King, 1
Hérigone, Pierre, 241, 292, 300
hermeneutics, 201, 204
Hero of Alexandria, 70–71, 176
Hevelius, Johann, 91–92, 120
Hin, Yang, 241
Hire, Philippe de la, 6
Hobbes, Thomas, 31, 330
Hooke, Robert, 167, 189
Horror, of the void, 44, 46, 47, 52, 53, 54, 58–59, 62, 85, 87, 105, 106, 107, 115–116, 118, 119, 120, 129, 150, 164, 165, 174–178, 239–240, 333, 336
humours, 66

349

Index

hydraulics, 129, 136, 141–145, 153
hydrostatic principle, 136, 141–142, 145, 333, 336
Huygens, Christiaan, 13, 15, 122, 316, 317, 320, 330, 331
Huygens, Constantin, 13, 14, 42, 108, 109, 122

Imagination, 223, 238
induction, principle of complete, 246–248, 260, 337

Jansenist, 206, 223, 283
Jaki, Stanley, 192
Jesuits, 3, 46, 65–71, 84–88, 333
Jesus Christ, 189, 195, 201, 204
Jews, 199–200, 203–204
Jones, Richard, 114
Jones, Richard Foster, 190
Jouanisson, Roland, 170

Kircher, Athanasius, 26, 28, 31
Koyanagi, Kimigo, 168–170
Koyré, Alexandre, 45, 231
Krailsheimer, Alban, 197, 199
Kurobe, Akira, 168

Ladislaw VII, King, 13
La Font de l'Arbre, 111, 121, 182–184
Lalouvère, Antoine, 325–328
Langres, 107
La Porte, Dr, 110
La Ville, Councillor, 110
Leavenworth, Isabel, 54, 64
Leibniz, Gottfried Wilhelm, 8, 121, 320, 322, 323, 324, 328, 330, 338
Le Pailleur, Jacques, 4, 5, 77–83, 86, 109, 159
Le Tenneur, Jacques, 48, 108, 126
lever, 130, 133, 136, 139, 141, 143, 149, 154, 328
Lhermet, J., 195
light, 62, 65, 67, 68, 75, 79, 92, 170, 213–214, 260, 334–335

Locke, John, 76–77
logic, 68, 209, 215, 219, 222, 235
London, 7
Louis XIII, 13
Louis XIV, 5
Louise Marie de Gonzague, Queen of Poland, 1, 13, 49
Lucan, 189
Lucretius, 160
Luynes, duc de, 223

Mach, Ernst, 138
Macôn, 121
Macrobius, 192
Maclaurin, Colin, 7
macrocosm, 66, 83
Magdeburg, 30
Magiotti, Raffaello, 24, 26, 28, 32, 36, 38
Magni, Valerio, 42, 49, 50, 84–85
Mahnke, Dietrich, 231
Maignan, Emmanuel, 27, 28, 30, 31
Mare, Father de la, 111
Mariotte, Adme, 102
Marolles, Michel de, 5
Martini, Raymond, 197
mass, conservation of, 173
mathematical mind, 219–221
Matton, Sylvain, 120
Mazauric, Simone, 120
Medici, Giovanni-Carlo de, 39
Menjot, Antoine, 188
Méré, Antoine Gombaud chevalier de, 203, 217, 218, 257–261, 284, 337
Mersenne, Marin, 3, 5, 7, 8, 13, 24, 26, 28, 31, 32, 36, 39, 41, 42, 45, 49, 51, 52, 85, 89, 91, 92, 102, 107, 108, 109, 122–126, 134, 135, 140, 141, 149, 151, 155–159, 167, 216, 248, 260, 319–322, 326
Mesnard, Pierre, 1, 5, 160, 181
Method, combinatorial, 261, 263, 269–271–289; direct probability, 261, 281, 282;

350

Index

expectations 261, 264–265, 269–270, 278, 283–289, 318
microcosm, 66, 83
Micanzio, Fulgenzio, 94
Middleton, W.E. Knowles, xi, 17, 27, 28, 30, 33, 34, 36, 37, 92
Milanesi, Vincenzo, xi
Milky Way, 192
Moivre, Abraham de, 241
moment, 138, 139
momentum, 139
Montaigne, Michel seigneur de, 189, 196, 202, 213, 224, 231, 283
Mont Blanc, 30
Mont Faron, 131
Montferrand, 1
Morin, Jean Baptiste, 241
Mosnier, Canon Claude, 110, 111, 120, 121
Mosnier, Pierre, 120
Mourlevat, Guy, 12
Mozart, Wolfgang Amadeus, vii
Mydorge, Claude, 5
Mylon, Claude, 316, 324

Nantes, 107
Nevers, 107
Newton, Isaac, 81, 154, 189, 221, 322, 323
Nicole, Pierre, 223
Nicomachus of Cerasa, 297
Noël, Etienne, x, 65–88, 97, 100, 102, 117, 122, 159, 188, 205, 209, 214, 239, 323, 334–336
number theory, 382–283

Oxford, 30, 325
orders, hierarchy of, 226–228

Pacioli, Luca, 260
Padua, University, xi
Pantin, Isabelle, 192
Pappus, 9
parabola, 6
Paris, 3, 13, 39, 41, 44, 49, 50, 51, 65, 95, 106, 114, 115, 120, 121,
123, 124, 157, 170, 171, 179, 182–184, 247, 257, 274, 283, 318, 328
parallel lines, 6
paradox, 6, 70, 132, 135, 197, 227, 240, 284, 315, 320
parsimony, 178
Pascal, Blaise: calculating machine, 9–15,95; childhood, 1–9; dreaming, 222–223,229; illness, 1–3,53,67,95, 333; experiments in childhood, 3; experiments on the vacuum, 41–47, 33–336; growth of mankind, 188–194; historical explanation, 187–194, 200; historian of science, 174–178; laboratory reports, 51–53; mastery of Euclid, 3; new way of teaching the alphabet, 15–16; night of fire, 282; Pascal's Principle, 148–149, 153, 333; Pascal's triangle 245–248, 337–338; wager, 313–316, 338–339
Account of the Great Experiment, 106,115,120, 126
Conicorum Opus Completum, 8
Discourse on the Condition of the Great, 223–224
Essay on Conics and *Mystic Hexagon*, 7–8
History of the Cycloid, 320
Introduction to Geometry, 211
Letter of Presentation and *Notice for the Calculating Machine*, 13, 55
Letter to de Ribeyre, 84, 123, 143, 192
Letter to Le Pailleur, 80–83,109, 159
Letter to Noël, 67–70, 205
Letter to Queen Christina, 14–15, 226–227
Letter to the Parisian Academy, 5, 9, 288

Pascal, Blaise: calculating machine (Cont.)
  New Experiments Concerning the Vacuum, 50, 53–64, 65, 74, 81, 83, 84,86,87, 99, 106, 107, 115, 116, 122, 174, 187, 202, 335
  On Figurate Numbers, 298, 304–305, 337
  On the Art of Persuasion, 207
  On the Boundaries of the Powers of Integers, 297–298
  On the Equilibrium of Liquids, 59, 69, 115, 129–154, 187, 221
  On the Geometrical Mind, 79, 207, 211, 213, 215–219
  On the Weight of the Mass of Air, 115,119, 129, 155–185, 187, 221, 336
  Pensées, ix 178, 195–206, 212, 214,215, 219, 220, 222, 223, 227–234, 236–240, 313–316, 319, 324, 334
  Potestatum Numericarum Summa (The Summing of the Powers of Intergers), 228, 282, 291–
  Preface to the Treatise on the Vacuum, 187–194
  Product of Consecutive Integers, 304–306
  Provincial Letters, 206–207, 283, 319, 324, 326
  Recognition of Multiple Numbers, 307–309
  Reply to Father Noël
  Treatise on the Arithmetical Triangle, 245–247, 265–268, 272, 284
  Treatise on the Sine, 328–330
  Treatise on the Vacuum, 129
  Use of the Arithmetical Triangle for Combinations, 255
Pascal, Etienne, 1, 2, 3, 4, 9, 10, 41–44, 65, 71, 83, 109
Pascal, Gilberte, 1, 2, 3, 4, 9, 10, 12, 95, 322

Pascal, Jacqueline, 1, 15–16, 51, 95–96, 283
Pascal, Marguerite, 2, 322– 323
Patrizzi, Francesco, 75
Paul, Saint, 201,204
Pecquet, Jean, 103
Peiresc, Claude Fabri de, 5
Pell, John, 321
Périer, Florin, 44, 48, 52, 59, 63, 80, 88, 96, 101, 105–115, 120, 121, 129, 133, 134, 166, 168, 174, 183
permutation, 252–253, 278–279
perspective, 6, 236
persuasion, art of, 235–239
Petit, Pierre, 5, 41–44, 50, 51, 52, 53, 57, 71, 179, 182
Philoponus, John, 72, 118
Pierius, Jacques, 45, 90, 91
Pintard, René, 92
Pisa, University of, 71
Place des Vosges, 5
Place Royale, 5
Plato, 117, 118, 213
Plutarch, 192
poetry, 225
Popkin, Richard H., 12
Port Royal, convent, 15, 283
Port Royal, grammar of, 15, 16
powers of integer, 296–298
Pozzo, Cassiano dal, 41
Prager, Frank D., 24
pressure, 130, 133, 135, 137, 141–142, 147–151, 158, 175, 214, 220–221
problem of points, 260–289
progress, 187–194, 205
projective geometry, 6, 337
Ptolemy, 68, 96, 205
pump, 17, 18, 20, 21, 23, 24, 35, 53, 54, 57, 71, 167, 171, 175, 176, 179
Puy-de-Dôme, 52, 88, 96, 97, 107, 108, 109, 110, 111, 119, 120, 121, 122, 123, 125, 126, 127, 160, 177, 178, 182–184, 333

Index

Pythagoras, 117

religion, 194–207, 315, 323, 324
Renaissance, 6
Reneri, Henri, 37, 157
Rey, Jean, 156, 176
Ribeyre, Antoine de, 42, 46, 50, 84, 123, 143, 192
Ricci, Michelangelo, 33–39, 105, 116, 122, 324, 325
Richelieu, Cardinal, 1, 14, 158, 241
Roannez, Duc de, 323–324, 328
Roberval, Gilles Personne de, 5, 13, 39, 45, 46, 47, 48, 49, 50, 5154, 57, 62, 73, 84–85, 88–97, 99, 102–103, 108, 122, 125, 126, 165, 182, 248, 263, 274–275, 278, 280, 286, 319, 320, 321, 323, 324, 325, 326, 327
Rohault, Jacques, 134, 170–171
Rome, 24, 28, 30, 35, 38, 39, 45, 49
Rouen, 1, 3, 9, 12, 30, 42, 44, 45, 50, 51, 108, 109, 188

Saci, Isaac Le Maître de, 283
Saint-Martin, Brulart de, 84
Sánchez Sorondo, Marcelo, v
Schickard, Wilhelm, 12
Schmitt, Charles B., 17
Schneider, Ivo, 297
Scholasticism, 42
scholarship, 187
Schott, Gaspar, 24
Scripture, 188, 195–207, 334
Séguier, Chancellor, 12, 13
Sellier, Philippe, 191, 199
Shakespeare, William, 17
Shapin, Steven (and Schaffer, Simon), 113, 114, 161
Shea, William, 61, 138
Simili, Raffaella, xi
Simplicius, 155
sin, original, 194, 197–198, 219
Sluse, François, René de, 38, 324, 325, 330

Smith, D.E., 261
Snell, Willebrord, 145
sound, 28–31
space, 70–71, 75–77, 81, 86, 217–219, 229, 231–232
sphere, infinite, 231, 233
Stevin, Simon, 139, 142, 145, 147
Stifel, Michard, 241
Stockholm, 14, 42
Struik, D.J., 7, 247, 329
sympathy, 116–117, 336
systematic variation, 177–178
Sweden, 41, 43

Tallemant des Réaux, Gédéon, 5
Tartaglia, Niccolo, 241, 285
Taton, René, 5, 6, 7, 12
Tega, Walter, xi
telescope, 192, 205, 218
thermometer, 115, 124, 134
theology, 77, 85, 187, 188, 335
Thirion, J, 100
time, 191, 211, 215–217, 229, 233, 236
thought-experiment, 22, 51, 146–148, 150–152, 161, 163, 168, 336
Torricelli, Eva, ngelista, x, 24, 31, 32–39, 41, 42, 43, 44, 51, 52, 53, 54, 63–64, 80, 83, 84, 87, 91, 99, 105, 116, 122, 136, 143, 153, 157, 174, 210, 221, 319, 320, 322, 326, 327, 335
Torrini, Maurizio, 33

Uchida, Masao, 168
Urban VIII, Pope, 24

vacuum, 5, 17, 21–39, 193, 202–203, 237, 238, 239
vacuum in a vacuum, 100–105, 119, 120, 122, 165–171, 177
Van Schooten, Frans, 9, 324
Viète, François, 9, 300
Ville, Antonio de, 94
virtual displacement, 136–141, 155, 220, 221

virtual velocity, 137–141
Viviani, Vincenzo, 32, 35
void: see vacuum

Waard, Cornelis de, 28, 37, 85, 157
Ward, Seth, 113
Wallis, John, 113, 324, 325, 326, 330
Warsaw, 13, 42, 49, 85, 88
Water, boiling of, 30, 50

Weather forecasting, 180, 182
Weaver, Warren, 248, 258
Westfall, Richard, 138, 141
work, 137, 154
Wren, Christopher, 113, 324, 325

Zizka, Jan, 117
Zucchi, Nicolò, 28, 31, 161